Economics, Law, and Institutions in Asia Pacific

The Asia Pacific region is expected to steadily enhance its economic and political presence in the world during the twenty-first century. At the same time, many serious economic and political issues remain unresolved in the region. To further academic enquiry and enhance readers' understanding about this vibrant region, the present series, Economics, Law, and Institutions in Asia Pacific, aims to present cutting-edge research on the Asia Pacific region and its relationship with the rest of the world. For countries in this region to achieve robust economic growth, it is of foremost importance that they improve the quality of their markets, as history shows that healthy economic growth cannot be achieved without high-quality markets. High-quality markets can be established and maintained only under a well-designed set of rules and laws, without which competition will not flourish. Based on these principles, this series places a special focus on economic, business, legal, and institutional issues geared towards the healthy development of Asia Pacific markets. The series considers book proposals for scientific research, either theoretical or empirical, that is related to the theme of improving market quality and has policy implications for the Asia Pacific region. The types of books that will be considered for publication include research monographs as well as relevant proceedings. The series show-cases work by Asia-Pacific based researchers but also encourages the work of social scientists not limited to the Asia Pacific region. Each proposal and final manuscript is subject to evaluation by the editorial board and experts in the field.

All books and chapters in the Economics, Law and Institutions in Asia Pacific book series are indexed in Scopus.

Han Phoumin · Rabindra Nepal · Fukunari Kimura ·
Farhad Taghizadeh-Hesary
Editors

Large-Scale Development of Renewables in the ASEAN

Economics, Technology and Policy

ERIA
Economic Research Institute for ASEAN and East Asia

Springer

Editors
Han Phoumin
Economic Research Institute for ASEAN
and East Asia (ERIA)
Jakarta, Indonesia

Fukunari Kimura
Economic Research Institute for ASEAN
and East Asia (ERIA)
Jakarta, Indonesia

Rabindra Nepal
Faculty of Business and Law, School
of Business
University of Wollongong
Wollongong, NSW, Australia

Farhad Taghizadeh-Hesary
School of Global Studies
Tokai University
Kanagawa, Japan

ISSN 2199-8620 ISSN 2199-8639 (electronic)
Economics, Law, and Institutions in Asia Pacific
ISBN 978-981-99-8238-7 ISBN 978-981-99-8239-4 (eBook)
https://doi.org/10.1007/978-981-99-8239-4

This Springer imprint is published by the registered company Springer Nature Singapore Pte Ltd.
The registered company address is: 152 Beach Road, #21-01/04 Gateway East, Singapore 189721, Singapore

Paper in this product is recyclable.

Preface

The Association of Southeast Asian Nations (ASEAN) has launched an ambitious program of much needed decarbonization. Excluding the Philippines, nine ASEAN governments have pledged to achieve net-zero targets by 2050, while Indonesia has set a target date of 2060. These states have also pledged to reduce greenhouse gas emissions in their nationally determined contributions (NDCs). The Paris Agreement acted as a catalyst for the ASEAN member states to revise their power sector development plans to include lofty commitments to energy sector decarbonization. The region has agreed to collectively increase its share of renewable energy installed power capacities to 35% by 2025.The International Renewable Energy Agency (IRENA) estimates that the region can meet its growing energy demand by replacing 75% of its energy-related CO_2 emissions by 2050 with renewables. This compares to almost halving emissions as of this writing, necessitating urgent action if this degree of decarbonization is to be achieved.

However, a successful transition to renewable energy sources is neither a quick fix nor merely a matter of greener technological substitution through adequate financing. Economic and policy aspects are equally significant to facilitating greater renewable deployment in ASEAN. This book is a response. It is unique and timely in documenting that achieving large-scale renewable deployment relies on combining the distinct albeit interrelated forces of economics, technology, and policy.

Section one focuses on the economic aspects of facilitating large-scale deployment of renewables. It consists of five chapters, which encompass regional and country-specific case studies alike. Chapter 1, "Electricity Market Design and Large Share of Renewables: Lessons for ASEAN," by Nepal et al., begins by reviewing the wholesale market design features of Singapore and the Philippines, which are ASEAN's existing liberalized electricity markets. The chapter then draws out market-based policy lessons for ASEAN from the case studies of the eastern Australian National Electricity Market, the Western Australia wholesale electricity market, and the UK electricity market. One proposal it offers is that liberalized electricity markets require balancing markets with government to achieve renewable energy and net-zero emission targets by aligning energy policy with climate policy.

Chapter 2, “Multi-objective Auctions for Utility-Scale Solar Battery Systems in ASEAN and East Asia,” by Toba et al., examines large-scale solar photovoltaic (PV) and battery energy storage systems auctions in East and Southeast Asia by revisiting their theoretical and conceptual frameworks. This chapter addresses the demand for Environmental, Social, and Governance (ESG) characteristics to be taken into account from the perspective of such key stakeholders as investors, government, bidders, and communities, as regards efficient allocations of risks, costs, and benefits. A key finding of this analysis is that integrating ESG in auction designs and business models is possible and can benefit business and sustainable development alike.

Chapter 3, “Power Trade and Hydroelectricity Development in the Greater Mekong Sub-region: Perspectives on Economic and Environmental Implications,” by Chang, examines how cross-border power trade affects the development of hydropower potential in the Greater Mekong Sub-region (GMS) within the ASEAN power trade model framework and shows economic and environmental implications thereof. Its findings strongly suggest that cross-border power trade will aid such development, as well as endorsing the view that such trade further promotes the development of other renewable energy sources.

Chapter 4, “Tradeable Renewable Energy Credits Market: Lessons from India,” by Sawhney, focuses on India. It analyzes the country’s renewable energy credit (REC) market experience over the preceding decade and examines the implications of changes in trading rules. Since auctions commenced in March 2011, REC market prices have steadily declined, falling as low as 3–6% between 2017 and 2121 as the inventory of unsold RECs accumulated, despite the renewable certification rate initially rising sharply from 2% in 2011–12 to 15% in 2014–15. While the certification rate has increased following a market design overhaul in 2022, the inventory of unsold RECs lingers. The chapter concludes that target underachievement and non-compliance with state renewable purchase obligations must be tackled with deep reforms in how power distribution companies function rather than the REC mechanism per se.

Chapter 5, “Rooftop PV with Batteries for Improving Self-consumption in Vietnam: A Cost-Benefit Analysis,” by Dan and Phoumin, is another country-specific case study. Vietnam must expand its use of renewables to achieve net-zero emissions by 2050 while meeting growing energy demand and facilitating such technological initiatives as energy storage. This chapter examine the costs and benefits of rooftop solar plus battery in a sample factory in Ha Tinh province, using some 115 MWh of grid-connected electricity annually in manufacturing building materials and installing 137 kWp solar with battery for self-sufficiency. The study also offers policy recommendations for Vietnam to meet its sustainable development targets.

Section two addresses the technological aspects of facilitating large-scale renewables deployment, with three country-specific studies and one regional case study. Chapter 6, “The Role of Battery Energy Storage Systems and Energy Market Integration in Indonesia’s Zero Emission Vision,” by Pramudya et al., ran simulations using the Balmoral energy model on 230 grid systems to estimate the impact of net-zero targets on optimal capacity expansion, electricity production mixes, emissions, and electricity supply costs. The results confirmed that zero emissions objectives would benefit significantly from integrating solar PV and battery energy storage, with emphasis on the importance of replacing phased-out coal-fired power plants with nuclear power by 2060.

Chapter 7, "Deployment of Renewable Energy and Utility-Scale Batteries in Australia: Lessons Learned and Policy Implications for Other Countries," by Grozev et al., reviews the underlying trends and outcomes of renewable energy utilization in Australia's National Electricity Market (NEM). The purpose of the chapter is to update available information with the most recent renewable energy and battery developments in the NEM, as well as describe the energy dynamics in South Australia, which remains the country's most advanced state in terms of penetration of wind and solar PV generation. The study also summarizes the cost projections of renewable generation technologies in Australia, including a summary of the main policy support schemes used in Australia to facilitate renewable energy investments.

Chapter 8, "Effects of Digital Technologies on Renewable Energy Development: Empirical Evidence and Policy Implications from China," by Zheng et al., investigates the effects of digital technologies on renewable energy development by estimating their influence on renewable energy market integration in China, finding that digital technologies have significantly bolstered such development in China. An entropy weight method is utilized to construct an index of digital technologies. These findings will provide valuable policy guidance to ASEAN countries regarding in achieving carbon-neutral energy transitions.

Chapter 9, by Phoumin et al., is "Potential Solar, Wind and Battery Storage Deployment to Decarbonize Emissions in ASEAN." It investigates the maximum contribution of solar and wind deployment together with energy storage potentials to change solar and wind deployment from intermittent to more stable loads by combining energy storage systems. Findings provide policymakers with a useful guide on how to scale up solar and wind with battery storage in order to facilitate profound decarbonization in ASEAN economies in the future. It also affirms that reaching carbon neutrality will require multiple approaches to decarbonize emissions in all sectors.

Section Three encompasses regional case studies focusing on policy aspects of large-scale renewables deployment. Chapter 10, "India's Cross Border Electricity Trade with South-Asian and BIMSTEC Countries," by Sharma et al., assesses the present status of the Cross Border Electricity Trade (CBET) regime among India and its energy trade partners, Nepal, Bhutan, Bangladesh, and Myanmar, and its effects on energy security. It develops a mathematical model based on simple energy balance in scenarios of without and with CBET, respectively, while integrating secondary storage for purposes of realism. The first scenario shows that if renewables are added to the mix, curtailment of these power sources to balance the grid could be unavoidable. The second scenario shows that the storage and generation capacity and curtailment period for renewables could be reduced by facilitating greater energy imports through an interconnected grid.

Chapter 11, "Toward a Coherent Policy Approach to Solar Uptake in Southeast Asia: Insights from Indonesia and Vietnam," by Yang et al., examines the experiences of Indonesia and Vietnam in adopting utility-scale solar power. It highlights the need to create coherent and effective policy frameworks capable of addressing both the emergence and wider adoption of niche electricity technologies and the reconfiguration of incumbent regimes. It suggests that a key strategy is to focus initial efforts on promoting clean technologies that have already played a significant role in the generation mix to reconcile the need for rapid transitions to address the climate crisis with usually prolonged transitions.

Chapter 12, "Impact of Policy on Solar PV Supply for ASEAN and Beyond," by Best et al., assesses the role of renewable energy policy in solar PV. It documents that such policy lags influence by up to six years on changes in solar energy supply per capita, based on a composite renewable energy policy index. Economic policy instruments such as carbon pricing and feed-in tariffs have the most robust impact on solar use. It concludes by recommending that expanded implementation of carbon pricing in ASEAN member states is an opportunity not to be missed.

This book is a compendium of important empirical studies providing pragmatic policy recommendations that will assist ASEAN in transitioning to sustainable energy sources through large-scale deployment of renewables. The focus of the book is on the environmental sustainability dimension of the energy policy trilemma by considering the role of economics, technology, and policy, in achieving energy transitions, while also providing a valuable resource for interested researchers.

Jakarta Pusat, Indonesia — Dr. Han Phoumin
Wollongong, Australia — Assoc. Prof. Rabindra Nepal
Jakarta Pusat, Indonesia — Prof. Fukunari Kimura
Tokyo, Japan — Assoc. Prof. Farhad Taghizadeh-Hesary

Acknowledgments

This study was produced by a working group of selected outstanding researchers from various academic backgrounds. We acknowledge support from all involved, including the many people who the authors met and interviewed to obtain country-specific information. Special thanks go to Hidetoshi Nishimura, former President of the Economic Research Institute for ASEAN and East Asia (ERIA); Tetsuya Watanabe, President of ERIA; Fukunari Kimura, Chief Economist, ERIA; Koji Hachiyama, Chief Operating Officer, ERIA; and Toru Furuichi, Director General for Research and Policy Design Administration, ERIA, for their encouragement, support, and advice in ensuring that the project would address the emerging energy issues that the region confronts, particularly the need to have appropriate energy transition while considering the acceleration of renewables and clean fuels and technologies. We are very grateful to Michael House of Word House Ltd. for copy editing. We also thank ERIA's publication team for kindly facilitating the outsourcing of the copy editing work. Finally, we are very grateful to Juno Kawakami, Shinko Mimura, Carolyn Zhang, and Sridevi Purushothaman for managing the publishing process.

Contents

About the Editors

Dr. Han Phoumin is Senior Energy Economist with the Economic Research Institute for ASEAN and East Asia (ERIA). He has more than 20 years of experience at various international and intergovernmental organizations and multidisciplinary research consortiums related to energy markets and technologies, the environment, integrated water resource management, governance, and economic development in the region of the Association of Southeast Asian Nations (ASEAN) and East Asia. He specializes in economic development and policy and applied econometrics. Much of his career over the past decade has involved the power sector, sustainable hydropower development, renewable energy research, energy efficiency, clean coal technology, energy security, and energy demand and supply forecasting.

Dr. Rabindra Nepal is Associate Professor of Economics at the School of Business of the Faculty of Business and Law of the University of Wollongong, Australia. Dr. Nepal is internationally recognized Academic and Researcher in energy, environment, and resources and has previously consulted for such international organizations as the World Bank, the European Commission, the International Centre for Integrated Mountain Development (ICIMOD), the Economic Research Institute for ASEAN and East Asia (ERIA), and the National Research Institute (NRI) of Papua New Guinea. Dr. Nepal has widely published more than 90 journal articles and book chapters, in addition to two co-edited volumes. With a strong policy focus in his research, he edits international journals with a similar emphasis, including Co-editor of the journal *Economic Analysis and Policy* and serving as International Advisory Board Member of the journal *Energy Policy*. He has taught diverse classes in Economics.

Prof. Fukunari Kimura was born in Tokyo in 1958. He received his Bachelor of Laws from the Faculty of Law, University of Tokyo, in 1982. He then received Master of Science and Ph.D. titles from the Department of Economics, University of Wisconsin-Madison, in 1990 and 1991, respectively. He worked for the Department of Economics, State University of New York, at Albany as Assistant

Professor in 1991–1994 and in the Faculty of Economics of Keio University as Associate Professor in 1994–2000. In particular, he has recently been active in writing academic/semi-academic books and articles on international production networks and economic integration in East Asia.

Dr. Farhad Taghizadeh-Hesary is Associate Professor of economics at Tokai University, Japan. He is also Co-founder and Vice-President of the International Society for Energy Transition Studies (ISETS), based in Australia. In 2022, he was listed on Clarivate's list of highly cited researchers, a global index of the top 0.1% of scientists. In 2023, he received the Medal and Commendation for Science and Technology from the Minister of Education, Culture, Sports, Science, and Technology (MEXT) of Japan for his outstanding research achievements in renewable energy financing, especially for designing the green credit guarantee scheme. He serves as Editor-in-Chief of the *Journal of Environmental Assessment Policy and Management* (The World Scientific) and Associate Editor and Board Member of several other journals. His research credits include authoring over 275 academic journal papers and book chapters and editing 20 books from such major publishers as Springer, Routledge, World Scientific, and the Asian Development Bank Institute. He holds a Ph.D. in economics from Keio University, Japan, sponsored by a scholarship from the Government of Japan (MEXT).

Chapter 1
Electricity Markets Design and Large Share of Renewables: Lessons for ASEAN

Rabindra Nepal, Han Phoumin, and Ashish Agalgaonkar

Abstract ASEAN economies such as Malaysia and Vietnam have ambitions of establishing liberalized and fully competitive wholesale electricity markets. However, skyrocketing natural gas prices have exposed the vulnerability that liberalized markets globally face from external energy price shocks. ASEAN also has a target of increasing the renewable energy share of its primary energy mix to 35% by 2025. This chapter examines how ASEAN can establish a competitive wholesale electricity market which delivers affordable and reliable electricity while lurching toward achieving greater sustainability. It begins by reviewing the wholesale market design features of Singapore and the Philippines, ASEAN's existing liberalized electricity markets. Next, it draws out market-based policy lessons for ASEAN from the case studies of the eastern Australian national electricity market, the Western Australia wholesale electricity market, and the UK electricity market. Then it argues that the liberalized wholesale electricity market model based on merit-order dispatch may not facilitate integration of large-scale renewables in the absence of appropriate supporting arrangements within wholesale market rules and design and public policy outside of markets. One option is alternative spot market design features such as one proposed in Greece based on market splitting and decoupling gas prices from electricity prices. Liberalized electricity markets require balancing the market with government to achieve renewable energy and net-zero emission targets, by aligning energy policy with climate policy.

R. Nepal (✉)
School of Business, University of Wollongong, Wollongong, Australia
e-mail: rnepal@uow.edu.au

H. Phoumin
Economic Research Institute of ASEAN and East Asia, Jakarta, Indonesia

A. Agalgaonkar
School of Electrical, Computer and Telecommunications Engineering (SECTE), University of Wollongong, Wollongong, Australia

H. Phoumin et al. (eds.), *Large-Scale Development of Renewables in the ASEAN*, Economics, Law, and Institutions in Asia Pacific,
https://doi.org/10.1007/978-981-99-8239-4_1

1 Introduction

Global electricity markets faced a looming energy crisis in 2022, primarily due to shortages of natural gas resulting from Russia's war with Ukraine. Liberalized electricity markets risked being unable to meet demand as supply withdrawals caused by power producers' non-bidding to avoid losses seemed likely due to non-recovery of costs. The reason is that wholesale electricity prices in liberalized markets are based on natural gas merit-order dispatch,[1] especially during peak load times. The merit order effect is how wholesale electricity market prices are set. Power stations supply electricity to wholesale markets, meeting aggregated demand in such a sequence that the cheapest offer made by the power station with the lowest operating costs, i.e., the lowest short-run marginal costs, instigates dispatch. In energy-only markets, such as wholesale electricity markets in Australia and Singapore, prices are based on providers' offers to supply electricity to the market at particular volumes and prices at set times versus demand at any given time. Increased supplies of renewable energy will thus eventually lower merit-order dispatch-based wholesale power prices at electricity exchanges.

Figure 1 shows that an increased supply of renewable energies, indicated by higher-capacity availability measured in gigawatts (GW), depresses wholesale electricity prices while changing marginal providers from hard coal-fired generators to lignite-burning generators. As the market clearing is based, however, on uniform-price auctions, where identical units of a homogenous commodity are sold for the same price (Khezr and Nepal 2021), such low-cost generators, other than the marginal providers, such as renewables, nuclear- and lignite-based power generators earn plant profits comprising 'infra-marginal profits,' the wholesale price difference arising when all generators except the marginal generators receive a higher price than their marginal costs and 'capacity rents,' the profit value created by owning scarce capacity. Wholesale market design based on merit-order dispatch should thus naturally reward investments in such low-cost power generation as renewables, allowing renewable generators to maximise profits. Aside from such markets as Germany, however, the merit-order wholesale price setting design has not delivered market-based incentives to drive renewable energy surges. Some of the distinct features of electricity market models implemented around the world and barriers within these market models limiting the energy mix transition are presented in a study by Johnathon et al. (2021).

On the other hand, the ability of liberalized wholesale electricity to deliver the energy policy trilemma of affordability, security, and sustainability has been increasingly questioned in recent times. The electricity markets of the Asia–Pacific region are rife with such examples (Phoumin et al. 2022). Imported natural gas comprises some

[1] Merit-order dispatch applies to bid- and cost-based wholesale electricity market designs alike. The only difference is that in the latter, quantity and, especially, price components of bids are regulated as implemented in many Latin American countries (McRae 2019). The price component is set at the marginal cost of the plant, based on fuel input prices and technical characteristics for thermoelectricity. The system operator solves its own dynamic programming model to determine the value of water for hydroelectricity.

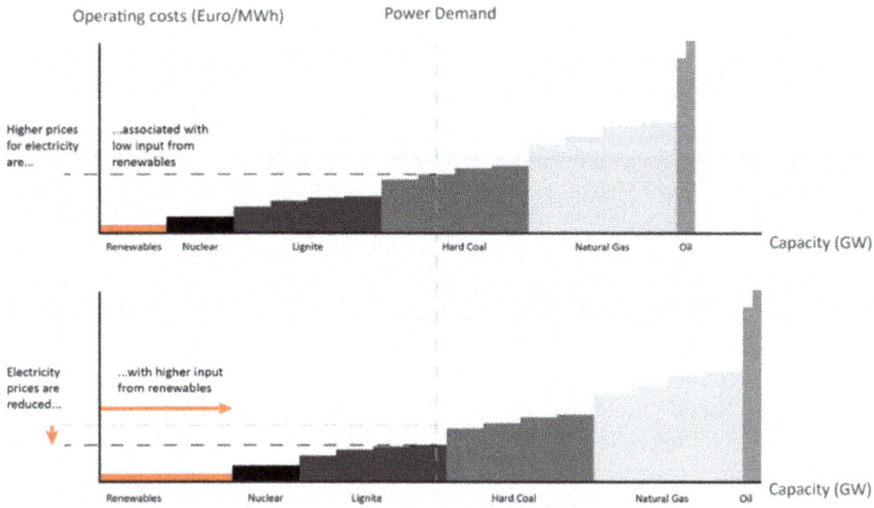

Fig. 1 Merit-order effects in energy-only markets. *Source* Appun (2015). https://www.cleanenergywire.org/factsheets/setting-power-price-merit-order-effect

95% of Singapore's electricity. Wholesale electricity prices spiked in Singapore in 2021, causing some retailers to exit the market. As electricity retailers have locked-in contracts with consumers while insufficiently hedging against such wholesale price spikes, the natural gas price increase further drove up wholesale electricity prices, crippling the market. South Australia experienced a 13-day market suspension in late 2016, the second such event since the national electricity market (NEM) commenced operations in 1998. Both markets faced higher volatility before Russia's invasion of Ukraine due to higher global liquefied natural gas (LNG) prices driven by gas prices more than doubling between 2015–2017, while coal prices also nearly doubled in the same period. Competition introduced to the Philippines' electricity markets failed to eliminate underlying system inefficiencies (Bacon 2019). In addition, the Philippines has some of the highest power tariffs in Asia, averaging between USD 0.18–USD 0.20 per kWh in its main grid, due to such factors as domestic taxes, regulatory incentives focused exclusively on generation capacity, and reliance on imported fuel (Ahmed 2020). Electricity prices in the liberalized markets of the Philippines and Singapore are amongst the highest in Southeast Asia (Ali et al. 2022).

Liberalized markets responded differently to the aforementioned skyrocketing global natural gas prices in 2022. In Australia, the NEM was suspended in June 2022 for about a week. In Singapore, the energy market authority (EMA) tightened licensing requirements for electricity retailers to protect consumers against provider failure risk as some retailers exited the market, as mentioned above. A novel hedge-based energy market model that allows renewable generators to secure hedge contracts from flexible generating technologies as insurance against weather-driven energy deficits is proposed in Johnathon et al. (2023). Globally, electricity markets are now more regulated than before as a consequence of Russia's war with

Ukraine (Nepal and Jamasb 2022), demanding electricity market design capable of facilitating greater integration of renewables into energy markets to deliver secure, sustainable, and affordable electricity in liberalized wholesale electricity markets in ASEAN. The idea is crucial, as ASEAN is one of the few regions of the world where coal-fired power has been expanding, suggesting that reducing its share thereof in its electricity mix remains essential to achieving a sustainable future (Ali et al. 2022). Singapore, accordingly, has set a target of deriving 5% of its peak electricity demand from renewables by 2020 and 4% of its total electricity from renewables by 2030, while the Philippines has a target of 15,234.3 MW renewables capacity, also by 2030. This chapter accordingly reviews the characteristics of liberalized electricity markets in Singapore and Philippines, and points out lessons learned from electricity market designs in Australia's eastern and western jurisdictions and the United Kingdom that could help facilitate greater renewables development and deployment in ASEAN while also liberalizing ASEAN's electricity markets.

Section 2 provides a comparative review of electricity markets in Singapore and the Philippines. Section 3 reviews the designs of Eastern Australia's NEM, the wholesale electricity markets of Western Australia and the UK. Section 4 addresses public policy. Section 5 concludes the chapter.

2 Wholesale Electricity Markets in ASEAN

The primary motives for the ASEAN economies implementing reform initiatives in their electricity sectors during the mid-to-late 1990s were improved productivity and attracting private sector investment (Sharma 2005). According to Phoumin et al. (2022), Singapore became the first country in Southeast Asia to launch a competitive electricity market in 2001, which the Philippines emulated in 2006. South Korea introduced competition in electricity generation in 2001. Malaysia amended an electricity reform law in 2001 and established an electricity regulator in 2002. Myanmar established an energy regulator in 1996 while enacting an electricity reform law in 2014. The Philippines and Vietnam enacted electricity reform laws and established regulators in 2001. ASEAN member states such as Malaysia and Vietnam aspire to establish competitive wholesale electricity markets. Nepal et al. (2022) investigated the socioeconomic impacts of power sector reforms, accounting for cross-sectional dependence in the 18 non-OECD Asian economies, a classification which includes all ASEAN member states. The ASEAN power sector reforms have helped improve economic outcomes, including social welfare, and reduce network losses, resulting in greater operational efficiency.

Many ASEAN member states have state-run vertically integrated electricity markets with a single-buyer model that includes participation by independent power producers (IPPs), as well as corporatized electricity sectors and regulators. Open and third-party access and distribution privatization remain limited, however. As mentioned above, only Singapore and the Philippines have fully liberalized electricity markets, discussed in detail hereinafter. Malaysia will be the third ASEAN member state with a fully liberalized electricity supply industry (ESI), following a second round of reforms which were announced in September 2018. The study by Aris et al. (2022) assess the performance of ASEAN's liberalized electricity markets, with emphasis on Singapore and the Philippines, with the study finding highlighting that in neither country has liberalization led to reductions in CO_2 emissions. Liberalization has, however, increased Singapore's renewable electricity generation share, though not that of the Philippines.

The Phillipines has greater installed generating capacity, approx. 20,000 MW in 2018, than Singapore, approx 13,650 MW in 2018. Singapore has achieved universal electricity access while the Philippines had 93% electricity access as of 2018. Singapore has no hydroelectric resources, implying dim prospects for tidal or wind energy due to low speeds thereof. Singapore generate only a small fraction of renewable electricity from biomass and solid waste, although it is exploring its geothermal potential. By contrast, the Philippines has extensive hydroelectric, geothermal, onshore wind, and solar power, even though coal-fired power is still the greater part of its existing energy mix, accounting for 57% in 2020, compared with approximately 21% for renewables. The Philippines has set the aforementioned target of 15,234.3 MW from renewables by 2030 (Ahmed et al. 2017).

2.1 Singapore

As previously mentioned, Singapore created the first liberalized electricity market in ASEAN (Ali et al. 2022). Its major characteristics include open access to transmission and distribution networks and vertical separation of contestable, i.e., competitive, and non-contestable, i.e., monopoly, market segments, including a wholesale bidding market and full retail competition. The retail market has progressively opened up to competition since 2001, with access extended to all consumers since November 2018.

Singapore commenced power sector reform in 1995 with the corporatization of the public utilities board (PUB) that had managed its gas and electricity sectors. The Singapore power (SP) holding company was created to stimulate electricity competition, holding stakes in all ESI segments: electricity generation, through Power Senoko and Power Seraya; electricity transmission and distribution, through Power Grid; and retail electricity supply, through Power Supply Limited. Temasek Holdings, another SP holding company, ran Tuas Power, an independent power generating company.

The Singapore electricity pool (SEP), a competitive wholesale market facilitating wholesale electricity trading, was launched in 1998. EMA is an independent regulatory body, was established in 2001 with the objective of delivering reliable and affordable electricity, in hopes of fulfiling expectations of more profound electricity liberalization. The energy market company (EMC), an EMA subsidiary, operates the national electricity market of Singapore (NEMS), which facilitates competitive wholesale and retail electricity trading. Being the market operator, NEMS matches

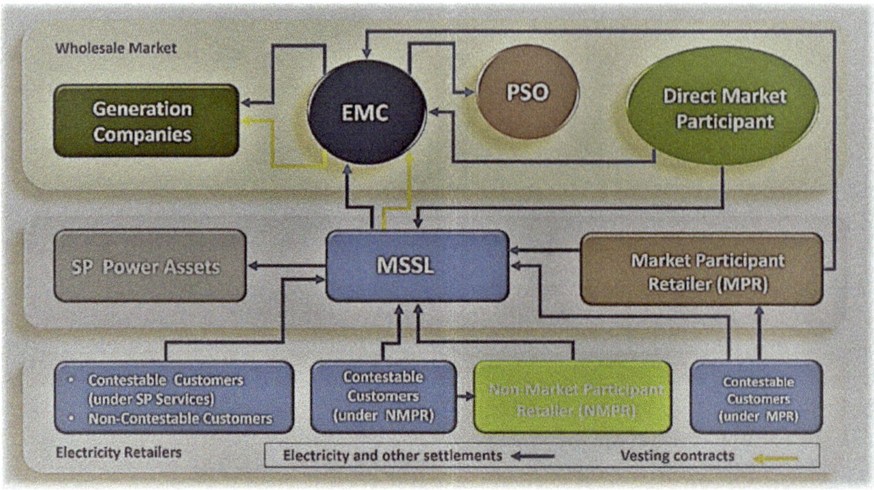

Fig. 2 NEMS governance in vertically unbundled Singapore ESI. *Source* Adapted from EMA (2010), Aris et al. (2022), Ali et al. (2022)

electricity demand and supply at 30-min intervals to determine the wholesale electricity rates at which electric companies are compensated. EMA also ensures secure power system operation. The half-hourly spot price determines (a) the dispatch quantity that each plant produces, (b) the reserve and regulation capacity that each plant is required to maintain, and (c) the corresponding wholesale spot market prices for energy, reserve, and regulation (EMA, 2010). Figure 2 shows the governance of NEMS and its chief stakeholders, which are also key market participants.

It was believed that deregulation would bring about lower electricity costs due to the various efficiency gains possible, with lower-bound estimates of as much as 8% of production costs in cost gains (Chang and Tay 2006). Chang (2007) showed that the generation market is fairly competitive and not much room remains to exercise market power given a low lerner index post-liberalization value, even though the NEMS generation market appears highly concentrated. As mentioned above, Singapore began opening its retail market in 2001, with consumers able to buy electricity either from chosen retailers or directly from the wholesale market at the half-hourly wholesale electricity rates. The open electricity market (OEM) was launched in April 2018, allowing Jurong area consumers, residential and commercial alike, to buy electricity from the retailer of their choice. It is anticipated that this market initiative will be gradually extended to all Singaporean consumers.

2.2 The Phillipines

The Phillipines embarked on market-oriented electricity reform in 2001 with the enactment of the electric power industry reform Act (EPIRA), inaugurating electricity restructuring and privatization. It facilitated vertically separating the previously vertically integrated national power corporation (NPC) by creating the national transmission company (TransCo.) beginning in 2003. The NPC assets were privatized while an independent energy regulator called the energy regulatory commission (ERC) was established. EPRIA is thus considered the most comprehensive legislative initiative in the Philippine power sector. Phillipine power sector reforms occurred in the context of broader macroeconomic reforms following the 1997 Asian financial crisis and high electricity prices providing sufficient incentive for said restructuring.

EPRIA also faciliated establishing the wholesale electricity spot market (WESM), a centralized platform for buyers and sellers to engage in spot electricity trading, where prices are determined based on demand, i.e., actual usage, and supply, i.e., availability. While officially commencing operations in March 2004, WESM has undergone a series of reforms since the Philippine electricity market corporation (PEMC), which is the autonomous group market operator and governing body, incorporated as a non-stock, non-profit corporation registered with the securities and exchange commission (SEC). WESM started commercial operation in Luzon in June 2006, while the Visayas grid was integrated into the WESM and began commercial operation therein in December 2010. The retail competition and open access (RCOA) arrangement was implemented in June 2013. PEMC instigated central scheduling and dispatch of energy, and contracted reserves in December 2015. In January 2016, preferential dispatch for renewable energy resources was integrated into the WESM, while WESM Mindanao was launched in June 2017. Figure 3 shows the governance of WESM and its chief stakeholders, which are also key market participants.

Control of WESM passed to the independent electricity market operator of the Philippines (IEMOP) in September 2018. Electricity customers with average monthly peak demand of at least 750 kW (down from the 1 MW threshold implemented in 2013) were also allowed to enter agreements for retail electricity supply. The RCOA threshold was further lowered to 500–749 kW level in February 2021 (Ali et al. 2022). The wholesale electricity spot price is used to settle traded quantities net of bilateral contracts, i.e., quantities not covered by bilateral contracts, in the Philippine wholesale electricity market (Rudnick and Velasquez 2019).

While the Phillipines has succeeded in implementing planned reform steps which were completed by 2013 (Bacon 2019), changes in market concentration after reforms did not significantly reduce high electricity prices in its wholesale electricity market (Poquiz 2015). Foster and Rana (2020) argued that despite the Philippines being an aggressive reformer in Southeast Asia, it actually lags behind in transmission network planning. Sharma et al. (2004) argued that the Phillipines should concentrate on competitive sourcing of new generation capacity and public ownership of transmission and distribution networks while regulating retail electricity supply, rather than pursuing electricity industry privatization outright.

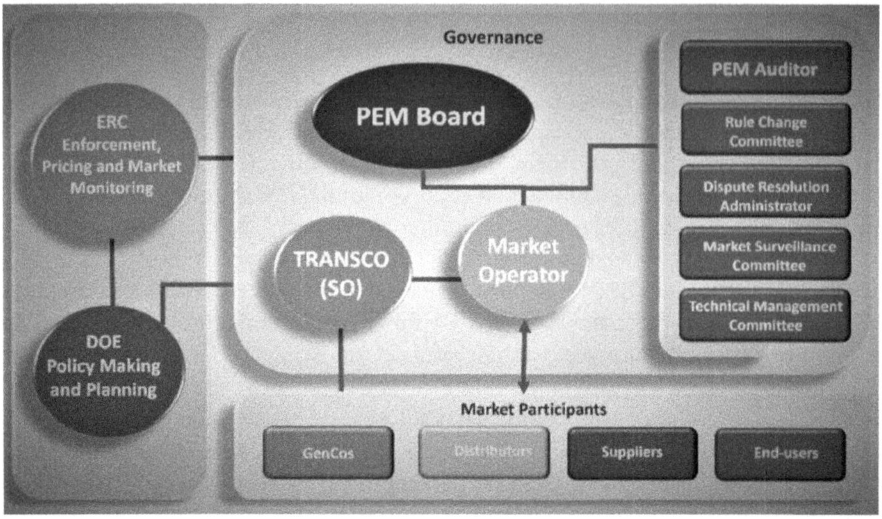

Fig. 3 WESM governance in vertically unbundled Philippine ESI. *Source* Adapted from Samantha (2019)

Table 1 shows the timeline of reforms in the Singaporean and Philippine electricity markets. In liberalized markets, the wholesale electricity market performs as a 'pool' under a centrally coordinated dispatch process, as electricity cannot be easily stored while supply needs to be matched instantaneously with demand. Luzon (which includes Manila), Visayas, and Mindanao are the three main islands of the Philippines, with Luzon and Visayas being physically interconnected (Rudnick and Velasquez 2019).

3 Review of Case Studies

In this section, we review the wholesale market arrangements in the eastern and western jurisdictions of Australia as well as the UK. Australia's case history is particularly relevant to ASEAN because the Sydney declaration of the Australian-ASEAN Summit 2018 provided a roadmap for expanding trade to renewable energy (Do and Burke 2022), and there is also a private-sector proposal to develop an Australia-Asia PowerLink connection to Singapore via Indonesian waters (Sun Cable 2021). Australia is also ASEAN's largest trading partner, and this relatively stable trade relationship is likely to grow through more renewable energy trade, given Australia's abundant solar and wind resources. Australia also operates different wholesale electricity markets in its eastern jurisdictions, which is an energy-only market, and its western regions, which is an energy capacity market, offering valuable insights into the performance of varying markets where the share of renewables is growing. The

Table 1 Timeline of wholesale electricity market reforms in the Phillipines and Singapore

	Singapore		Phillipines
1995	SP assumes responsibility for electricity and piped gas functions from PUB	2001	EPIRA enacted; ERC and Transco created
1998	SEP launched	2002	WESM rules promulgated
2001	EMA Act and Electricity Act enacted, establishing EMA	2003	PEMC incorporated
2002	NEMS established, replacing SEP	2004	PEMC designated WESM autonomous group market operator (AGMO)
2003	NEMS opened for contestable consumers with more than 20,000 kWh average monthly consumption	2006	WESM started commercial operation in Luzon grid
2006	Contestable consumers redefined as having more than 10,000 kWhaverage monthly consumption	2010	Visyas grid incorporated into WESM
2009	Temasik holdings divests three power companies	2013	RCOA implemented for electricity customers with at least 1 MW monthly peak demand
2010	Private electricity generators commence trading	2016	Contestable consumers with at least 750 kWh monthly peak demand are mandated to sign contracts with electricity retailers
2014	Contestable consumers redefined as having more than 4000 kWh average monthly consumption	2017	WESM Mindanao launched
2015	Contestable consumers redefined as having more than 2000 kWh average monthly consumption		
2018	Commencement of open electricity market extension to all consumers in stages		

UK, on the other hand, operates a contracts wholesale market, day-ahead spot market, and capacity market. In 2014, Nord Pool spot took ownership of the UK wholesale market, i.e., the day-ahead market, and the power exchange (Nord Pool in the UK). The UK also has one national wholesale electricity price at any given moment, unlike the eastern jurisdictions of Australia, which has five regional electricity prices determined every five minutes, i.e., zonal pricing. The UK and Australia have also comprehensively reformed their power sectors, including vertical separation, competition in generation and retail, incentive regulation of transmission and distribution networks, independent regulation, and privatization.

3.1 The National Electricity Market

The national electricity market (NEM) in the eastern jurisdictions of Australia commenced operations as a wholesale electricity spot market in December 1998, comprising five interconnected states that also act as price regions: Queensland (QLD), New South Wales (NSW) (including the Australian Capital Territory (ACT)), South Australia (SA), Victoria (VIC), and Tasmania (TAS). While one objective for it was to integrate these regional markets into one (Nepal and Foster 2016), the NEM integration process is ongoing, owing to regional electricity rate differences arising from such factors as underlying network constraints, lack of adequate interconnector capacity, extreme weather and other events, and regulatory sanctions (Do et al. 2020a, b; Naeem et al. 2022).

Generators offer supply bids with details specifying amounts of electricity offered at specified prices for set time periods, and may re-submit amounts offered at any time (AEMO 2021). The Australian energy market operator (AEMO) then deploys generators to produce electricity, with the cheapest generator put into operation first, suggesting that demand, i.e., consumption, is met in the most cost-efficient way, i.e., the merit-order dispatch. The dispatch price for wholesale electricity delivery is determined every 5 minutes beginning October 1, 2021, as opposed to the prior 30-minutes (by means of aggregating 6 prices determined at every 5 minutes) wholesale electricity spot market settlement. The objective of the 5-minutes price settlement period is to provide a better price signal for investment in peaking generation technologies such as batteries and gas peaking generators (AEMO 2021). In fiscal year 2020–21, renewable generation as a percentage of total generation was as follows: wind, 10.45%; hydroelectric, 7.21%; grid-scale solar, 3.85%; distributed photovoltaic (PV), 7.09%; battery energy storage systems, 0.05%; and biomass, 0.09%. According to AEMO (2021), there could be sufficient renewable resources available to meet 100% of baseline consumer demand during certain periods in the NEM by 2025. NEM also has approximately 14 GW of distributed solar, as of December 2021, and is now the largest generator in the region. As the NEM continues to retire fossil-fuel plants, approximately 60 GW of new grid-scale renewables will be required by 2040 to take their place, as per the 2020 integrated system plan (ISP) forecasts released by AEMO.

As mentioned, NEM is an energy-only market, which implies inevitable price volatility. The maximum spot price, also known as the maximum price cap, is set at $15,000/MWh, while the market price floor is − $1000/MWh. AEMO recovers costs from customers to pay generators, as most customers purchase their electricity through a retailer, rather than participating directly in the NEM. Retail electricity prices are still regulated in Victoria, the Australian Capital Territory, Tasmania, and regional Queensland, however. Financial risks arising from said price volatility during trading periods are managed through financial derivatives, including swaps or hedges, options, and futures contracts. NEM also experienced a 13-day market suspension in South Australia during late 2016, the second such suspension since NEM commenced operations in 1998. It originated from severe weather conditions that damaged transmission and distribution assets, followed by decreased wind farm output and a loss of synchronism distrupting the Heywood interconnector, which interlinks South Australia and Victoria, resulting in a supply–demand mismatch in SA (AER 2018).

As per the 2020 ISP, renewable energy zones (REZs) will meet NEM's need for large-scale renewable generation, as these areas offer high-quality renewable energy resources. Some 14 REZs were announced in 2020 by the state governments, including five in New South Wales, six in Victoria, and three in Queensland. Connecting new renewable energy resources to the grid by implementing the REZ reform framework remains a challenge, however, necessitating investments in transmission network expansions and augmentations. Simshauser (2021) explored funding REZs through a consumer-funded regulatory model as well as a renewable generator-funded market model. The Australian government also has a renewable energy target (RET) policy scheme, which encourages renewables to make electricity generation less emissions-intensive. It offer large-scale generation certificates for large scale power generators and small-scale technology certificates to owners of small-scale systems for every MWh of power generated, which electricity retailers purchase. The retailers submit their credits to the clean energy regulator to meet RET legal obligations (Australian Government 2022). NEM also has declining feed-in tariffs (FITs) for renewable energy, which pays for excess electricity that small-scale solar PV or wind power systems generate and sell back to the grid.

3.2 *Western Australia Electricity Market*

The wholesale electricity market (WEM) of Western Australia commenced operations in 2006, supplying electricity to over 1.1 million Western Australian households and businesses annually in the south west interconnected system (SWIS). WEM has such objectives as providing economically efficient, safe, and reliable electricity, as well as encouraging competition among SWIS generators and retailers. Much work remains to achieve these objectives, however, as outlined in Simshauser and Wild (2009) and Khezr and Nepal (2021).

Retailers buy electricity from generators in this wholesale market and sell to consumers. Generators comprise scheduled generators, including baseload coal- and gas-fired generators and intermittent wind and solar generators, with a total capacity of 6 GW. WEM market customers are retailers, large scale consumers, and demand-side participants. The latter are a load or group of loads which reduce consumption, thereby satisfying supplemental generation requirements (AEMO 2022). As of February 2022, the annual renewables mix as a percentage of total generation was as follows: wind, 19.2%; and grid solar, 2.1%. The largest SWIS generator is rooftop solar, with installations on one in three households, and it is anticipated that commercial and residential solar will provide 45% of total expected generation capacity by 2030–2031 (AEMO 2022). Western Power is the network operator responsible for working with plant owners to maintain secure and reliable power system operations.

WEM is distinguished from NEM in that WEM design comprises a wholesale electricity trading component and a capacity component. The reserve capacity mechanism (RCM), which is ultimately funded by the market customers, ensures that sufficient generation is available to meet demand during peak periods. AEMO, the market operator, thus has to operate and settle the RCM and also buy and sell electricity from the following markets under wholesale arrangements: the short term energy market, the load following ancillary service (LFAS) market, and the balancing market. WEM is a relatively new capacity-energy market where electricity providers are remunerated for making capacity available, alongside a wholesale market where market participants interact to supply and purchase electricity on a half-hourly basis. Simshauser and Wild (2009) have argued that WEM market regulation succeeds in eliminating the possibility of infra-marginal rents events while keeping the administratively determined price cap low as well.

The higher uptake of rooftop solar in WA is associated with such government financial incentives as FITs and falling PV system prices. Ma et al. (2016) found the PV systems' capitalization effect is an estimated 2.3–3.2% property value premium associated with PV systems, allowing homeowners to fully recover the costs of PV investments when properties are sold. Nor should the 'warm glow' aspects of these installations be ruled out (Ma and Burton 2016). The distributed energy buyback scheme (DEBS) replaced the renewable energy buyback scheme (REBS) on September 8, 2020 for the installation of new and upgraded distributed energy resources in WEM. DEBS offers eligible customers a time of export payment, distinguishing between peak versus off-peak rates, for electricity exported to the grid from such distributed sources as rooftop solar PV systems, batteries, and electric vehicles. FIT recipients may also qualify for DEBS by upgrading their systems. WA's large coal and gas reserves have delayed the transition to renewables, however, despite the state having some of the best solar and wind resources in the world. Coal and gas accounted for the largest shares of the WEM generation mix as of February 2022, at 44.7% and 33.5% respectively. WEM generators were also unable to meet the large–scale renewable energy target (LRET) of 23.5% of wholesale electricity purchases from renewables in 2020.

3.3 The UK Wholesale Electricity Market

The UK electricity market consists of generators and suppliers, with the latter purchasing electricity from the former at wholesale prices. Therefore, the UK wholesale electricity market operates under a self-dispatch system where buyers, i.e., suppliers, and sellers, i.e., generators, of electricity enter into contracts ahead of time for anticipated demand at prices either bilaterally negotiated, i.e., under forward/future contracts, or determined through demand and supply matching on public exchanges such as spot markets, i.e., the day-ahead spot market (Liu et al. 2022). This arrangement arose in 2005 when the new electricity trading arrangements (NETA) changed its name to the British electricity trading transmission arrangements

(BETTA), paving the way for establishing a single market for electricity of England, Wales and, Scotland, i.e., Great Britain (less Northern Ireland). The BETTA reforms replaced the 'pool' arrangements, which were based on a centrally dispatched process where the system operator determined the least cost way of clearing the market by matching supply and demand and eventually communicating a planned running order to each participant. The contracts, which may now be signed years ahead of fulfillment in the forward Market, include forward contracts, future contracts, and options (Liu et al. 2022).[2] The spot market is also used for buying and selling of electricity ahead of realtime, with primary contract types being day ahead auctions and intraday trading, including half-hourly trading. Nordpool and Epexspot operated spot electricity trading prior to 2014, whereupon Nord Pool Spot took over the UK wholesale market and power exchange (Nord Pool in the UK).

The UK currently has a capacity market which was introduced per the electricity market reform (EMR) policy of 2013 under the Energy Act of 2013. Its purpose is to ensure electricity supply security by providing payments for reliable electricity sources. There are thus large incentives for new investments in capacity as well as making existing capacity available, ensuring sufficient supply to meet demand at all times. The EMR also replaced the Renewables Obligations System with a "contracts for differences" (CfDs) scheme in October 2014. CfDs are awarded for 15-year terms, so as to support deployment of large-scale renewable projects bigger than 5 MW. Eligible technologies include onshore and offshore wind, solar PV, geothermal, hydropower, ocean power (tidal and wave), landfill gas, sewage gas, anaerobic digestion, biogas, biomass, and CHP plants (IEA 2019). The mechanism is based on the difference between an agreed strike price and the market price. If the market price exceeds the strike price, the renewable provider would be required to pay the difference to the CfD counterparty, whereas if the strike price exceeds the market price, would mean the CfD counterparty would be required to pay the difference to the renewable provider. CfDs thus provide incentives for renewables investment by offering renewable project developers, facing high upfront costs and long lifetimes, direct protection from volatile wholesale prices (UK Government 2022), while also protecting consumers from increased support costs for high electricity rates.

The UK wholesale electricity market has experienced 'the decline of coal' and the 'ascent of wind' in its electricity generation mix (National Grid 2022). Great Britain went a full day without any power generation from coal on April 21, 2015, while the market recorded a week without power generation from coal May 1–8, 2019. British wind farms averaged a record 20.90GW on November 2, 2022. In 2021, fossil fuels (coal, oil and gas) accounted for 43% of total generation, followed by renewables (solar, wind, hydroelectric) at 34.2%, and other sources (nuclear, biomass, transfers and storage, and others) at 23.7%.

Table 2 summarizes the major wholesale market design characteristics of the markets examined in this chapter. They vary in pricing models, dispatch processes,

---

[2] As defined in the study by Liu et al. (2022) study, forward contracts specify pre-contract tariffs and delivery schedules, while future contracts allow trading of contracts. Options are rights to buy and sell electricity during specific periods at specified tariffs, and may also be traded.

Table 2 Major wholesale market design characteristics[3]

Market feature/ market	NEM (South-Eastern Australia)	WEM (Western Australia)	UK	NEMS (Singapore)	WESM (Phillipines)
Wholesale electricity pricing model	Zonal pricing	Zonal pricing (the market being one zone in itself)	National pricing	Nodal pricing	Nodal pricing
Wholesale market rules: self-dispatch versus centralised dispatch	Bid-based centralised dispatch (pool)	Bid-based centralised dispatch (pool)	Self-disptach	Bid-based centralised dispatch (pool)	Bid-based centralised dispatch (pool)
Spot market settlement period	Every 5 min	Half-hourly	Half-hourly	Half-hourly	Hourly nodal prices
Energy only markets versus energy capacity market	Energy-only market (bilateral contracts; spot market)	Energy-capacity markets (bilateral contracts; spot market, capacity market)	Energy-capacity market (contracts market, spot market and capacity market)	Energy-only market (bilateral contracts such as vesting contracts and spot market)	Energy-only bid-based power pool (bilateral contracts and spot market)
Support for renewable energy	FiTs; REZs	FiTs	CFDs	No subsidies; streamlining registration for self-consumption from renewables; implementing regulatory sandbox framework[4]	Net metering; FiTs; preferential dispatch for renewable electricity in wholesale spot market
Gross versus net pool	Gross pool	Gross pool	Net pool	Gross pool	Net pool

market settlement periods, and capacity markets. The Great Britain (GB) and WEM markets have capacity markets to avoid the 'missing money' problem that results from price caps in energy-only markets such as NEM and Singapore limiting rents on scarce capacity in peak periods, reducing incentives to invest. It occurs when prices for electricity in competitive wholesale electricity markets fail to adequately reflect the value of investment in the resources needed for reliable electricity supply (Hogan 2017).

[3] Please refer to the IEA/IRENA policies database to learn more about the 'in-force' and 'ended' policies.

[4] This framework allows regulations to be relaxed in a sandbox capable of accommodating new products and services for testing within defined parameters.

4 Policy Lessons and Discussions

The case studies of Australia and the UK'S wholesale electricity market design characteristics together with renewable energy development therein provide important lessons for ASEAN in rethinking electricity market design to accommodate the rising penetration of renewables and wholesale markets.

First, renewables development in wholesale electricity markets can be pursued regardless of whether the market is physically interconnected to another cross-border market as in the UK, which is connected to Ireland through the East–West Interconnector, a 500 MW high-voltage DC submarine and subsoil power cable), or isolated, i.e., no cross-border connection, as in Australia. Notwithstanding, the need for adequate interconnector capacity and the mitigation of underlying network constraints thereby is an enabler rather than a barrier for integrating renewable energy in wholesale electricity markets (Nepal and Foster 2016; Do et al. 2020b). Do et al. (2020b) have further proposed supporting renewable generation by setting an appropriate carbon price in interconnected wholesale markets. Removing such barriers as transmission access and energy curtailment alongside transmission expansion in planning can accelerate renewable energy development, as countries such as the Philippines strive to benefit from competitive renewable energy zones (CREZ) (Lee et al. 2020).

Second, relying on incentives of profiting from the least-cost dispatch mechanism in wholesale electricity markets will be insufficient to drive adequate renewable energy development in liberalized markets. NEM has identified REZs for large-scale grid-based renewable development, and supports decentralized renewable energy adoption through FiTs and other subsidies. The GB electricity market, on the other hand, has established the CFDs scheme to support renewable energy. In ASEAN, Singapore has no subsidies for renewables, while the Philippines has implemented net metering, FiTs, and preferential dispatch for renewable electricity in its wholesale spot market. Other ASEAN member states, instead operating primarily single-buyer models in the absence of full-fledged competitive wholesale electricity markets, should follow the examples of Singapore and the Philippines in supporting renewable energy development through such policies as the foregoing.

Third, those ASEAN member states aiming to establish competitive wholesale electricity markets, such as Malaysia and Vietnam, should take care to accommodate wholesale electricity market design features conducive to renewable energy development. The experience of the Philippines and Singapore suggest that nodal pricing, or locational marginal pricing (LMP), of wholesale electricity is effective, as LMP divides national networks into hundreds or even thousands of nodes, each with their own unique wholesale prices. Price signals such as nodal pricing thus provides are truly cost-reflective, such that wholesale prices typically vary from node to node in each trading period thereby providing level playing field for distributed generators (Agalgaonkar et al. 2004). The current underlying institutional governance of ASEAN electricity markets also suggest that other ASEAN economies should embrace a bid-based wholesale power pool based on a centralized dispatch as in

Singapore, the Philippines, and Australia, rather than self-dispatch as in the UK. Capacity markets as in the UK and Western Australia can effectively support renewable energy development and avoid the 'missing money problem' if there are not already adequate and targeted arrangements to support renewable energy outside of the wholesale market. Countries like the Philippines could also benefit from introducing shorter market settlement and dispatch periods such as half-hourly rather than hourly, or even every 5 minutes as in the NEM, to provide better price signals for investment in peaking generation technologies.

Fourth, the crisis in the gas markets arising from Russia's invasion of Ukraine, and the resulting negative supply shocks, have exposed the fragilities of liberalized electricity markets relying on natural gas as peaking plants and determining wholesale electricity prices under the merit-order dispatch mechanism. Decoupling electricity prices from natural gas prices is essential (Maurer et al. 2022). Greece considered wholesale market splitting, where a mandatory pool for low-variable cost technologies, including wind and solar, but also nuclear, run-of-river hydro, and fossil fuel cogeneration will be established. Under this scheme, electricity provided will be remunerated via CFDs arrangement based on their full costs. A second wholesale market would then be established per convention for such other providers as selected fossil fuel condensing plants. ASEAN economies pursuing competitive wholesale electricity markets have the opportunity to consider these market designs to provide better price signals for renewable energy technologies from the outset.

5 Conclusions

This chapter reviewed the design features and performance of liberalized bid-based wholesale electricity markets operating under a merit-order dispatch mechanism while needing to decarbonize through large-scale renewable energy integration. We informed our understanding of such liberalized electricity markets in ASEAN as the Philippines and Singapore by undertaking a comparative case study review of the wholesale electricity markets operating in eastern and western Australia, as well as the UK. Important policy implications were drawn for ASEAN, where other member states including Malaysia and Vietnam are aiming to establish competitive wholesale electricity markets and increase the share of large-scale renewable energy in their power generation mixes.

Almost all ASEAN member states have set renewable energy targets as well as net zero emissions targets, providing a level starting point for large-scale renewable energy development. We conclude, however, that such development in wholesale electricity markets requires aligning national energy policy with climate policy. Australia has recently agreed to blend emissions reduction into its national energy objective as the government aims to reduce greenhouse gas emissions by 43% below 2005 levels by 2030, as well as achieve net-zero emissions by 2050. The national energy objective guides rule-making and other energy policy decisions concerning electricity and gas generation and transmission, as well as retail energy, and therefore allows the government to undertake a custom approach to doubling Australia's renewables capacity every decade toward meeting said net zero emissions target. Such large-scale renewable energy development in wholesale electricity markets also requires balanced public- and private-sector participation from the outset.

References

AEMO (2021) The national electricity market—fact sheet, December. Available at https://aemo.com.au/-/media/files/electricity/nem/national-electricity-market-fact-sheet.pdf

AEMO (2022) The wholesale electricity market—fact sheet, March. Available at https://aemo.com.au/-/media/files/electricity/wem/wholesale-electricity-market-fact-sheet.pdf?la=en&hash=ED1512DAF6230ABBA3008B1954AB46A5

AER (2018) The black system event compliance report, Australian Energy Regulator, December. Available at https://www.aer.gov.au/news-release/aer-releases-investigation-report-into-south-australias-2016-state-wide-blackout

Agalgaonkar AP, Kulkarni SV, Khaparde SA, Soman SA (2004) Placement and penetration of distributed generation under standard market design. Int J Emerg Electr Power Syst 1(1)

Ahmed SJ (2020) Philippines power sector can reach resilience by 2021, Institute for Energy Economics and Financial Analysis, June. Available at https://ieefa.org/wp-content/uploads/2020/06/Philippines-Power-Sector-Can-Reach-Resilience-by-2021_June-2020.pdf

Ahmed T, Mekhilef S, Shah R, Nadarajah M, Seyedmahmoudian M, Horan B (2017) ASEAN power grid: a secure transmission infrastructure for clean and sustainable energy for South-East Asia. Renew Sustain Energy Rev 67:1420–1435

Ali H, Phoumin H, Suryadi B, Farooque AA, Yaqub R (2022) Assessing ASEAN's liberalized electricity markets: the case of Singapore and the Philippines. Sustainability 14:11307

Appun K (2015) Setting the power price: the merit order effect. Clean Energy Wire—Journalism for the Energy Transition, January 23, Available at https://www.cleanenergywire.org/factsheets/setting-power-price-merit-order-effect

Aris H, Mohd Zawawi IS, Jørgensen BN (2022) The Philippines' and Singapore's journeys towards liberalised electricity supply industries—takeaways for Malaysia. Energies 13(14):3514

Australian Government (2022) Large-scale renewable energy target market data, renewable energy target, clean energy regulator. Available at https://www.cleanenergyregulator.gov.au/RET/About-the-Renewable-Energy-Target/Large-scale-Renewable-Energy-Target-market-data

Bacon R (2019) Learning from power sector reform: the case of the Philippines. Policy Research Working Paper No. 8853. World Bank, Washington, DC

Chang Y (2007) The new electricity market of Singapore: regulatory, framework market power and competition. Energy Policy 35(1):403–412

Chang Y, Tay TH (2006) Efficiency and deregulation of the electricity market in Singapore. Energy Policy 34(16):2498–2508

Do TN, Burke P (2022) Is ASEAN ready to move to multilateral cross-border electricity trade? Asia Pacific Viewpoint. https://doi.org/10.1111/apv.12343

Do H, Nepal R, Smyth R (2020a) Interconnectedness in the Australian national electricity market: a higher-moment analysis. Econ Rec 96(315):450–469

Do H, Nepal R, Jamasb T (2020b) Electricity market integration, decarbonisation and security of supply: dynamic volatility connectedness in the Irish and Great Britain Markets. Energy Econ 92

EMA (2010) Introduction to the national electricity market of Singapore; technical report; Energy Market Authority: Singapore. Available at https://www.ema.gov.sg/cmsmedia/handbook/nems_111010.pdf

Foster V, Rana A (2020) Rethinking power sector reform in the developing world. Sustainable Infrastructure. World Bank, Washington, DC

Hogan M (2017) Follow the missing money: ensuring reliability at least cost to consumers in the transition to a low-carbon power system. Electr J 30(1):55–61

IEA (2019) Contracts for difference (CfD), IEA/IRENA renewables policies database, 25 October, Retrieved 16 Nov 2022, from Website https://www.iea.org/policies/5731-contract-for-difference-cfd

Johnathon C, Agalgaonkar AP, Kennedy J, Planiden C (2021) Analyzing electricity markets with increasing penetration of large-scale renewable power generation. Energies 14(22)

Johnathon C, Agalgaonkar AP, Planiden C, Kennedy J (2023) A proposed hedge-based energy market model to manage renewable intermittency. Renew Energy 207:376–384

Khezr P, Nepal R (2021) On the viability of energy-capacity markets under decreasing marginal costs. Energy Econ 96

Lee N, Dyreson A, Hurlbut D, McCan M, Isabel N, Edward V, Reyes NCR, Capongcol MC, Cubangbang HM, Agustin B, Paul Q, Bagsik J, Leisch J (2020) Ready for renewables: grid planning and competitive renewable energy zones (CREZ) in the Philippines. United States.https://doi.org/10.2172/1665863

Liu J, Wang J, Cardinal J (2022) Evolution and reform of UK electricity market. Renew Sustain Energy Rev 161

Ma C, Burton M (2016) Warm glow from green power: evidence from Australian electricity consumers. J Environ Econ Manag 78(C):106–120

Ma C, Polyakov M, Pandit R (2016) Capitalisation of residential solar photovoltaic systems in Western Australia. Aust J Agric Resour Econ 60(3)

McRae SD (2019) Market power in cost-based wholesale electricity markets: evidence from Mexico, December, American Economic Association 2020 conference, Available at https://www.aeaweb.org/conference/2020/preliminary/paper/r5H6kA6i

Maurer C, Schlecht I, Hirth L (2022) The Greek market design proposal would be the end of electricity markets as we know them, July 28. Available at https://www.euractiv.com/section/electricity/opinion/the-greek-market-design-proposal-would-be-the-end-of-electricity-markets-as-we-know-them/

Naeem M, Karim S, Rabbani MR, Nepal R, Uddin GS (2022) Market integration in the Australian National Electricity Market: fresh evidence from asymmetric time-frequency connectedness. Energy Econ 112(C)

National Grid (2022) National grid: live, data. Retrieved 16 Nov 2022, from https://grid.iamkate.com/

Nepal R, Foster J (2016) Testing for market integration in the Australian national electricity market. Energy J 37(4):215–237

Nepal R, Jamasb T (2022) Global electricity reforms—what/where next? International Association for Energy Economics (IAEE) Webinars, 12 Oct. Available at https://www.iaee.org/en/webinars/webinar_jamasb.aspx

Nepal R, Phoumin H, Musibau H, Jamasb T (2022) The socio-economic impacts of energy policy reform through the lens of the power sector—does cross-sectional dependence matter? Energy Policy 167

Phoumin H, Nepal R, Kimura F, Uddin GS, Taghizadeh-Hesary F (2022) Revisiting electricity market reform: lessons for the ASEAN and East Asia. Springer, Singapore, pp 1–194

Poquiz JLS (2015) A decade of reforms in the Philippine power sector: a price-concentration analysis on the whole sale electricity spot market. DLSU Bus Econ Rev 25:95–118

Rudnick H, Velasquez C (2019) Learning from developing country power market experiences : the case of the Philippines (English). Policy research working paper; no. WPS 8721; The World Bank Group, Washington D.C. Available at http://documents.worldbank.org/curated/en/428331548771494859/Learning-from-Developing-Country-Power-Market-Experiences-The-Case-of-the-Philippines

Samantha L (2019) Liberalization of electricity markets in selected places, Research Office Legislative Council Secretariat, IN19/18-19.2019. Available at https://www.legco.gov.hk/research-publications/english/1819in19-liberalization-of-electricity-markets-in-selected-places-20190823-e.pdf

Simshauser P (2021) Renewable energy zones in Australia's national electricity market. Energy Econ 101(C)

Simshauser P, Wild P (2009) The Western Australian power dilemma. Aust Econ Pap 48(4):342–369

Sharma D, Madamba SE, Chan MRL (2004) Electricity industry reforms in the Philippines. Energy Policy 32(13):1487–1497

Sharma D (2005) Electricity reforms in the ASEAN: a panoramic discourse. Econ Pol Wkly 40(50):5318–5326

Sun Cable (2021) Australia Asia power link. Retrieved 16 Nov 2022, from https://suncable.sg/australia-asia-power-link

UK Government (2022) Policy paper—contracts for difference, Department for Business, Energy and Industrial Strategy, Updated, 13 May. Available at https://www.gov.uk/government/publications/contracts-for-difference/contract-for-difference

Chapter 2
Multi-objective Auctions for Utility-Scale Solar Battery Systems: Lessons for ASEAN and East Asia

Natsuko Toba, Tooraj Jamasb, Luiz Maurer, and Anupama Sen

Abstract Auctions are an increasingly popular means of competitively promoting and procuring renewable energy to meet energy, social, and climate change objectives. To succeed, the technology designs need to accommodate technological progress, declining costs, and increasing Environmental, Social and Governance (ESG) demand. This analysis examines international experiences with large-scale solar photovoltaic (PV) and battery energy storage systems (BESS) auctions, which may be useful for East and Southeast Asia. It revisits auctions' theoretical and conceptual frameworks while concentrating on the ESG aspect from the perspective of such key stakeholders as investors, government, bidders, and communities, regarding efficient allocations of risks, costs, and benefits. It then relates this framework to real-world practices and international evidence on solar PV with and without BESS. The analysis shows that integrating ESG in auction designs and business models is possible and can benefit business and sustainable development. This analysis' focus on the ESG and solar PV plus BESS in auctions are nearly non-existent in the existing academic literature according to the review by del Río and Kiefer in Energy Policy 173 (2023).

Financial support from the Copenhagen School of Energy Infrastructure (CSEI) is acknowledged. CSEI functions are funded by CBS and energy sector partners.

N. Toba (✉)
Energy Policy Research Group, University of Cambridge, Cambridge, UK
e-mail: ntoba@ifc.org

T. Jamasb
Copenhagen School of Energy Infrastructure, Department of Economics, Copenhagen Business School, Frederiksberg, Denmark

L. Maurer
LM Engineering and Consulting, Richmond, USA

A. Sen
Smith School of Enterprise and the Environment, University of Oxford, Oxford, UK

H. Phoumin et al. (eds.), *Large-Scale Development of Renewables in the ASEAN*, Economics, Law, and Institutions in Asia Pacific,
https://doi.org/10.1007/978-981-99-8239-4_2

1 Introduction

East and Southeast Asia (ASEAN) are dynamic regions undergoing transitions into sustainable growth pathways, especially concerning energy. The International Monetary Fund (IMF), as of October 2022, forecasts Asian economy to expand much more slowly than in the preceding two decades while Asia's economic performance remains relatively sound in an increasingly sluggish global economy (IMF 2022). Among the 16 Least Developed Countries (LDC) in the United Nations' category of being on the path to graduation, ten are World Trade Organization (WTO) members, including ASEAN members Cambodia, Lao PDR, and Myanmar. The phasing-out of international support measures associated with LDC status may present challenges to graduating LDCs attempting to integrate into the global economy, such as stricter compliance with climate and other environmental, social, and governance (ESG) regulations. Six global brands that source garments and footwear from Cambodia wrote to its government in August 2020, stating that its proposed increase in coal-fired electricity could reduce the country's prospects for attracting future investment (Voice of America 2020).

According to the International Energy Agency (2022), Southeast Asia will see rapid growth in energy demand. In its Stated Policies Scenario (STEPS), based on a business-as-usual assumption, the region's oil-dominated demand rises more than 3% annually from 2021 to 2030, faster than in the previous decade. Renewables, natural gas, and coal demand all rise rapidly, with coal continuing to dominate, although its share of power generation declines from 42% today to 39% by 2030.

The International Renewable Energy Agency (IRENA) has estimated average annual investment needs for renewable energy and energy efficiency in East and Southeast Asia totaling US $582 billion under its Planned Energy Scenario (PES) and US $830 billion under the Transformative Energy Scenario (TES) during 2016–2050 (IRENA 2020a; base year for US$ prices unavailable). These needs are despite decreasing renewables costs, as seen in IRENA reporting that total installed costs for utility-scale solar PV plants fell 81% between 2010 and 2020, from US $4731 per kilowatt (kW) to US $883/kW (IRENA 2022; information on nominal or real prices unavailable).

IEA reports utility scale lithium-ion battery prices falling from US $4285 per kilowatt-hour (kWh) in 2010 to US $1568/kWh in 2017 (IEA 2020; information on nominal or real prices unavailable).Notwithstanding, the S&P Global-owned IHS predicts that a battery module price increase of 5% in 2022 amid fierce demand for lithium-ion phosphate batteries in electric vehicles (EV) will drive up the overall cost of stationary battery projects by some 3%. IHS Markit forecasts that lithium-ion battery prices will not fall before 2024, thanks to rising metal prices, EV demand, and China's near-monopoly on the sector (Hall 2022). Solar PV system prices have also increased in 2021–2022, due chiefly to supply chain constraints (Stevens 2022).

In the wake of fossil fuel prices soaring from 2021 to 2022, solar power has helped meet electricity demand and enhance energy security in Asia. In China, India, Japan, South Korea, Vietnam, the Philippines, and Thailand, solar electricity generation reduced potential fossil fuel expenditures by approximately US $34 billion from January to June 2022, equivalent to 9% of total fossil fuel costs those countries incurred during that time (Edianto et al. 2022).

Power systems aspiring to high renewables penetration rates with mostly variable renewable resources will probably require a variety of storage technologies, whose owner should procure through a competitive process to meet the power system least-cost objective. As the renewable energy sector progresses, policies must take changing market conditions and new technical and socioeconomic challenges into account to ensure a just and inclusive transition encompassing the energy sector and more. Falling costs of new technologies, expanding growth in variable renewables, i.e., solar and wind, and greater emphasis on climate and other ESG objectives by policymakers and stakeholders have altered the conditions for new market entrants and new power generation projects. One instrument on the rise is auctions to promote competition for the market as policymakers seek to procure renewable electricity at the lowest possible price while fulfilling other social or economic objectives. While enough data for statistical analysis are unavailable, general auction price trends might better reflect technology cost trends than earlier feed-in tariff schemes with government-set prices, per Fig. 1.

Morality in competitive markets is increasingly important for investors, shareholders, and consumers (Tirole 2017, 2021; Dewatripont and Tirole 2022). Financiers' demand for return on ESG is on the rise, with global debt issued for ESG purposes forecast to reach US $1.3 trillion in 2022 (Institute of International Finance 2022) from the approximately US $30 billion in 2013 reported by Bloomberg New

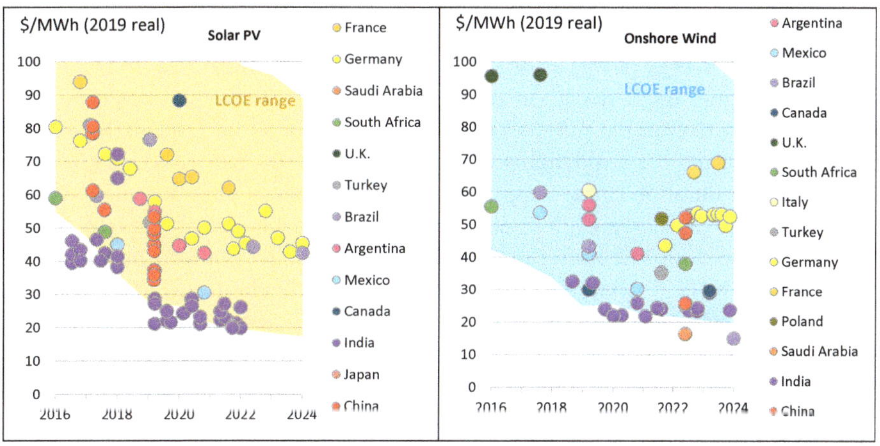

Fig. 1 Levelized bids for auctions across G-20 by project commissioning year, 2016–2024. *Source* Bloomberg New Energy Finance (BloombergNEF 2021). *Note* To make the winning auction tariffs comparable across countries, BloombergNEF levelizes the capacity-weighted average winning tariff, estimating the average inflation-indexed tariff for the lifetime of the project. BloombergNEF removes the effect of subsidies, standardizes inflation, and adds a merchant tail for the lifetime of the project after the auction tariff expires. Levelized bids are shown by their commissioning dates

Energy Finance (BloombergNEF). The European Union (EU) will require funds to disclose information about how they reduce potential negative impacts of their investments beginning in 2023.

According to Theobald (2022) major impediments to institutional investments in emerging and frontier markets are that institutions and fund managers are increasingly applying ESG considerations in their investment strategies that exclude or down-weight emerging and frontier markets. However, some investors use an active ESG approach in addition to, or instead of, ESG screening, in which they identify investment opportunities to improve ESG outcomes using the Sustainable Development Goals (SDGs) as their targets (Theobald 2022). For the following reasons, this study concentrates on auctions for procuring utility-scale solar photovoltaics (PV) and battery energy storage systems (BESS) with long-term power purchase agreements (PPA) on the order of 15–25 years or other sufficient cost recovery periods:

First, some countries in ASEAN and East Asia, such as Japan, Korea, and Singapore, have wholesale electricity markets based on auctions in energy markets, e.g., the day-ahead and real-time markets, and capacity markets, which include forward markets. Other ASEAN countries, such as Vietnam, Cambodia, Indonesia, and Lao PDR, retain a state-owned single buyer model, i.e., centralized agents which purchase power from generators, with electricity purchased from private independent power producers (IPPs) with PPAs often combined with power generated by state-run providers. While the latter countries may lack competitive electricity markets, auctions for procuring contracted amounts of electricity provide opportunities for bidders to compete for specific market segments under the PPAs.

Second, while corporate renewable energy PPA volumes are increasing as companies aspire or need to decarbonize their activities, they face challenges in delivering 24/7 renewable energy power as of 2022 (LDES Council, McKinsey and Company 2022). Achieving 100% decarbonization with variable renewables requires long-duration energy storage (LDES) technologies. As shown in Fig. 2, technologies with low energy capacity costs and high power capacity costs (the blue area) are most suitable for longer duration storage applications on the order of days at a time and less frequent charge discharge cycles. Examples include metal-air batteries, hydrogen, thermal storage with low round-trip efficiency (RTE), and pumped hydro storage with medium RTE. Technologies with intermediate capabilities, including redox flow batteries (RFBs) with medium RTE, are in the green area. Technologies in the brown area, including lithium-ion battery high RTE, are better suited to shorter duration applications on the order of hours and more frequent cycling. EV battery development has significantly improved short-duration electricity storage prospects, while long-duration storage technologies have not experienced similar levels of help from other market drivers (MIT 2022).

Small-scale renewable energy and storage systems, such as small islands, tend to approach the 24/7 renewable energy target more closely, as shown in Fig. 3 use of lithium-ion batteries for longer durations in larger systems to complement wind power, such as in Ireland, is assessed as too expensive (Newbery 2020). Competitive auctions improve the transparency of renewable energy PPAs by enabling investments

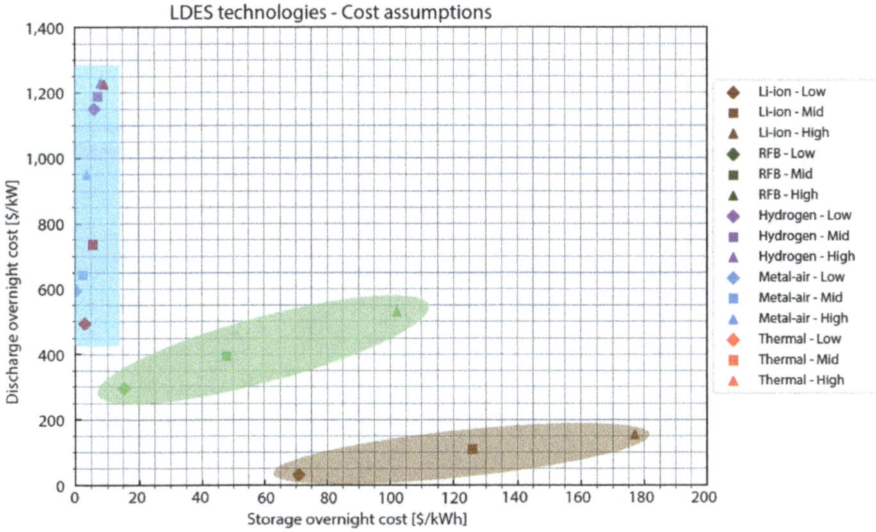

Fig. 2 Three classes of energy storage technologies, grouped by discharge power and storage overnight capital costs in 2050 (US $2020 prices). *Source* Massachusetts Institute of Technology (2022). *LDES* Long-duration energy storage. *RFB* Redox flow battery

in clean, dispatchable capacity that drives down costs, and more precise climate and ESG compliance.

This study's focus on ESG aligns with and is more comprehensive than the Paris Agreement on climate change. It particularly examines renewable energy installations, which tend to be located in ecologically and socioeconomically sensitive areas. Climate, being an aspect of the "E" for "environment" in ESG, is an abiotic factor of ecosystems. Thus, the global community must consider the impact of its investments on the ecosystem beyond climate mitigation, adaptation, and resilience if we are to achieve sustainability. According to a UN report (UNEP 2022), climate, biodiversity, and land degradation goals will be out of reach unless investments into nature-based solutions reach US $384 billion/year by 2025, more than double the current US $154 billion/year as of 2022. Annually, private capital represents only an estimated 17% (US $26 billion) of total investments into nature-based solutions. Private sector actors will have to combine net-zero with being nature-positive, complying with Task Force on Climate-related Financial Disclosures (TCFD) and Nature-related Financial Disclosures (TNFD).

A review of 607 academic publications renewable electricity auctions identified in March 2022 (del Río and Kiefer 2023) finds that study's focus on multi-criteria auctions and auctions on solar PV plus BESS, i.e., dispatchable renewable energy sources (RES) electricity generation, are almost non-existent in their reviewed academic literature. This review's finding is consistent with this study and the facts that in April 2023, the government of United Kingdom issued a call for evidence on introducing non-price factors into the contracts for difference scheme, such as ESGs

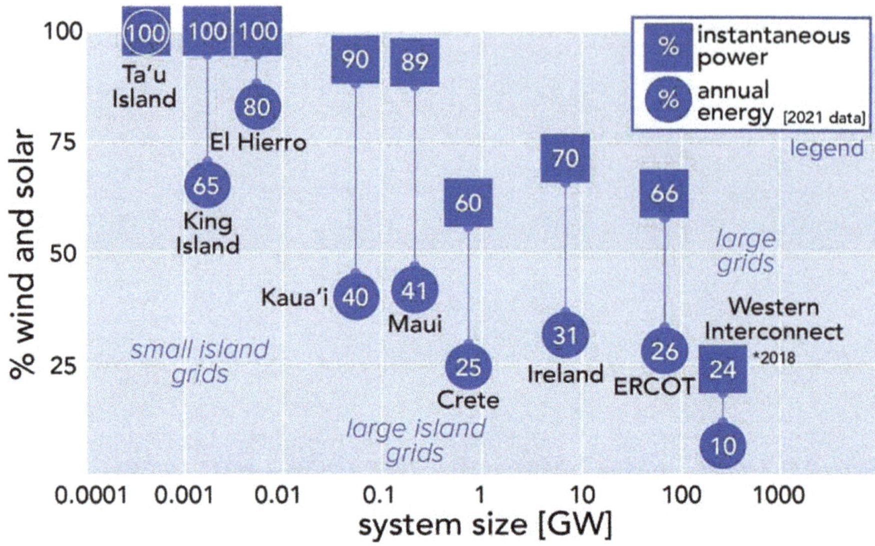

Fig. 3 Instantaneous power versus annual energy by grid system size. *Source* Kroposki (2022). *ERCOT* Electric reliability council of Texas

(Government of United Kingdom 2023) and that the United States Federal Energy Regulatory Commission (FERC) has issued only broad electric storage rulings that are not yet specific to hybrid resources such as solar PV and BESS as of May 2023.

Section 2 begins with a theoretical and conceptual framework of auction markets where demand and supply have their own ESG objectives, before assessing risks and providing case histories of measures to mitigate said risks, including the complementary role of auctions, among other market instruments. Section 3 briefly reviews concerned auction methods and their contractual forms. Section 4 discusses several business models with case histories. Section 5 is a literature review. In Sect. 6, the conclusion, we note that broader policy support might facilitate integrating ESG into competition and better environmental outcomes.

2 Conceptual Frameworks

2.1 Static, Dynamic, and Incentive Frameworks

2.1.1 Static Framework

Auction design's main objectives include efficiency, fairness, transparency, and simplicity, subject to the firms', i.e., bidders', incentive compatibility, individual rationality, and participation constraints. This analysis uses a simplified framework building on Tirole (2017, 2021) and Dewatripont and Tirole (2022), which assumes a unit-demand, i.e., an official selecting a bidder on behalf of consumers and the public interest, and n sellers, i.e., bidders, $i \in \{1, \ldots, n\}$. To compete, sellers select a price p_i and an ESG choice a_i, both in $\mathbb{R}^+$. Higher a_i values signify higher ESG choice levels, at least in the relevant range $\left[0, \widehat{a_i}\right]$ where $\widehat{a_i} \leq +\infty$. a_i has a welfare impact $W_i(a_i)$, with $W''_i < 0$ and $W_i'(0) = +\infty$. Thus, there exists $\widehat{a_i}$ such that $W_i'(a_i) > 0$ if and only if $a_i < \widehat{a_i}$. Let $\mathbf{a} \equiv (\mathbf{a_1} \ldots \mathbf{a_n})$ denote the vector of ESG choices.

The vector $\left\{p_{i,}a_i\right\}$ determines the net price $\widehat{p_i}$ perceived by the buyer. Seller i faces a demand function $D_i\left(\hat{\mathbf{p}}\right)$ where $\hat{\mathbf{p}} \equiv \{p_1, \ldots p_n\}$ denotes the vector of net prices, and also refers to $D_i\left(\widehat{p_i}, \hat{\mathbf{p_{-i}}}\right)$, where $\hat{\mathbf{p}}_{-i}$ denotes the vector of net prices charged by seller i's rivals. The buyer's cost or benefit of ESG is a function $\phi_i(a_i)$ with $\phi_i'' \geq 0$ such that.

$$\widehat{p_i} \equiv p_i + \phi_i(a_i). \tag{1}$$

When the buyer is ESG-irresponsible, then $\phi_i'(a_i) > 0$, as demand decreases with the morality of the firm's offer. Conversely, ESG-responsible buyer demand increases with the morality of the firm's offer: $\phi_i'(a_i) < 0$, while ESG-neutral buyer demand remains unchanged regardless of morality: $\phi_i'(a_i) = 0$.

Seller i's unit cost c_i may depend on her ESG choice $a_i : c_i(a_i)$ with $c_i'(a_i) \gtrless 0$. The sellers are substitutes, and hence, demand elasticity is $(\partial D_i/\partial \hat{p}_i < 0 < \partial D_i/\partial \hat{p}_j)$, and marginal revenue is decreasing in price ($(p_i - c_i)D_i\left(\hat{\mathbf{p}}\right)$ is concave in p_i). $\eta_i\left(\hat{\mathbf{p}}; \sigma\right) \equiv \left(-\partial D_i/\partial \hat{p}_i\right)/(D_i/p_i)$ denotes price elasticity of demand for supplier i's services.

Assumption 1 (*elasticity of demand*): Seller i's elasticity of demand increases with competitive pressure: $\frac{\partial \eta_i}{\partial \hat{p}_j} < 0$.

Objective functions. Sellers care about profit and ESG impact, as ESG is part of requirements to bid in the auction and/or requirements that the seller's, i.e., the firm's, investors impose. Let $\alpha_i \geq 0$ denote seller i's intrinsic ethics, that is, the weight on welfare relative to weight on profit.

Assumption 2 (*consequentialism*). As net prices determine demand, seller i's social welfare perception depends on net prices and ESG choices: $\mathbf{W}_i\left(\hat{\mathbf{p}}, \mathrm{a}\right)$. Perceived welfare impact scales with actual impact, making it proportional to demand: non-increasing function $\Gamma_i(a_i)$ *such that* $\Gamma i(0) = +\infty$ and $\lim_{a_i \to \hat{a}} \Gamma_i(a_i) = 0$, *and* $\frac{\partial \mathbf{W_i}}{\partial a_i} = \Gamma_i(a_i) D_i\left(\hat{\mathbf{p}}\right)$.

Seller i maximizes the sum of profit and internalized perceived social welfare as ESG impact; letting $\alpha_i \geq 0$ denote the intensity of her social preferences, her utility function is:

$$U_i \equiv [p_i - c_i(a_i)]D_i\left(\hat{\mathbf{p}}\right) + \alpha_i \mathbf{W}_i\left(\hat{\mathbf{p}}, a\right) \equiv \left[p_i - (c_i(a_i) - \alpha_i \mathbf{W}_i\left(\hat{\mathbf{p}}, a\right)\frac{1}{D_i\left(\hat{\mathbf{p}}\right)})\right]$$

$$D_i(\hat{\mathbf{p}}) \equiv \left[p_i D_i(\hat{\mathbf{p}}) + \alpha_i \mathbf{W}_i(\hat{\mathbf{p}}, a)\right] - c_i(a_i) D_i(\hat{\mathbf{p}}). \tag{2}$$

That $\frac{\partial \mathbf{W_i}}{\partial a_i}$ is proportional to demand D_i is consistent with consequentialism. ESG choices are uniform over seller i's demand and so their impact is proportional to demand.

The following is a simplified equilibrium behavior illustrating the foregoing in a first-price (pay-as-you-bid) auction with incomplete information. Each of n bidders' private value v (parameter) is drawn from distribution F, denoted as $v_i \equiv p_i(v_i) - c_i(a_i) D_i(\hat{\mathbf{p}})$ from Eq. (2) where $p_i(v_i) \equiv p_i D_i(\hat{\mathbf{p}}) + \alpha_i \mathbf{W}_i(\hat{\mathbf{p}}, a)$. Bidder will bid at bidding price $p_i\ (v_i)$ (decision variable), and the expected utility is:

$$\mathbb{E}(\mathrm{u}(\mathrm{p_i}(\mathrm{v_i}), \mathrm{v})) = (\mathrm{p_i}(\mathrm{v_i}) - \mathrm{v_i})\mathrm{Pr}(\mathrm{Win}|\mathrm{p_i}(\mathrm{v_i})) \tag{3}$$

By the envelope theorem, $\frac{\mathrm{d}u}{\mathrm{d}v} = \frac{\partial u}{\partial p_i(v_i)} \frac{\partial p_i(v_i)}{\partial v} + \frac{\partial u}{\partial v} = \frac{\partial u}{\partial v}$, then, $\frac{\mathrm{d}u}{\mathrm{d}v} = \Pr(Win|p_i(v_i)) = \Pr(lowest\ bid) = \Pr(lowest\ value) = F(v)^{n-1}$. Utility is rewritten as $u(v) = u(0) + \int_0^v F(v)^{n-1}\mathrm{d}v = \int_0^v F(v)^{n-1}\mathrm{d}v$, which substituted into Eq. (3) results in: $p_i(v_i) = \frac{u(p_i(v_i), v)}{\Pr(Win|p_i(v_i))} + v_i = F(v)^{-(n-1)} \int_0^v F(v)^{n-1}\mathrm{d}v + v$. For example, where $v \sim U$ on $[0, 1]$, then $F(v) = v$, and $p(v) = \frac{v}{n} + v = \frac{v(1+n)}{n}$. Given that the optimal bid converges to the value as $n \to \infty$, in the limit the buyer can extract the bidder's full surplus. In equilibrium, the bidder bids the expected value of the second lowest value, given that the bidder has the lowest value.

The buyer will select the seller who bids at the lowest price p. While the ESG-responsible buyer may consider social welfare impact $\hat{p} \equiv p + \phi(a)$ in selecting the bidder, they will weigh the bid offer price p higher than $\phi(a)$. As the auctioned quantity (demand) is fixed, seller i tries to minimize the offer price p_i. Rearranging Eqs. (1) and (2) results in:

$$p_i \equiv \widehat{p_i} - \phi_i(a_i) \equiv \left[U_i - \alpha_i \mathbf{W}_i(\hat{\mathbf{p}}, a)\right] * \frac{1}{D_i(\hat{\mathbf{p}})} + c_i(a_i) \tag{4}$$

In these equations, the seller i's controllable cost is $c_i(a_i)$. Hence, the seller tries to reduce cost c_i and/or ESG concerns a_i, either by increased efficiency or cutting corners. Examples of the latter include, but are not necessarily limited to, choosing lower-quality and thus cheaper inputs, and reducing ESG performance and/or quality. As cost c_i depends on ESG efforts a_i, however, cutting corners might incur greater costs than the bid offer p_i. Less effort in social and environmental impact assessment, mitigation and management measures, and benefit sharing with local communities could delay contract execution, leading to cost overruns and penalties. Low-quality equipment may cost more in maintenance, repair, and replacement.

2.1.2 Dynamic Framework

In a dynamic intertemporal setting where auctions are held over the years, sellers may choose not to participate in an auction and wait for subsequent auctions, when more information about the auction process may be available. Sellers who do participate, however, will glean more information from participation than those who decline. Assuming an initial auction where all bidders have the same prior information, participating bidders would gain additional information by their participation, resulting in more posterior information. In the next auction, those bidders who participated in the earlier auction thus have updated prior information that those who did not participate perforce lack, leaving the latter at a potential disadvantage.

In the above setting, based on Bergemann and Juuso (2010) and Bergemann and Välimäki (2019), the flow marginal contribution to welfare $m_i(\theta_t)$ of seller i is: $m_i(\theta_t) = M_i(\theta_t) - \delta M_i(\theta_t, h_t^*)$, where M_i is the marginal welfare contribution of seller i, time $t = 0, 1 \ldots$, common discount factor $\delta \in (0, 1)$, allocation $h_t \in H$, Markovian state $\theta_t = \left(\theta_{1,t,\ldots,}\theta_{It}\right) \in \Theta$, private (Markovian) signal $\theta_{i,t+1}$ of i generated by conditional distribution function $\theta_{i,t+1} \sim P_i(\cdot|h_t, \theta_{it})$ and socially efficient allocation rule (after all histories C_t; the histories are bidders reporting state θ_t and allocation):

$$a_t^* : H_t \to [0, 1]'. \tag{5}$$

Expanding the flow term with respect to time gives: $m_i(\theta_t) = (W(\theta_t) - W_{-i}(\theta_t)) - \delta\left(W\left(\theta_{t+1}|h_t^*\right) - W_{-i}\left(\theta_{t+1}|h_t^*\right)\right)_i$, where the first bracket indicates M_i starting at t and the second bracket indicates M_i starting at $t + 1$ and h_t^* on the right-hand side. Further expending the flow term with respect to identity (rearranging) gives: $m_i(\theta_t) = (W(\theta_t) - \delta W\left(\theta_{t+1}|h_t^*\right)) - (W_{-i}(\theta_t) - \delta W_{-i}\left(\theta_{t+1}|h_t^*\right)$, where the first bracket indicates current value with bidder i and the second bracket indicates current value without bidder i but with h_t^* in the right hand side. Given the marginal contribution to welfare is $M_i = v_i - p_i$, and by rearranging, price bidder i is:

$$p_i = v_i - M_i \tag{6}$$

By adjusting Eq. (6) into an intertemporal setting, the socially efficient allocation rule (5) satisfies ex post incentive and ex post participation constraint with payment

$$p\colon p_{i,t}\left(h^*(\theta_t), \theta_{-i,t}\right) = v_i\left(h^*(\theta_t), \theta_{-i,t}\right) - m_i(\theta_t) \tag{7}$$

2.1.3 Incentive Framework

The average age of coal-fired power plants in East and Southeast Asia is on the order of 10–15 years (World Bank 2022), despite the need for renewable power in these regions. It is thus crucial to plan the retirement of such plants to ensure a smooth and just transition over the medium- and long-term. In some cases, electricity resource planning and adequacy requirement and/or tightening ESG and climate regulations as incentives toward 24/7 green power, especially by corporations, necessitate additional renewable energy, such as solar, to replace the retiring coal, which often provides baseload. A combination of solar PV and BESS is thus one technology option for replacing retired coal-fired power plants such as the foregoing. As a means of early coal power retirement, Germany has been holding one-sided subsidized compensation auctions to purchase the capacity of coal-fired power plants during 2020–2027 with a price cap per capacity (Reuters 2021; World Bank 2022).

Coordinated arrangements include staged product-matching auctions. The first stage thereof, building on the radio spectrum reallocation incentive auctions by the United States Federal Communications Commission (FCC) in 2016–2017 (Leyton-Brown et al. 2017; Royal Swedish Academy of Sciences 2020), is a reverse auction to determine a price at which coal-fired power producers voluntarily relinquish their coal power capacity and indicate the amount of carbon dioxide equivalent (CO_2e) emissions avoided by said retiring coal-fired power capacity. The second stage is a forward auction for avoided CO_2e emissions, which may be repeated until the supply prices of avoided CO_2e equal the purchase prices, or the difference is reduced enough for the host government or donors to make up the remaining shortfall.

Figure 4 illustrates the first- and second-stage auctions, repeated over four rounds until demand, i.e., carbon buyers, and supply, i.e., coal-fired power being retired, align. In the third stage of the auction, the corresponding freed-up coal-fired power capacity will be matched by reverse auctions of solar PV and BESS, while such backup generators as gas turbines may be required, as solar PV and BESS alone remain as yet unable to provide 24/7 dispatchable power or replace the baseload, as shown in Figs. 2 and 3 storage retrofit strategies for coal and other thermal power plants may also play a part in the not-too-distant future when the costs and implementations of same become clearer.

Fig. 4 First-stage reverse and second-stage forward auctions. *Source* Author

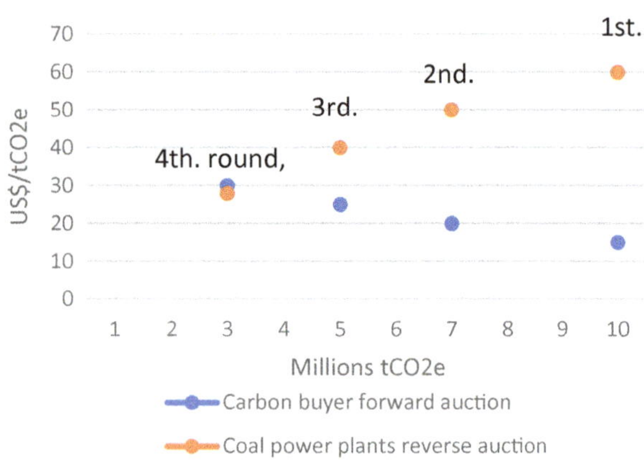

2.2 *Risks*

As in other auctions for renewable energy resources, competitive procurement of paired solar PV and BESS is subject to certain risks, hence returns for investors and economic and social impact; see, e.g., Maurer et al. (2020), Cote et al. (2022), and Roth et al. (2022). Market designs of said auctions must therefore ensure that the benefit of market competition outweighs the cost. They should mitigate and manage risks for the markets to provide incentives and signals for the right investments, in terms of type, amount, timing, and externalities, to deliver affordable quality electricity to consumers. Non-market alternatives, such as non-transparent bilateral contracts negotiated with unsolicited power providers, are likely to result in suboptimal welfare outcomes. Following is a summary of key risks, formats, and measures to mitigate and manage risks, concerning ESG pertaining to the solar PV and BESS auctions.

2.2.1 Bidding

Bidders have the allocation risk of not winning. The resources they expend in applying and preparing, and meeting the physical prequalification criteria of the auction are sunk costs if they lose. Such a risk is significant if the auctioned items are limited, i.e., fixed demand, and if such costs are large relative to the bidder's financial resources and project portfolio. Thus, smaller companies and local community organizations may be at a disadvantage, undermining auctions' diversity, equity, and equality, as well as ESG objectives (Eberhard et al. 2014; Amazo et al. 2021; Cote et al. 2022). As a rule of thumb, sunk costs should not exceed 3–5% of capital expenditures (Haufe and Ehrhart 2018). Expenditures on ESG-related prequalification criteria may reduce overall costs if ESG issues prove too costly and/or time-consuming for project realization. Examples include environmental and social impact assessments (ESIA) or proof of community engagement (Amazo et al. 2021), which may have significant monetary and non-monetary costs, such as political economy, time, and effort. However, less efforts on ESG related prequalification, such as inadequate community involvement may slow or halt renewable energy projects, as in canceled wind farms in Mexico and Kenya (Business and Human Rights Resource Centre 2018, cited in IRENA 2019). ESG-related prequalification criteria may also make timely commissioning more likely because bidders can account for enhancing, mitigating, and managing expected ESG impact in their bids, thereby reducing ESG uncertainties.

One design option is for the auction planner to pay costs common to all bidders, being more resource efficient than requiring each bidder to individually pay such costs. China provides a case study of this approach, to be discussed hereinafter. For example, if the auction planner identifies a site for solar PV and BESS in advance, the planner should pay for ESIA, community engagement, and land and other permits and authorizations, which each bidder can adjust to reflect their circumstances. A second option is for the auction planner to reduce research costs and information asymmetry among bidders, e.g., large or small, international or local electric utilities, community-based organizations or private companies, etc., by sharing indicative costs and information when soliciting bids. Hawaiian Electric Company, Inc. (Hawaiian Electric) included such indicative costs of Supervisory Control and Data Acquisition (SCADA) communications, security system interconnection, and station services (e.g., overhead lines and transformers) when soliciting bids for renewable dispatchable generation and storage on O'Ahu (Hawaiian Electric 2019). They could also include indicative ESG-related costs in like manner, as local bidders may have more local ESG information. A third option is to limit prequalification requirements to preliminary social impact evaluations and evidence of community engagement, to be finalized after the bidder wins the award, with penalties for non-finalization or tying granting of licenses to successful finalization (Amazo et al. 2021). Design strategies may include any or all of these.

As suggested above, design should encourage diverse participation by smaller actors and investors who are less able than larger ones to cope with auctions' complexity and competitiveness. Strategies such as (i) reduced prequalification, (ii) different pricing rules, and (iii) quotas, may significantly affect and even distort outcomes. A lack of clear taxonomy of protected groups may result in unintended consequences, as happened in Germany in 2017, where preferential rules led to artificial citizen energy communities for onshore wind that were awarded more than 90% of the auction volume (Kitzing et al. 2019; Cote et al. 2022). In Australia, qualifications for the state of Victoria's 2017 renewable energy auction scheme included proof of community engagement and benefit sharing. Community projects and other small-scale actors could not compete against larger and more established players, however, due to (i) nascent community initiatives at the time of the auction, (ii) technology-neutral auction schemes, (iii) high up-front costs for proposal preparation, and (iv) lack of economies of scale. Thus, Victoria had to employ other support schemes, such as grant funding (Renewable Communities Program), to support community energy initiatives (IRENA 2019). A study of South African renewable energy auction program during 2011–2015 finds (i) some market concentration did not undermine project pricing or market development, (ii) preferential conditions for small, local players has been more effective at counteracting market concentration than lowering of entry barriers and (iii) policy certainty and predictability seem more important to counteract market concentration than any auction design measures (Kruger et al. 2021).

2.2.2 Awarding and Contracting

Bidders, i.e., suppliers or sellers, tend to have differing information about true demand and may have varying cost profiles of the bid item, i.e., solar PV and BESS. They may also have different financial profiles to diversify risk and take more strategic approaches. Winning an auction may also mean that other parties and the demand have better information than the winner about the bid item's value, and as the lowest bid wins, bidders also try to underbid each other, including trying to shade their bids. Doing so, however, may cause them to inflate their bids or bid below what would be financially viable. Such was observed in multi-item auctions under uniform pricing rules in Germany where several bidders submitted bids below €0.01/kWh, in Spain when an auction in 2015 resulted in a clearing price of zero, and in the British Contract for Difference (CfD) auctions in 2015 where two solar projects were withdrawn for submitting bids at irrationally low prices (Tongsopit et al. 2017). On the other hand, in first price, or pay-as-you-bid, single-item auctions, the bidders' strategy is to bid just below the second lowest bidder, as described in Sect. 2.1.1.

Irrational underbidding risk is relevant given (i) declining costs of solar PV and BESS, and (ii) uncertainties in financing and materials costs of same. In August 2022, Malaysia extended power purchase agreements from its fourth large-scale solar (LSS4) tender for large-scale PV from 21 to 25 years because of concerns about project bankability, due to rising material prices and fears of rising interest rates. Several project owners asked the Malaysian Energy Commission to review electricity bids, which it rejected. The LSS4 program awarded 823 MW of capacity across 30 projects. Out of a total of 2457 MW awarded, only 1160 MW were operational by the second quarter of 2022 (Table 1; Santos 2022a).

Table 1 shows that barely a quarter of the capacity awarded by auction in India since 2017 had been commissioned as of early 2022, and several companies that had been awarded PPAs surrendered capacity, due chiefly to low tariffs and rising costs. Indian turbine manufacturers are turning to exports, while developers are moving from auctions and long-term PPAs to options that fetch better prices through direct sales to commercial and industrial customers and sales via the Indian Energy Exchange. Other longer-term challenges in India include the high cost of capital, grid connection, permitting, and land acquisition. Large wind and solar power projects require large amounts of land, often leading to development on local communal lands. Land rights issues are thus becoming more contentious around the world (REN21 2022).

Delays and underbuilding may arise from factors beyond the developer's control. Significant causes of construction underperformance include obtaining environmental and social permits and grid access. It is therefore essential to allocate such responsibilities fairly between bidder and auctioneer (Diniz et al. 2023). Alternatively, qualification requirements may include permits, although doing so may constrain the pool of participants. Many jurisdictions should streamline and make permitting processes more transparent. In Mexico, social impact permits have become a bottleneck in deploying awarded projects, especially due to unclear and lengthy institutional processes. While Mexico made such permits as prequalification, instead of a post-award requirement, in the fourth auction round, Mexico ultimately cancelled the auction (IRENA 2019).

Reducing uncertainty is one design strategy for mitigating underbidding risk if bidders are rational. Each bidder would revise its bid if they had information about other bidders. Such information might be inferred by competitors' bids in open, though not sealed bid auctions. Thus, the reverse clock auction yields lower bids, theoretically. An auction planner can set time limits on project completion to

Table 1 Auctioned project realization rates

	India	Malaysia	Brazil	United Kingdom	The Netherlands	France	Ireland	California, US	China
Auction years	1997–2022	2016–2022	2009–2010	1990–2001	2011	2011	1995–2003	2011–2015	2003–2004
Technology	Wind	Solar PV	Biomass, wind, small hydro	Technology neutral	Technology neutral	Solar PV	Wind, hydro, biomass, and CHP	Technology neutral	Wind
Realization rate	25% by 2022	47% by 2022	~ 30% by 2014	~ 30%	68% by 2015	< 50%	~ 30% by 2005	> 75%	100% by 2007

CHP Combined heat and power
Source Wigand et al. (2016), Kreiss et al. (2017), REN21 (2022), and Santos (2022a)

reduce underbidding. Maurer et al. (2020) notes reverse clock auctions are likely to become the industry standard as business models for standalone and co-located or hybrid BESS facilities mature. Disclosures could, however, invite bidders to implicitly collude, especially with large multi-project bidders in an environment with low competition, while setting a reserve price could mitigate same (Haufe and Ehrhart 2018).

A Vickrey auction or a Vickrey-Clarke-Groves (VCG) mechanism will induce bidders to bid their true values (no shading) as their dominant strategy, because the winning bidder would be awarded the opportunity cost, regardless of the bidder's own value. For a single item, the mechanism is referred to as a second-price sealed-bid auction, or simply a Vickrey auction where bidders simultaneously submit sealed bids. While the highest bidder wins the item, unlike standard sealed-bid tenders, the winner pays the amount of the second-highest bid. In reverse auctions, the buyer instead pays the second-lowest bid. This second-price sealed bid is de facto equivalent to the English clock auction. While economists have been extensively researching VCG, including in Sect. 2.1.2, it is rarely applied in practice. Ausubel and Milgrom (2004) discuss several possible weaknesses of VCG, including possibilities of very low revenues (in reverse auction, very high revenues) or vulnerability to collusion.

Under-contracting risk is an auction outcome where the amount of capacity or generation contracted is less than expected, which may be high if an auction has a low participation rate and/or does not impose penalties on winning bidders who do not sign PPAs. The design strategy for mitigating under-contracting is to require bid bonds or impose other penalties to make it costly for selected bidders to walk away without signing contracts. A selected bidder may still choose not to sign a PPA because the financial penalties are usually larger for breach of contract than turning down a contract (Maurer et al. 2020).

If a firm bids to supply more than the contracting capacity required at auction, it faces a risk of over-contracting and having to buy on the spot market to honor the contract. A study of the Chilean experience from 2006 to 2011 finds that a higher cost of over-contracting for entrants, especially smaller ones, than for incumbents may pose a barrier to entry (Bustos-Salvagno 2015). The study finds that incumbents are on average presenting lower bids than entrants, due in part to a significant difference in the cost of over-contracting, which is directly related to their level of risk-aversion. Incumbents with diversified portfolio of generating technologies have an advantage over entrants, especially the smaller ones. Consequently, entrants are asking for a risk premium that influences competition, as their bids do not represent a serious threat for incumbents (Bustos-Salvagno 2015). One strategy for mitigating risk and increasing competition is to design auctions to cater to technology profiles, e.g., variable renewable energy. In its electricity auction of November 2014, Chile allowed renewable bidders to bid for eight-hour blocks. This rule allowed solar and wind generators to bid more aggressively since they could bid when their over-contracting cost was at a minimum. While the historical average was around four generators, there were seventeen bidders in this instance (Bustos-Salvagno 2015).

2.2.3 Construction and Operation

Nonrealization risk is the failure of auction winners to implement their contracted projects. Selected bidders may opt out before signing contracts. As seen in Table 1, projects often have low realization rates for such reasons as underbidding, low prequalifications, cost increases, missing deadlines, permits, ESG impacts, unavailability or more remote location of grid connections than expected, and premature commencement (Kreiss et al. 2017; Tongsopit et al. 2017; Kitzing et al. 2019; Szabó et al. 2022). In China, the government secured the land and procured environmental permits, and most of the bidders were state-owned companies that could cross-subsidize their wind projects and bid low prices (Wigand et al. 2016). In Germany, deadlines can be extended once when a lawsuit has been filed against a project, and lawsuits against onshore wind construction are not uncommon there (Tongsopit et al. 2017). Conflicting policy objectives in designs, e.g., lowest price versus local content, might also result in a low realization rate of the winning projects, as in Indonesia (Tongsopit et al. 2017).

A design strategy for improving realization rates might include high financial prequalifications and adjusted physical prequalifications relative to sunk costs, penalties covered by financial prequalifications (e.g., bid bonds), and increased competition (Kreiss et al. 2017; Kitzing et al. 2019; Haufe and Ehrhart 2018; Matthäus 2020). While stricter prequalifications and penalties might increase bids and reduce participation, increased competition may offset same. A study based on 250 observations from 220 auctions taken place in 16 European countries from 2012 to 2020, suggests that policymakers should either strive for short realization periods with financial prequalifications or for long realization periods with no financial prequalifications (Anatolitis et al. 2022). In Germany, increased competition and decreased public support may improve project realization rates (Haufe and Ehrhart 2018). By contrast, a 2021 survey found that developers tend to be less willing to participate in highly competitive auctions (Cote et al. 2022). Kremer (2022) notes a low ratio of private to social return with low barriers to entry.

Auctions are also used to allocate grid connections. Portugal held two large-scale tenders in 2019 and 2020 to resolve a glut of grid permit requests for solar projects. Some 52% of awarded grid-connection capacity went to PV or PV and BESS projects. The projects gain full access to the wholesale and ancillary services market and the option to sign a PPA with a utility or corporate off-taker. All projects under this merchant option will pay the system operator €5–40/MWh for 15 years for lifetime grid access (BloombergNEF 2021).

A theoretically and empirically proved design with low complexity for the bidders might facilitate appropriate bidding strategies to optimize outcomes, including ESG. Auctions should minimize incentives for strategic supply reduction, possibly with markets diversified into forward and wholesale segments. A long-term auction schedule ensures a degree of certainty, helping investors avoid risk. Ad hoc auctions undertaken without future auctions scheduled might force bidders to underbid to limit their losses of projects already at the advanced development stage.

A study shows that continuity in auction rounds, rather than ad-hoc auctions, increases long-term certainty for participation, as in California, and further finds that auction frequency depends on context and technology (Kitzing et al. 2019). In general, lower auction frequencies are appropriate for technologies with fewer bidders and larger projects, e.g., offshore wind, and more frequent rounds for technologies with more potential participants, e.g., solar PV (Kitzing et al. 2019). In China, solar PV auctions were held annually between 2019 and 2021, while renewable energy auctions have been held biennially in the UK since 2015, and quarterly in Italy between 2019 and 2022 (BloombergNEF 2021).

2.3 *Role of Auctions Among Other Policy Instruments*

As each policy instrument has its own strengths, selecting and designing a complementary mix of instruments may better mitigate and manage risks in scaling up solar PV and BESS in the electricity market. Kwon analyzes (2020) the effects of South Korea's policy mix of auctions, feed-in tariffs (FiT) for small solar PV producers, and renewable portfolio standards (RPS), summarized as follows. The country's long term contract auction scheme with sliding premiums is capable of (i) alleviating price risk for renewable electricity suppliers under RPS by fixing remuneration over long periods, (ii) counteracting lowered competitive pressures brought about by FiTs with the intense market competition of auctions, and (iii) reducing asymmetric information by influencing renewable energy certificates' (REC) spot prices and providing a reference price for FIT rates. A weekly REC spot market may mitigate sales risks arising from the long-term contract auction scheme holding only two rounds of bidding opportunities annually. FiTs can lower RPS price risks and mitigate transaction costs and sales risks of long-term contract auctions. Intense RPS market competition may also counteract the reduced competitive pressures that FiTs may engender. The following example demonstrates these complementary circumstances. The adoption of long-term contract auctions in 2017 resulted in falling REC spot prices due to rapid increases in small and medium solar PVs. Re-introducing FiTs for small solar PV suppliers in 2018 drove REC spot prices lower than long-term contract auction prices, implying that current REC spot prices may be lower than break-even prices for small solar PV. Hence, it may be necessary to raise the RPS target to reverse the falling trend in REC prices.

Having more than one policy instrument providing diverse market opportunities is particularly relevant for solar PV and BESS, given the ability of BESS to complement or substitute for other power system elements, including generation, transmission, distribution, and demand response. With climate change having uncertain impact on electricity demand and supply, sophisticated markets and analysis may help better plan, operate, and regulate future power systems, and ensure that these systems are reliable and efficient. In Australia, the Hornsdale wind power and BESS plant participated in an auction for frequency regulation and uses part of its storage for price arbitrage in the wholesale market, which may allow revenue and risk diversification based on a complementarity between price arbitrage (MWh) and frequency regulation (MW).

Forward auction markets can mitigate potential prices significantly above marginal costs in wholesale spot markets, e.g., day-ahead, real-time, etc. As shown in Fig. 5, forward price higher than wholesale spot market (Cramton and Stoft 2008; Ausubel and Cramton 2010; Cramton 2020), and in Fig. 6, forward price lower than same, a dominant wholesale market player might have less incentive to bid much higher than their marginal cost in the spot market as their forward sales secured through long-term contract auctions place them in a more balanced position (Cramton and Stoft 2008; Ausubel and Cramton 2010; Cramton 2020). A large electricity supplier with many projects in its portfolio or a supplier building a larger capacity than the auction requires, may behave like that. In the latter case, bidders may bid some of their capacity to anchor some of their revenues and sell the remainder on the spot market or to corporate off-takers.

The winner of the July 2022 Chilean auctions, a 253 megawatt-peak (MWp) solar and 1 gigawatt-hour (GWh) BESS project, will sell a portion of the electricity

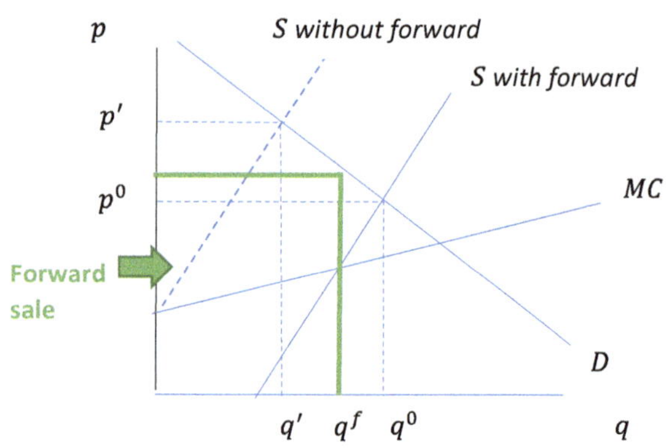

Fig. 5 Forward auction contracts may mitigate market power in spot market, case 1. *Source* Cramton (2020)

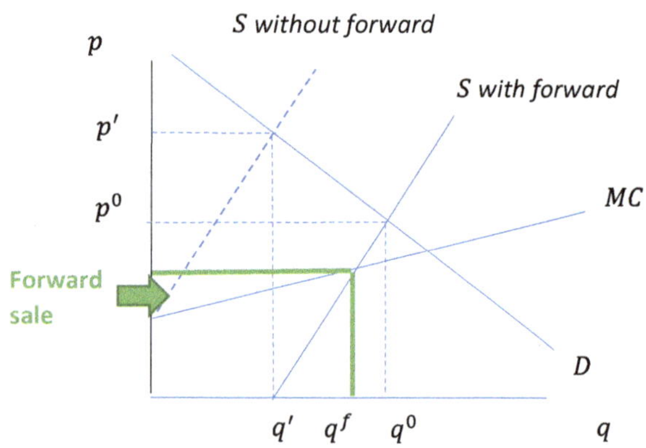

Fig. 6 Forward auction contract may mitigate market power in spot market, case 2. *Source* Author building on Cramton (2020)

generated to distribution companies under 15-year PPAs and the rest to private off-takers (NS Energy 2022). In such instances, the auction price could be lower than a solar PV/BESS wholly dedicated to the auction could achieve, by diversifying risks and achieving economies of scale. Brazil held multiple auctions for hydropower plants, where the developers sell most of the energy in the regulated market through the auction scheme and part of the remaining energy via corporate forward contracts. Solar PV/BESS suppliers' behavior across these different markets need to be closely monitored and audited. Regulators need to mitigate anti-competitive behavior and unreasonable cross-subsidies, as well as evading ESG obligations outside contract under auctions.

3 Modalities of Auctions and Contractual Agreements

3.1 Technology-Specific or Technology-Neutral Auctions

Auction designers must decide whether technology-neutral or technology-specific auctions better suit their objectives, per Table 1. The advantage of technology-neutral auctions over technology-specific auctions is lower costs especially large-scale projects, through encouraging diverse participants and competition (Anatolitis et al. 2022). In December 2017, a technology-neutral auction was held in Colorado in the US. Although storage capacity was not explicitly solicited, 105 of the 430 proposals included storage components with the median solar PV and BESS bid price being 20% lower than the cheapest prices under PPA in the US at the time (Lackner et al. 2019). The disadvantage of technology-neutral auctions versus technology-specific auctions is that they restrict diversification in such conditions as technology types, locations, and companies. For example, a study of European multi-technology auctions 80% of all multi-technology auction rounds from 2011 to 2020 were skewed, strongly or exclusively favoring one technology, while the dominant technologies of individual rounds vary (Melliger 2023) Different technologies have different planning, cost, construction, and operations characteristics. Thus, prequalification criteria and realization periods may affect them differently, potentially complicating ensuring a level playing field in such auction design aspects as ceiling prices, material and financial prequalification, penalties, and deadlines. While holding several auctions by technology category, rather than holding a single technology-neutral auction, might simplify auction design, it might also reduce competition for technologies that have limited application or are relatively new. Highly competitive technology-specific auctions such as those for ground-mounted solar PV in Germany are possible, with the influence of such sector characteristics as preexisting support (Wigand et al. 2016). Technological neutrality has been especially popular in Latin America (IRENA 2019).

3.2 Auction Contract Types

The major auction contract types are PPAs in most developing countries and contracts for differences. e.g., in the UK and Italy. Examples of PPAs are (i) blended tariffs including solar PV plus BESS, e.g., Malawi; Arizona, US; and Israel, (ii) solar energy tariffs and BESS capacity payments, e.g., Nevada, US; Portugal; and Uzbekistan, (iii) time variant tariffs, e.g., Chile, Nevada and Arizona, and India, and (iv) monthly lump-sum payments based on theoretical maximum PV output minus penalties for BESS unavailability or underperformance, e.g., Hawaii, US. While type (i) is the simplest, it does not offer different benefits, hence the values of the multiple services that BESS provides. Type (iv) is for small systems requiring long-term firm energy.

4 Solar PV and BESS Business Model

This section briefly describes the business model of solar PV and BESS business model, which includes either co-located plants that pair two or more generators and/or that pair generation with storage at a single point of interconnection, and full hybrids that feature co-location and co-control. Systematic empirical data and analysis on the business model and solar PV and BESS and PPAs are scarce, not to mention integrated ESG, especially in academic literature. The following is a summary of four business models in the United States (Seel et al. 2022), where the hybrid and co-located plants, dominated by solar PV and BESS, are growing rapidly at scale in many configurations and are distributed broadly across the United States where each state or region has distinct characteristics in terms of energy resources, regulations, markets, climate, etc.

4.1 Merchant Plant

The merchant plant business model is applicable for those countries with wholesale electricity markets, such as South Korea, the Philippines, Singapore, etc. The plant operator or IPPs maximize profit by responding to competitive price signals in organized electricity markets. The merchant solar PV + BESS plants earn revenue through (i) energy markets through energy arbitrage by charging the battery when wholesale electricity prices are low and selling when they are high, (ii) forward capacity markets and (iii) ancillary service markets. Even if wholesale electricity market prices do not always reflect system needs precisely, they provide a more dynamic dispatch signal to plants than regulated tariffs or incentive program rules and requirements.

4.2 Peak Load Reducer

The peak-load reducer business model generates value by reducing the load of a load-serving entity during peak times. The solar PV + BESS peak-load reducer primarily uses the battery to reduce load-serving entity costs. For example, utilities in Independent System Operator of New England (ISO-NE) pay for transmission service via a regulated peak-load pricing schedule and pay for capacity based on the forward capacity market price. The avoided costs from lower transmission-related and capacity related demand charges can be significant. The peak-load reducer business model forecasts monthly and annual peak hours, assesses the demand charges and dispatches the solar PV + BESS plant to reduce its reliance on the transmission network during those hours. Also, energy from the solar PV + BESS plant can lower the energy charges for the load-serving entity at the wholesale energy price. When not being utilized to lower peak load, the solar PV + BESS plant provides ancillary services sold directly to ISO-NE rather than indirectly reducing load-serving entity costs. The billing determinants based on coincident peaks become the primary dispatch signal. To the extent that system conditions coincide with the operator's expectations of the annual and twelve-monthly peak load events, the dispatch signal is dynamically responsive to grid needs, though not as directly as the merchant plant.

4.3 Incentive Program Participant

At an early stage of deployment of solar PV + BESS plants, an option of business models is to earn revenue by participating in government's incentive programs, such as feed in tariff, energy attribute certificates (EACs) such as renewable energy credits (RECs), tax credits and grants. The incentive program participant operates solar PV + BESS plants to comply with incentive program rules and regulations, such as a demand-response program, discharge at specific time periods, charging requirement from its paired solar PV, etc. As these incentive rules can deviate from direct wholesale market signals, solar PV + BESS plants that maximize revenue from such programs will operate differently from merchant plants and will yield a lower market value, while still likely being privately profitable. The United States offers a private owner of a solar PV + BESS plant an investment tax credit (ITC) for the BESS investment if it charges 75–100% of the time from the co-located solar PV unit. Despite the ITC support to the high capital costs of BESS, qualifying batteries charging at least 75% from the PV unit may limit the value these plants can provide to the grid. A solar PV + BESS plant operator may forgo charging from the grid, even if electricity costs are near-zero or negative, because doing so would reduce the share of the ITC the project can claim. Similarly, the operator may choose not to provide regulation-down service outside of hours when the solar PV is generating because doing so could reduce its ITC eligibility. Solar PV + BESS plants usually use this business model to complement the other primary business models and are typically IPPs.

4.4 Large Energy Consumer

Under a large energy consumer business model, private end-user characteristics are a major determinant of the dispatch of a solar PV + BESS and not bulk power system needs. This business model includes, but is not limited to, large manufacturing, industrial or commercial facilities, water treatment plants and mining operations. The large energy consumer typically places a premium on the ability to ride out multi-day outages and shorter outages lasting several hours. To meet these criteria, the battery unit may be kept at full state-of-charge during most hours and cycled only infrequently in the event of an outage. This operating strategy does not straightforwardly benefit the electric grid, although it can provide significant benefits to the end-user and possibly the local community in the event of a natural disaster or other form of major outage.

Large energy consumers are typically enrolled in industrial electricity tariffs, and a solar PV + BESS can reduce end-customer bills. In the United States, some large energy consumer faces a noncoincident peak demand charge and the solar PV + BESS discharges to reduce its monthly maximum demand, irrespective of whether it lines up with system demand. Lowering customer demand can reduce local congestion along the utility's distribution system, but the dispatch of the solar PV + BESS may provide less market value than if it directly responded to wholesale electricity market price signals. Industrial electricity tariffs may also include a coincident peak demand charge, which then provides a dispatch signal comparable to that of the peak-load-reducer business model.

This business model may be useful for export oriented large commercial and industrial firms in ASEAN and East Asian counties. Those firms need to meet increasing climate, 24/7 clean energy and other ESG regulations, and often grid electricity generation mix includes fossil fuels in many countries in the region (Table 2).

5 Toward Sustainable Development and 24/7 Clean Energy Transition

Table 3 summarizes key ESG risk mitigation and management costs versus avoided ESG costs and achieved benefits. As shown in Sect. 2, if ESG risk mitigation and management costs exceed avoided ESG costs and achieved benefits, bidders would be willing to transfer their private benefits, i.e., their net revenue, to ESG risk mitigation and management costs. Such transfer payment between private costs/benefits and externalities costs/benefits (welfare) is the concept that this analysis introduced to

Table 2 Summary of solar PV/BESS business model example

Circumstance or objective	Suitable business model	Ownership
Meet the real time electricity system needs, capacity adequacy and ancillary services needs	Merchant	IPPs
Reduce transmission and capacity costs and meet ancillary services needs	Peak load reducer	Load serving entities
Transform from an early development stage to full commercialization and improve demand response with feed in tariff, EACs such as RECs, tax credits, grants, etc.	Incentive participant	IPPs
Reduce electricity payments, increase resilience, and meet climate, 24/7 clean energy and other ESG requirement	Large energy user	Large manufacturing, industrial and commercial firms

Table 3 Key ESG risk mitigation and management costs versus avoided ESG costs and achieved benefits

ESG risk mitigation and management costs	Avoided ESG costs and achieved benefits
Environmental and social impact assessments and stakeholder engagement Benefit sharing Local employment, content, industry, and participation 24/7 clean power arrangements	Higher financing costs due to project delays Increased capital and operations and management costs Penalties Bid bonds (securities) and sunk costs due to project cancelation Greenwashing or non-compliance Local development benefits

incorporate ESG in competitive auctions. The effect can be seen in using cheaper ESG bonds and equity than non-ESG alternatives and grants (Kenway 2021; Lamdouar et al. 2022; Leonard Energy 2022). El Salvador's 2014 tender for solar and wind power required developers to invest 3% of their revenue in community social projects (IRENA 2019). The following are key findings of selected reviews of literature in addition to references already discussed in the previous sections.

5.1 Toward Sustainable Development

ESG goals should be embedded in project definitions for renewable auctions, or in other words, project qualification preconditions. Qualified bidders would then move to the next phase, where awards are based solely on price. Other current practices are to establish (i) one formula selection method including price and non-price factors with their weights in criteria, as in South Africa, Uganda, and Taipei, and (ii) merit adjustments to bid price, as in Malaysia (IRENA 2019; Amazo et al. 2021). Mixing monetary, i.e., price, and non-monetary, i.e., non-price, values run the risk of subjective judgement or loss of nuance. While some projects may have low social and environmental scores, and hence high risk, the low prices they offer as compensation result in their having the highest overall scores. Therefore, such projects are likely to win, which however could result in nonrealization of the projects due to the negative social and environmental issues that the low scores signify. Making price the only award criterion is more transparent, ensuring that only qualified bids are awarded contracts. If multi-criteria auctions are implemented, those criteria should be specific, quantitative, and similarly transparent to bidders.

ESG measures in auction designs should not expect too much from one project to generate local economic and social development, and thus need policy support. Examples of such support include local development, such as local factories, industry, research and development (R&D) facilities, supply, ownership, and employment. For example, the Chinese government provided significant policy support to develop the local solar industry, including several supply-side tools, such as grants, subsidies and low-cost loans, for more than a decade before it combined them with demand-side policy tools linked to performance requirements in cell manufacturing, such as cell efficiency in the Chinese top runner program (Münch and Scheifele 2023).

Local content is a blessing if capacity exists or is easy to build, or a curse if capacity hardly exists or preconditions for same are absent, such as regulatory frameworks and market potential. An initial step is understanding material and human resource requirements of various renewable technologies, assessing these requirements in the context of existing domestic resources and capabilities, and identifying ways to maximize domestic value creation by leveraging and enhancing local industries. Some countries, including Brazil, Russia, Malaysia, Argentina, Saudi Arabia, and Turkey, have imposed strict local-content requirements, which may cause auctions to fail consequently. South Africa discovered that creating a domestic manufacturing sector requires more than local content requirements, namely, a convincing government commitment to renewables and visibility about future demand. The government's years-long delay in signing PPAs with renewable auction winners was thus damaging. The lack of predictability and a small local market did little to encourage developing local industry, and most players that built factories have since shut down (BloombergNEF 2021). One reason for delays in early projects in Brazil was that its nascent domestic wind industry was not yet capable of supplying the equipment for developers to fulfil their local content quotas (IRENA 2019).

An India case study provides the first causal estimate of local content on firm-level innovation and production of solar PV auctions (Münch and Scheifele 2023). The Indian government simultaneously held solar auctions with and without local content from 2013. The study digitizes the results from the 41 auctions worth US $8.65 billion in solar module demand and collects annual revenue and solar patents of the 113 participating firms between 2004–2020. For causal identification, the study compares winners of local content with similar open auction winners in a staggered difference-in-difference estimation. Overall, the study finds winning local content auctions does not significantly increase firms' solar patents or sales. The key reasons why the policy did not create sustainable effects that local content are that (i) the small size and irregular frequency of auction, which neither allowed continuous and scale up of production to enable learning by doing nor generate sufficient revenue for re-investment in R&D and (ii) the reduction in competition in auctions due to lack of performance requirement that resulted in no incentive for the bidders to innovate.

Other emerging potential storage technology options might provide more feasible local supply opportunities. The currently dominant lithium-ion batteries that support solar PV, with their large-scale effects, are hard to produce locally. Such potential technology choices for LDES as flow batteries, compressed air storage, or thermo-electric storage might prove easier to localize because they involve a large portion of mid-technology local assembly.

The challenge of designing auctions to create long-term higher-skilled employment opportunities necessitates broad long-term systematic enabling policies. Long-term auction schedules and volumes signal longer-term market and job opportunities through project pipelines. In Uganda, staggered rather than simultaneous project development created learning curves that extended employment terms, reducing costs in time and resources on later projects. Quantitative employment targets in auctions should be accompanied by such benchmarks as quality, sustainability, and diversity. While South Africa's auctions exceeded job creation targets, most of the labor provided by its citizens was unskilled and short-term, leaving training, education, and development needs to fall by the wayside. Short-term, low-paid, unskilled jobs are not a lasting solution to poverty nor a path to sustainable development. In Senegal and Uganda, skilled construction workers for renewable energy projects were mostly expatriates, while the local community held mostly unskilled positions (IRENA 2019). The Noor-Ouarzazate concentrated solar power (CSP) complex in Morocco offered a wide range of employment opportunities to women, who represented only 4% of its workforce (IRENA 2019). Labor skill level development paths need broader policies in such areas as education and skills development to build local capacities as the sector evolves, which requires long-term planning. Attracting and retaining skilled workers is challenging in rural areas which are the sites of large renewable energy projects that could contribute to local economic development.

Local communities with high rates of poverty and inequality usually expect more from electricity supply projects than they can deliver. Engaging communities and maximizing benefits on the local level are crucial for project sustainability and can enable just and inclusive transitions. At the Morocco Noor-Ouarzazate CSP complex, local communities opted for as infrastructure and social services to benefit everyone, including women and children, rather than cash compensation for land use, which would benefit only male landowners (IRENA 2019). The South African Renewable Energy Independent Power Producer Procurement Programme (REIPPPP) required a community trust or a company that represents local communities, and as project shareholders, communities earn dividends to be invested in community development initiatives. In Namibia, NamPower, the national utility, included disadvantaged Namibians in auctions by such measures as 30% shareholding, management positions, skills and entrepreneurship development, community investments, and local hiring (IRENA 2019). Local community engagement can be a lengthy process, and as indicated, often involves land issues and political economy. Despite support from donors and international financial institutions, many initially promising projects, such as Guajira in Colombia and Turkana in Kenya, continue to face challenges (Mbugua 2021; Azzopardi 2022). In other instances, engagement may take the form of community power initiatives in Germany and Japan, onsite participatory planning with indigenous communities in Mexico, Latin America, and the Caribbean (IRENA 2019). Assessments of local stakeholder engagement are among the requirements for environmental and social impact assessments and governance, especially for projects financed by international institutions of the abovementioned kind.

A Canadian renewable electricity auction program case study demonstrate success in Indigenous equity participation and privately financed development (Hastings-Simon et al. 2022). The Alberta's Renewable Electricity Program in Canada implemented a series of reverse auctions for contracts-for-differences (CfD) between 2015 and 2019. It contracted for new renewable generation at prices in the range of CA $30 to CA $43/MWh (US $23 to US $33/MWh), well below expectations and among the lowest costs globally at the time, resulted in the government revenue of CA $75.5 million (US $60 million). The program steered new entrants into Alberta's power market, including through mandated Indigenous equity participation in one round of auctions. The price discovery and the incentive to develop new projects under the program spurred privately-financed development.

Auction designs could cope with land constraints, which are common in renewable energy projects, as described above. In Malaysia's Large-Scale Solar PV auction, plans to use land for economic activities besides solar generation, e.g., agriculture, might work significantly to the bidder's advantage. Germany's solar PV auctions cap the number of sites for ground-mounted projects on arable land, providing incentives to deploy in industrial zones rather than use land having agricultural or other alternative uses (IRENA 2019).

Auction planners could integrate geospatial least-cost electrification roll-out plans in auction designs, which can help exploit synergies between the energy sector and the broader economy to optimize energy transition benefits. Such plans, which have been applied in such countries as Kenya, Rwanda, Myanmar, and Papua New Guinea, represent the principle of one goal with many partners, which helps the government in policymaking and working with donors and partners, as well as serving as an investment prospectus. They coordinate off-grid and on-grid electricity alike, integrating demographic and geographic information system mapping techniques that combine technical, economic, demographic, and demand and supply data. Such plans constitute an inexpensive, dynamic planning platform capable of undertaking rapid updates to adapt to changes in key parameters, as with said geographic information systems (Independent Evaluation Group 2016). Designs thus based on geospatial electrification planning may mitigate projects being concentrated in resource-rich regions, resulting in more even regional distributions capable of spreading the socio-economic benefits of renewable energy projects, while also facilitating grid integration. The plans also help maintain the balance between achieving socio-economic objectives and procuring electricity at low prices, by aligning deployment policies with enabling and integrating polices. They increase project realization rates as they also coordinate auction schemes with permitting, e.g., the Netherlands, spatial planning, e.g.,

Ireland and the Netherlands, and grid availability, e.g., Brazil and Portugal (Wigand et al. 2016).

Auctions designs that integrate ESG and just and inclusive energy transitions may require policy support and grants which recipients win competitively with monitoring, reporting, and verification (MRV) requirements. They may include (i) industrial policies that enhance domestic capabilities, such as business incubation, research and development, supplier development, and support for small and medium enterprises in key sectors, (ii) education and training policies to increase technical, business and environmental management, and socioeconomic development capacities, (iii) labor market and social protection policies, including such employment services as job matching, on- and off-job training and labor mobility, and (iv) financial policies to ensure just transitions, including carbon pricing, green bonds, and revenue recycling schemes (IRENA 2019).

The US played an important role in introducing competitive bidding for energy procured by regulated utilities to serve their customers. A key piece of legislation was the Public Utility Regulatory Policies Act (PURPA) of 1978, which, while originally intended to increase conservation and foster co-generation, indirectly also provided a roadmap for regulators to mandate that utilities seek the most effective way to meet their customer needs, whether building new power plants or acquiring energy competitively from emerging IPPs. Different states adopted different methodologies, with the initially prevailing approach being establishing competitive tenders or requests for proposals, and basing award on both price and non-price factors, the most important of the latter being flexibility and dispatchability, while also taking ESG objectives into account (Plummer and Troppmann 1990). Over the years, many states have moved to pure auctions, where the price is the only factor in awarding contracts. In 2002, New Jersey pioneered an auction process for procuring most of its electric needs through an Internet-based auction whose winners were responsible for fulfilling all requirements, i.e., capacity, energy, ancillary services, etc., and the state's renewable portfolio standards (Fox 2005; BGS Undated).

5.2 *Clean Energy Transition*

As previously discussed, as of 2022, solar PV and BESS can approximate 24/7 clean energy supplies only in small and/or isolated systems. At the same time, many enterprises confront increasing pressures to use clean energy and report same, especially those associated with multinational concerns. Such large corporations are accordingly shifting from offsetting energy emissions, mostly by buying Renewable Energy Certificates (REC), to time-location tracked energy procurement. Several initiatives aim to accelerate the transition to 24/7 clean energy. One of these is EnergyTag, with more than 100 global participants including such tech giants and energy companies as Statkraft and Vattenfall. Google and Microsoft have also created partnerships that make their data centers more sustainable through hourly energy monitoring and matching with carbon-free sources from their clean-energy portfolios. The UN also launched the 24/7 Carbon-Free Energy Compact in 2021.

The transition to 24/7 clean energy might also drive higher ESG scores, which could facilitate access to cheaper capital in financial markets eager for green investment portfolios. An initial step for enterprises in ASEAN and East Asia to achieve 24/7 clean energy of allowing enterprises to trade RECs or equivalent and monitor and report types and amounts of energy used would help remain in global value chains and become more competitive. Rooftop solar regulations also need updating. In Cambodia, enterprises have challenges in installing solar PV on rooftops and cannot trade the resulting power among consumers. The regulator needs to rationalize electricity tariffs, for example, to set a fair level of capacity charge based on the highest amount of energy each consumer estimated to consume from the grid, i.e., excluding consumption from consumer's own generation such as rooftop solar PV, to operate, maintain, and invest in systems to ensure that electricity remains available at all times to all consumers and help reduce unnecessarily high peak demand. Carefully designed tariffs are becoming even more important as the system needs to integrate variable renewable energy, consumers, buildings, and EV within the system to achieve 24/7 green power as closely as possible.

5.2.1 Case Study: Thai Partial-Firm Renewables Auction

In 2017, Thailand conducted its third renewable energy-exclusive auction, as part of a new Small Power Producers (SPP) Hybrid Program. It had a ceiling of 300MW capacity from 10 to 50MW plants, and a starting (ceiling) price of Thai Baht (B) 3.66 (US $0.11) per kilowatt hour (kWh). Bidders proposed their maximum percentage discount from the ceiling price (IRENA 2019; O'Mealy et al. 2020).

Thailand became the first country in Asia to require developers to supply partial-firm power generation, i.e., delivering electricity at full capacity during peak hours, rather than merely installing new capacity. It also held the first auction in Asia to allow bids based on either a single technology, or a hybrid combining two or more technologies, to allow consistent feed-in to the grid. PPAs required that providers deliver between 100 $\pm$ 2% of specified capacity during peak periods, defined as 9AM-10PM on weekdays, and limit output at other times to 65 $\pm$ 2% of capacity (IRENA 2019; O'Mealy et al. 2020). The Thailand Energy Regulatory Commission (ERC) reported that 42 of 85 bids submitted had passed the pre-qualification stage and announced 17 projects with accepted bids. Of these, 14 were for biomass and the other three were hybrids with solar PV and BESS. The accepted bids ranged from 15.6 to 99.99% of the ceiling price, with net prices of B1.85–3.38/kWh (US $0.06–0.11/kWh) (IRENA 2019; O'Mealy et al. 2020).

In March 2018, the Minister of Energy announced that the Government of Thailand (GoT) would not buy additional power from new renewable energy projects for the next five years due to a high reserve power margin. The GoT later stated, however, that it might consider procuring new renewable energy projects that could sell electricity below the Electricity Generating Authority of Thailand's (EGAT) wholesale price. It also noted that it would tie new renewable energy procurement to its new Power Development Plan. These announcements left domestic and international renewable energy developers and investors alike uncertain about the Thai renewable energy development policy and regulatory environment, with some shifting their plans and investments elsewhere in the region (O'Mealy et al. 2020). The 2017 auction participants noted low winning prices for project realizations (O'Mealy et al. 2020). As of October 2022, the GoT reports many uncompleted projects in past programs (Santos 2022b).

6 Conclusion

This study offers the following conclusions. First, theoretically and empirically proved auction market design with low levels of complexity for bidders may facilitate bidding strategies intended to optimize outcomes, including ESG. A design strategy intended to improve realization rates might include high financial prequalification and adjusted physical prequalification relative to sunk costs, penalties covered by financial prequalification, and increased competition. Designs incorporating multiple select policy instruments rather than one policy instrument would enable said instruments to complement each other, e.g., PPAs awarded through long-term contract auctions, wholesale markets, etc.

Second, ESG goals in renewable auctions should be part of project definition and as such should be preconditions for project qualification, allowing awards based solely on price. Auction planners might integrate auction designs within geospatial least-cost electrification roll-out plans, which could facilitate exploiting synergies between the energy sector and the broader economy to optimize the benefits of green transitions. Designs that integrate ESG and just and inclusive energy transitions may require policy support and grants that recipients win competitively and adhere to MRV requirements.

Third, the transition to 24/7 clean energy may drive higher ESG scores, which might facilitate access to cheaper capital in financial markets eager to greenify investment portfolios. An initial step for enterprises in ASEAN and East Asia to build 24/7 clean energy would be allowing enterprises to trade RECs or equivalent, and building capacity for monitoring and reporting types and amounts of energy used, would help such firms remain in global value chains and make themselves more competitive.

Acknowledgements We acknowledge financial support from the Economic Research Institute for ASEAN and East Asia (ERIA). We thank Peter Mockel, Jonathan Walters, Pasquale Lucio Scandizzo and Sarah Lawson for their comments.

References

Amazo A, Lotz B, F Wigand, Lawson S, Monteforte A, Eisendrath A, Gutierrez J, Paz A (2021) Auction design and the social impact of renewable energy projects. Scaling up renewable energy (SURE). The United States Agency for International Development (USAID)

Anatolitis V, Azanbayev A, Fleck AK (2022) How to design efficient renewable energy auctions? Empirical Insights from Europe. Energy Policy 166:112982. https://doi.org/10.1016/j.enpol.2022.112982

Ausubel LM, Milgrom P (2004) The lovely but Lonely Vickrey auction. Stanford Institute for Economic Policy Research

Ausubel LM, Cramton P (2010) Using forward markets to improve electricity market design. Utilities Policy 18(2010):195–200

Azzopardi T (2022) Colombia opens new wind farm amid indigenous protests. Windpower Monthly. 19 Jan 2022. Available at https://www.windpowermonthly.com/article/1737843/colombia-opens-new-wind-farm-amid-indigenous-protests. Accessed 1 Oct 2022

Bergemann D, Välimäki J (2019) Dynamic mechanism design: an introduction. J Econ Lit 57(2):235–274. https://doi.org/10.1257/JEL.20180892

Bergemann D, Juuso V (2010) The dynamic pivot mechanism. Econometrica 78(2):771–789. https://doi.org/10.3982/ecta7260

BGS Undated (2022) New Jersey Statewide basic generation service electricity supply auction. Available at https://www.bgs-auction.com/bgs.auction.overview.asp. Retrieved 29 Dec 2022

BloombergNEF (2021) Renewable energy auctions. NetZero Pathfinders. Bloomberg new energy finance (NEF). Available at https://www.bloomberg.com/netzeropathfinders/best-practices/renewable-energy-auctions/#example-2. Retrieved 1 Oct 2022

Business and Human Rights Resource Centre (2018) Renewable energy and human rights. Business and Human Rights Resource Centre. Retrieved from www.business-humanrights.org/en/renewable-energy-human-rights. Cited in International Renewable Energy Agency (IRENA) 2019

Bustos-Salvagno J (2015) Bidding behavior in the Chilean electricity market. Energy Econ 51:288–299. https://doi.org/10.1016/j.eneco.2015.07.003

Cote E, Đukan M, de Brauwer CPS, Wüstenhagen R (2022) The price of actor diversity: measuring project developers' willingness to accept risks in renewable energy auctions. Energy Policy 163:112835

Cramton P (2020) Auctions and market design. Presentation. 5 November 2020. Available at https://cramton.umd.edu/auctions-and-market-design/. Retrieved 2 Oct 2022

Cramton P. Stoft S (2008) Forward reliability markets: less risk, less market power, more efficiency. Utilities Policy 16:194e201

del Río P, Kiefer CP (2023) Academic research on renewable electricity auctions: taking stock and looking forward. Energy Policy 173. https://doi.org/10.1016/j.enpol.2022.113305

Dewatripont M, Tirole J (2022) The morality of markets. 24 March 2022

Diniz BA, Szklo A, Tolmasquim MT, Schaeffer R (2023) Delays in the construction of power plants from electricity auctions in Brazil. Energy Policy 175.https://doi.org/10.1016/j.enpol.2023.113467

Eberhard A, Kolker J, Leigland J (2014) South Africa's renewable energy IPP procurement program: success factors and lessons. World Bank Group, Washington, DC. World Bank. https://openknowledge.worldbank.org/handle/10986/20039 License: CC BY 3.0 IGO

Edianto AS, Suarez I, Waite N (2022) The sunny side of Asia. Ember, the Centre for Research on Energy and Clean Air (CREA) and the Institute for Energy Economics and Financial Analysis (IEEFA)

Fox S (2005) New Jersey's BGS auction: a model for the nation internet procurement may be used in other states. Fortnightly Magazine—September 2005. Public Utilities Fortnightly Available at https://www.fortnightly.com/fortnightly/2005/09/new-jersey%E2%80%99s-bgs-auction-model-nation. Retrieved 29 Dec 2022

Government of United Kingdom (2023) Call for evidence on introducing non-price factors into the contracts for difference scheme. April 2023. Department for Energy Security and Net Zero
Hall M (2022) IHS Markit: battery prices won't fall until 2024. 7 March 2022. PV Magazine. Available at https://www.pv-magazine.com/2022/03/07/ihs-markit-battery-prices-wont-fall-until-2024/. Retrieved 1 Oct 2022
Hastings-Simon S, Leach A, Shaffer B, Weis T (2022) Alberta's renewable electricity program: design, results, and lessons learned. Energy Policy 171.https://doi.org/10.1016/j.enpol.2022.113266
Haufe MC, Ehrhart KM (2018) Auctions for renewable energy support—suitability, design, and first lessons learned. Energy Policy 121:217–224. https://doi.org/10.1016/j.enpol.2018.06.027
Hawaiian Electric Company, Inc. (2019) Request for proposal of variable renewable dispatchable generation and energy storage Island of O'Ahu. 22 Aug 2019. Docket No. 2017–0352
Independent Evaluation Group (2016) Reliable and affordable off-grid electricity services for the poor: lessons from the World Bank Group experience. World Bank, Washington, DC. World Bank. https://openknowledge.worldbank.org/handle/10986/25391 License: CC BY 3.0 IGO
International Energy Agency (IEA) (2020) Energy technology perspectives. Special Repo on clean energy innovation. Specific price values are available in https://www.iea.org/data-and-statistics/charts/evolution-of-li-ion-battery-price-1995-2019. Retrieved 7 Nov 2022
International Energy Agency (IEA) (2022) World Energy Outlook (WEO)
International Monetary Fund (IMF) (2022) Regional economic outlook. Asia and Pacific: sailing into headwinds. Washington, DC: International Monetary Fund, 2022. Identifiers: 9798400220517 (paper), 9798400221200 (ePub), and 9798400221224 (PDF)
International Renewable Energy Agency (IRENA) (2019) Renewable energy auctions: status and trends beyond price. International Renewable Energy Agency, Abu Dhabi
Institute of International Finance (2022) Sustainable debt monitor. Financing the transition. 3 Nov 2022
IRENA (2020a) Global renewables outlook: energy transformation 2050. Abu Dhabi
IRENA (2022) Renewable technology innovation indicators: mapping progress in costs, patents and standards. International Renewable Energy Agency, Abu Dhabi
Kenway N (2021) ESG investors are getting 'something for nothing. 28 Oct 2021. ESG clarity. Available at https://esgclarity.com/esg-investors-are-getting-something-for-nothing/, Retrieved 1 Oct 2022
Kitzing L, Anatolitis V, Fitch-Roy O, Klessmann C, Kreiss J, Río P, Wigand F, Woodman B (2019) Auctions for renewable energy support: lessons learned in the AURES Project. IAEE Energy Forum/Third Quarter 2019. International Association for Energy Economics (IAEE)
Kreiss J, Ehrhart KM, Haufe MC (2017) Appropriate design of auctions for renewable energy support—Prequalifications and penalties. Energy Policy 101:512–520
Kremer M (2022) Innovation, experimentation, and economics. Presentation at eighth Richard Goode lecture. 30 Nov 2022. International Monetary Fund. https://www.imf.org/en/Videos/view?vid=6316368113112
Kroposki B (2022) UNIFI consortium overview. The Universal Interoperability for Grid-forming Inverters (UNIFI)
Kruger W, Nygaard I, Kitzing L (2021) Counteracting market concentration in renewable energy auctions: lessons learned from South Africa. Energy Policy 148.https://doi.org/10.1016/j.enpol.2020.111995
Kwon T (2020) Policy mix of renewable portfolio standards, feed-in tariffs, and auctions in South Korea: are three better than one? Utilities Policy 64:101056
Lackner M, Koller S, Camuzeaux JR (2019) Policy brief using lessons from reverse auctions for renewables to deliver energy storage capacity: guidance for policymakers. Rev Environ Econ Policy 13(1):140–148. https://doi.org/10.1093/reep/rey019
Lamdouar S, O'Connell P, Wong T (2022) Assessing ESG-Labeled bonds. May 6, 2022. Harvard Law School Forum on Corporate Governance. Available at https://corpgov.law.harvard.edu/

LDES Council, McKinsey and Company (2022) A path towards full grid decarbonization with 24/7 clean power purchase agreements. Long duration energy storage (LDES) Council in collaboration with McKinsey and Company

Leonard Energy (2022) How auction design affects the financing of renewable energy projects. H2020 project: auctions for renewable energy support. Webinar Leonard Energy 26 Jan 2022. Available at https://www.youtube.com/watch?v=u75HR7TKYKU. Retrieved 1 June 2022

Leyton-Brown K, Milgrom P, Segal I (2017) Economics and computer science of a radio spectrum reallocation. In: Proceedings of the National Academy of Sciences of the United States of America, vol 114, Issue 28. National Academy of Sciences, pp 7202–7209. https://doi.org/10.1073/pnas.1701997114

Massachusetts Institute of Technology (MIT) (2022) The future of energy storage. The MIT Energy Initiative

Matthäus D (2020) Designing effective auctions for renewable energy support. Energy Policy 142.https://doi.org/10.1016/j.enpol.2020.111462

Maurer L, Doyle P, Hyman E, Bauer L, Torres P (2020) Creating a level playing field for battery energy storage systems through policies, regulations, and renewable energy auctions. Crown Agents USA and Abt Associates, Prepared for USAID, Washington, DC

Mbugua B (2021) Lake Turkana wind power faces storms in title tussle. People daily. Available at https://www.pd.co.ke/business/lake-turkana-wind-power-faces-storms-in-title-tussle-101655/. Retrieved 1 Oct 2022

Melliger M (2023) Quantifying technology skewness in European multi-technology auctions and the effect of design elements and other driving factors. Energy Policy 175.https://doi.org/10.1016/j.enpol.2023.113504

Münch FA, Scheifele F (2023) Nurturing national champions? Local content in solar auctions and firm innovation local content requirements solar energy auctions green industrial policy staggered difference-in-difference propensity score matching. https://doi.org/10.7910/DVN/UW

Newbery D (2020) Implications of the national energy and climate plans for the single electricity market of the island of Ireland. Energy policy research group working paper, Cambridge Working Paper in Economics 2072, University of Cambridge

NS Energy (2020) Canadian Solar wins large solar plus BESS project in Chile's power auction. 2 Sept 2022. Available at https://www.nsenergybusiness.com/news/canadian-solar-zaldivar-solar-plus-bess-project-chiles/. Retrieved 1 Oct 2022

O'Mealy M, Sangarasri T, Khalid E, Hyman E (2020) Private sector recommendations for renewable energy auctions in Thailand and Malaysia. Crown Agents-USA, and Abt Associates, Prepared for USAID, Washington, DC

Plummer JL, Troppmann S (eds) (1990) Competition in electricity: new markets and new structures. Palo Alto, Calif. Public Utilities Reports, Inc.; QED Research Inc., Arlington Va

Renewable Energy Policy Network for the 21st Century (REN21) (2022) Renewables 2022 global status report. Renewable energy policy network for the 21st century (REN21)

Reuters (2021) German auction agrees terms to close 533 MW of coal power. 14 Dec 2021. Available at https://www.reuters.com/markets/commodities/german-auction-agrees-terms-close-533-mw-coal-power-2021-12-15/. Retrieved 1 Oct 2022

Roth A, Wigand F, Blanco A (2022) Policy Brief, January 2022, De-risking and scaling-up renewables through market-based policies, Practices from EU and the world—COP26 AURES II side event. Auctions for Renewable Energy Support II (AURES II)

Royal Swedish Academy of Sciences (2020) Scientific background on the sveriges riksbank prize in economic sciences in memory of alfred nobel 2020. Improvements to Auction Theory and Inventions of New Auction Formats. The Committee for the Prize in Economic Sciences in Memory of Alfred Nobel.

Santos B (2022a) Malaysia to grant 4-year PPA extensions to bidders in large-scale solar tender. 28 Oct 2022. PV Magazine. Available at https://www.pv-magazine.com/2022/10/28/malaysia-to-grant-4-year-ppa-extensions-to-bidders-in-large-scale-solar-tender/. Retrieved 1 Nov 2022

Santos B (2022b) Thailand introduces FIT scheme for solar, storage. 31 Oct 2022. PV Magazine. Available at https://www.pv-magazine.com/2022/10/31/thailand-introduces-fit-scheme-for-solar-storage/. Retrieved 1 Nov 2022

Seel J, Warner C, Mills A (2022) Influence of business models on PV-battery dispatch decisions and market value: a pilot study of operating plants. Adv Appl Energy 5.https://doi.org/10.1016/j.adapen.2021.100076

Stevens P (2022) Solar costs jumped in 2021, reversing years of falling prices. 10 Mar 2022. CNBC. Available at https://www.cnbc.com/2022/03/10/solar-costs-jumped-in-2021-reversing-years-of-falling-prices.html. Retrieved 1 Oct 2022

Szabó L, Bartek-Lesi M, Dézsi B, Diallo A, Mezősi A (2022) D2.2, December 2020, Auctions for the support of renewable energy: lessons learnt from international experiences—synthesis report of the AURES II case studies. Auctions for Renewable Energy Support II (AURES II)

Theobald S (2022) Drivers of investment flows to emerging and frontier markets. Mobilising institutional capital through listed product structures (MOBILIST). The United Kingdom Government's Foreign, Commonwealth and Development Office

Tirole J (2017) Economics for the common good. Princeton University Press

Tirole J (2021) Markets and morality. The second sustainable finance center conference keynote presentation. 2 Dec 2021. Available at https://www.youtube.com/watch?v=62zcKL3RlSY. Retrieved 1 June 2022

Tongsopit S, Amatayakul W, Saculsan PG, Nghia VH, Tirapornvitoon C, Favre R, Amazo A, Tiedemann S, Afanador A (2017) Designing renewable energy incentives and auctions: lessons for ASEAN. 4 Sep 2017. The United States Agency for International Development

United Nations Environment Programme (UNEP) (2022) State of finance for nature. Time to act: doubling investment by 2025 and eliminating nature-negative finance flows. Nairobi. https://wedocs.unep.org/20.500.11822/41333

Voice of America (VOA) (2020) Global brands say future orders at risk given Cambodia's increasing coal power. 12 Aug 2020. Available at https://www.voacambodia.com/a/global-brands-say-future-orders-at-risk-given-cambodia-increasing-coal-power/5540674.html. Retrieved 1 Oct 2022

Wigand F, Förster S, Amazo A, Tiedemann S (2016) Auctions for renewable energy support: lessons learnt from international experiences. Report D4.2, June 2016. Auctions for Renewable Energy Support: Effective use and efficient implementation options (AURES)

World Bank (2022) The use of auctions for decommissioning coal power globally. The World Bank, Washington, DC

Chapter 3
Power Trade and Hydroelectricity Development in the Greater Mekong Sub-region: Perspectives on Economic and Environmental Implications

Youngho Chang

Abstract This study examines how cross-border power trade affects the development of hydropower potential in the Greater Mekong Sub-region (GMS) in an ASEAN power trade model, drawing economic and environmental implications. Although utilization rate estimates of potential GMS hydropower capacity range from full utilization to under-utilization, the findings of this study strongly suggest that cross-border power trade will promote GMS hydropower and other renewable development, in turn reducing the amount of carbon dioxide emitted in the region as a whole, with associated positive environmental implications.

Keywords Power trade · Hydroelectricity · Cross-border interconnections · Electricity market integration · GMS

1 Introduction

The Association of Southeast Asian Nations (ASEAN) has been working to utilize energy resources across the Southeast Asian region through such integrated grid networks as the ASEAN Power Grid (APG) and the Trans ASEAN Gas Pipelines (TAGP). The Great Mekong Sub-region (GMS) is a successful pioneer of regional power trade.

ASEAN member states, especially those which comprise the GMS, namely Cambodia, Lao PDR, Myanmar, Thailand, and Vietnam, have tremendous hydropower potential (Chang and Li 2013; Li and Chang 2015; ASEAN Centre for Energy 2022). Table 1 presents actual hydroelectricity capacities as of 2018 and potential hydroelectricity capacities for the GMS states.

Y. Chang (✉)
School of Business, Singapore University of Social Sciences, 463 Clementi Road, Singapore 599494, Singapore
e-mail: yhchang@suss.edu.sg

H. Phoumin et al. (eds.), *Large-Scale Development of Renewables in the ASEAN*, Economics, Law, and Institutions in Asia Pacific,
https://doi.org/10.1007/978-981-99-8239-4_3

Table 1 Actual (2018) and potential GMS hydroelectricity capacity (MW)

Capacity/countries	Cambodia	Lao PDR	Myanmar	Thailand	Vietnam	Total
Actual	1330	5472	3259	3103.4	20,170	33,334.4
Potential	26,417.27	48,949.64	108,000	12,431.65	95,568.35	291,366.91
Utilization (%)	5.03	11.18	3.02	24.96	21.11	11.44

Source Chang and Li (2015), ASEAN Centre for Energy (2022)

As shown in Table 1, utilization rates of hydroelectricity potential in the GMS range from approximately 3% in Myanmar to approximately 25% in Thailand, and 11.44% in the aggregate, showing the tremendous hydroelectricity potential for the area, as mentioned above, while highlighting that these states have been slow to develop that potential, due chiefly to a lack of interconnections between sources and end users. As of April 2020, existing ASEAN cross-border bilateral interconnections capacity stands at 7720 MW, capacity under construction ranges from 555 to 625 MW, and capacity to be built from 18,369 to 21,769 MW (ASEAN Centre for Energy 2021).

There have been some suggestions for how to design a market for multilateral trade of electricity in ASEAN. The International Energy Agency (IEA) has proposed a four-stage model for ASEAN, comprising a bilateral stage, a secondary trading stage, a primary trading mode, and a fully integrated regional market (International Energy Agency 2019). More practically, a two-phase market development is proposed before transitioning to an integrated electricity market (Li et al. 2020). Apart from constructive suggestions for integrating electricity markets, there are some cautious hopes and doubts alike. Cross-border power sales between Singapore and Malaysia might set a good example for harmonizing and liberalizing the ASEAN electricity market (Trowers and Hamlins LLP 2021). A limited institutional capacity may hinder further hydroelectricity development in Lao PDR (Tran and Suhardiman 2022). It may be premature for ASEAN to implement multilateral cross-border electricity trade, chiefly due to bilateral power agreements and huge investment costs (Do and Burke 2022).

Most cross-border interconnections within the GMS are dedicated connections between exporters and importers with no third-party access (Ricardo Energy and Environment 2019), which may hinder the development of full-fledged open access grid interconnections in the GMS.

The purpose of this study is to demonstrate how GMS power trading will promote hydroelectricity development in the region, and to draw justifications of cross-border bilateral and open-access interconnections that will help the GMS states cooperate on driving such development. This study collects and evaluates information relating to the status of the aforementioned cross-border bilateral interconnections in the GMS to construct a model incorporating such interconnections. It then solves the model and derives solutions and policy implications, using the General Algebraic Modeling System (GAMS).

The remainder of this chapter is structured as follows. Section 2 reviews how electricity markets can be integrated and what benefits such integrated markets will bring. Section 3 presents the data, the model, and two scenarios, while Sect. 4 presents results, discussions, and policy implications. Section 5 concludes.

2 Literature Review

As with market integration in general, the key drivers of energy market integration are cooperation and coordination rather than control and confinement (Chang 2021). The International Energy Agency (IEA) has suggested a four-stage model for multilateral electricity trade in ASEAN, comprising a bilateral first stage, a second stage of secondary trading, a primary trading mode at the third stage, and a fully integrated regional market at the fourth and final stage (International Energy Agency 2019). A two-phase scheme is also proposed as being more practical before ASEAN fully integrates its electricity markets (Li et al. 2020).

Apart from the methodological aspects of electricity market integration, there is room for some cautious hope and doubt alike. Cross-border power sales between Singapore and Malaysia may provide an example of how to harmonize and liberalize ASEAN electricity markets (Trowers and Hamlins LLP 2021). Most cross-border interconnections within the GMS are dedicated connections between exporting and importing states, with no third-party access (Ricardo Energy and Environment, 2019), preventing GMS from developing full-fledged open access grid interconnections. As such, bilateral agreements and huge investment costs might make multilateral cross-border electricity trade in ASEAN premature (Do and Burke 2022). In Lao PDR, limited institutional capacity might forestall further hydroelectric development there (Tran and Suhardiman 2022).

European nations offer examples of successfully integrating state or national electricity markets (Jamasb and Pollitt 2005). ASEAN can similarly succeed in doing so (Chang and Li 2015; Chang et al. 2016). African states may also successfully integrate their national electricity markets provided they meet certain preconditions for trading, institutional arrangements, setting practical timetables, and identifying future prospects (Oseni and Pollitt 2016). Grid interconnection in Northeast Asia might result in an economically efficient power system and similarly efficient renewable energy utilization in the region despite large initial outlay requirements (Otsuki et al. 2016). While South Asia has very low-level cross-border power trade, short- and medium-term bilateral electricity cooperation there might build confidence in such trade, eventually promoting electricity market integration (Singh et al. 2018).

There have been studies of whether and how cross-border power trade benefit nations or regions. Regional power market integration in ASEAN demonstrated institutional and policy aspects of regional development in association with energy cooperation (Yu 2003; Yu et al. 2005; Watcharejyothin and Shrestha 2009; Economic

Consulting Associates 2010). Linking resource-rich and resource-poor ASEAN member states may reduce overall costs of meeting its growing electricity demand (Chang and Li 2013).

It is anticipated that hydropower will promote cross-border power trade between the US and Canada that may aid the US low-carbon economy transition (Yuan et al. 2021), although cross-border power trade in North America has instead increased natural gas-fired power generation (Siddiqui et al. 2020). As Table 1 shows, hydropower is the most abundant renewable resource in ASEAN (BP 2022), followed by wind. It is anticipated that ASEAN cross-border power trade will promote wind energy development as well as hydropower (Chang and Phoumin 2021). Excess electricity from such renewables, unless otherwise curtailed, might also be tapped to generate hydrogen that may replace fossil fuels being currently used to generate electricity (Chang and Phoumin 2022).

Transmission grids are essential for cross-border power trade. As mentioned above, infrastructure investments to build integrated electricity markets incur tremendous upfront outlays, which might be offset by benefits accruing from cross-border power trade, resulting in net positive gains, albeit small ones (Li and Chang 2015). Constructing cross-border energy trade infrastructure in the Central Asia Regional Economic Cooperation Energy Corridor (CAREC) appears to reduce the region's carbon emissions per GDP, increase GDP per energy use, and promote renewable electricity (Qadir and Dosmagambet 2020), proving that promoting power trade through connecting grids in integrated electricity markets bring net benefits to participating states.

As noted above, there have been studies on power development and economic benefits in integrated electricity market. There have not, however, been many studies on whether and how cross-border power trade promotes hydropower and other renewables development. As mentioned above, the purpose of this study is accordingly to examine whether cross-border power trade will accelerate ASEAN hydropower development and estimate possible gains after accounting for regional grid connection costs.

3 Data, Model, and Scenarios

3.1 Model

This study builds a cross-border power trade model following Turvey and Anderson (1977), Chang and Tay (2006) and Chang and Li (2013). The key innovation is including cross-border grid construction cost in the model, which is solved using the General Algebraic Modeling System (GAMS).

3.2 Descriptions

The objective of the cross-border power trade model is to minimize the cost of meeting demand for electricity in ASEAN for the study period of 2018–2040. Said cost chiefly comprises capital, operation, transmission, and carbon costs. The capital expenditure (CAPEX) of a given type of power generation capacity at a given time is expressed as follows:

$$\sum_{i=1}^{I}\sum_{v=1}^{T}\sum_{m=1}^{M} c_{miv} * x_{miv}, \tag{1}$$

where x_{miv} is the capacity of plant type m, vintage v, in country i. Vintage indicates when a given type of capacity is built and made operational. c_{miv} is the corresponding capital cost per unit of capacity of the plant. A time dimension is added to the equation besides the vintage dimension for simulation purposes and consistency in presentation with other cost terms, allowing capital cost amortization using a capital recovery factor.

The operational expenditure (OPEX) of said given type of power generation capacity at the given time is expressed as follows:

$$Opex(t) = \sum_{i=1}^{I}\sum_{j}^{J}\sum_{v=-V}^{t}\sum_{p=1}^{P}\sum_{m=1}^{M} F_{mitv} * u_{mijtvp} * \theta_{jp}, \tag{2}$$

where u_{mijtvp} is the power output of plant m, vintage v, in year t, country i, block p on the load, and exported to country j. F_{mitv} is the corresponding operating cost that varies with v, and θ_{jp} is the time interval of load block p within each year in the destination country.

The amount of carbon emissions of different power generation types or technologies is expressed as follows:

$$\sum_{m=1}^{M}\sum_{i=1}^{I}\sum_{j=1}^{J}\sum_{v=-V}^{T} u_{mijtvp} * \theta_{jp} * ce_m \tag{3}$$

and the carbon cost in year t is expressed as follows:

$$CC(t) = cp_t * \left(\sum_{m=1}^{M}\sum_{i=1}^{I}\sum_{j=1}^{J}\sum_{v=-V}^{T} u_{mijtvp} * \theta_{jp} * ce_m\right), \tag{4}$$

where ce_m is the carbon emissions per unit of power plant capacity of type j plant, and cp_t is the carbon price per unit of carbon emissions in year t.

Cross-border transmission costs comprise tariffs and transmission loss. The tariff is paid to recover grid line capital investments and operational costs. Transmission loss may be significant over long transmission distances. To model tariffs, let tp_{ijv} be the amount of new transmission capacity added between country i and j at year v. Then ct_{ijv} and co_{ijv} are the annualized CAPEX, with a 30-year contract and stipulated IRR embedded, and OPEX of the new transmission capacity, respectively. The total cost of cross-border power transmission in year t is expressed as follows:

$$TC(t) = \sum_{i=1}^{I} \sum_{j=1}^{J} \sum_{v=-V}^{T} \left(\boldsymbol{ct}_{ijv} + \boldsymbol{co}_{ijv} \right) * \boldsymbol{tp}_{ijv}. \tag{5}$$

As mentioned above, the objective of the power trade model is to minimize the total cost of electricity during the period being studied. The objective function is written as:

$$obj = \sum_{i=1}^{I} \sum_{v=1}^{T} \sum_{m=1}^{M} c_{miv} * x_{miv} + \sum_{t=1}^{T} \{Opex(t) + CC(t) + TC(t)\} \tag{6}$$

Several constraints are required to optimize the above objective function. This study makes some key assumptions to ensure meeting domestic demand and surplus electricity trading. First, total installed regional power generation capacity should be greater than or equal to total regional electricity demand. Second, total electricity output in each country is constrained by the load factor of each installed capacity of all types of electricity generation in the county. Third, the electricity supply of all countries in the region to a given country should be greater than or equal to said country's electricity demand. Fourth, total electricity supply from one country to all countries in the region, including said country itself, must be less than or equal to the country's total available supply capacity at a given time.

Equation (7) shows a first set of constraints, which requires total power capacity to meet total power demand in the region. Q_{itp} is the power demand of country i in year t for load block p:

$$\sum_{i=1}^{I} \sum_{j=1}^{J} \sum_{m=1}^{M} \sum_{v=-V}^{t} u_{mijtvp} \geq \sum_{i=1}^{I} Q_{itp} \tag{7}$$

The second constraint, shown in Eq. (8), states the constraint of load factor lf_{mi} of each installed capacity of power generation. kit_{mi} is the initial vintage capacity of type m power plant in country i:

$$u_{mijtvp} \leq lf_{mi} * (kit_{mi} + x_{miv}) \tag{8}$$

The third constraint, shown in Eq. (9), states that all countries' power supply to a given country must be greater than the country's demand. $tl_{i,j}$ is the ratio of transmission loss in cross-border electricity trade between country *i* and country *j*:

$$\sum_{j=1}^{J}\sum_{m=1}^{M}\sum_{v=-V}^{t} u_{mijtvp} \cdot tl_{ij} \geq Q_{itp} \tag{9}$$

The fourth constraint, shown in Eq. (10), states that total supply of power of one country to all countries, including itself, must be less than the aggregate of the country's available power capacity at the time:

$$\sum_{j=1}^{J} u_{mijtvp} \leq \sum_{m=1}^{M}\sum_{v=-V}^{t} lf_{mi} * (kit_{mi} + x_{miv}) \tag{10}$$

The fifth constraint, shown in Eq. (11), is capacity reserve constraint. pr is the rate of reserve capacity as required by regulation. $p = 1$ represents the peak load block:

$$\sum_{i}^{I}\sum_{m=1}^{M}\sum_{v=-V}^{t} lf_{mi} * (kit_{mi} + x_{miv}) \geq (1 + pr) * \sum_{i}^{I} Q_{it,p=1} \tag{11}$$

Hydroelectric facilities have a so-called energy factor constraint, as shown in Eq. (12). ef_{mi} is the energy factor of plant type m in country i. Other facilities have $ef = 1$:

$$\sum_{p=1}^{P}\sum_{j=1}^{J} u_{mijtvp} \leq ef_{mi} * (kit_{mi} + x_{miv}) \tag{12}$$

Last, development of power generation capacity faces the resource availability constraint, per Eq. (13). $XMAX_{mi}$ is the type of resource constraint of plant type m in country i:

$$\sum_{v=1}^{T} x_{miv} \leq XMAX_{mi} \tag{13}$$

3.3 Scenarios

This study establishes the following broadly defined scenarios. The first, which serves as a reference case, is constructed by replicating the current cross-border transmission network in which no cross-border power trade takes place. This study provides

Table 2 Scenarios

Scenarios	Maximum import %	Remarks
No trade	No cross-border imports	Reference case
50% trade	Up to 50% imports	Alternative case

Source Author's compilation

a cost estimate lower bound by assuming that the planned ASEAN Power Grid operates on schedule, as well as that transmission costs and losses over distances between countries are accounted for in calculating the total cost of meeting the region's electricity demand. This base scenario, or business-as-usual case, replicates current GMS system-to-system interconnections where no third-party or open access is allowed, to reflect current GMS cross-border interconnections. Table 2 provides a summary of these scenarios.

The alternative scenario of 50% trade is constructed by assuming that planned transmission connections are built and operational. Under this scenario, the minimum electricity supplied from domestic sources is 50%, and thus, up to 50% of a country's electricity demand can be met with imports. Table 3 provides planned interconnections with completion dates and capacities as of 2018.

Table 3 Ongoing and planned cross-border power transmission line projects (APG+)

Country A	Country B	To be completed in	Capacity (MW)
Cambodia	Vietnam	2019	465
Cambodia	Thailand	2025	1800
Indonesia	Malaysia	2020	200
Brunei	Malaysia	2020	100
Laos	Thailand	2018	1169
Laos	Thailand	2019	1535
Laos	Thailand	2020	630
Laos	Thailand	2023	1040
Laos	Vietnam	2020	100
Singapore	Malaysia	2018	600
Singapore	Indonesia	2020	600
Philippines	Malaysia	2020	500
Myanmar	Thailand	2022	1190
Myanmar	Thailand	2025	10,150

Source Chimklai (2013), Zhai (2010), ADB (2013), APERC (2004), Bunthoeun (2012)

3.4 Data

This study covers the ASEAN member states of Brunei, Cambodia, Indonesia, Lao PDR, Malaysia, Myanmar, the Philippines, Singapore, Thailand, and Vietnam. Technologies for power generation discussed in this study are coal, coal CCS, diesel, natural gas, natural gas CCS, hydro, small hydro, geothermal, wind, solar PV, and biomass. The period studied by this optimization simulation model is 2018–2050.

Key data are existing capacities of power generation types, CAPEX and OPEX thereof, load factor and life expectancy of each vintage thereof, resources available for power generation in each country, peak and non-peak demand and duration of demand in each country, projected demand growth rates, and cross-border power trade transmission cost and loss. This study takes data used in Chang and Li (2013) and updates initial capacities given in Chang and Li (2013) based on data taken from ASEAN Centre for Energy (ACE), 2020 and IRENA (2019). Figure 1 shows ASEAN's initial installed capacity in by plant type as of 2018.

The load factors of various generation technologies are as shown in Table 4.

Refer to the Appendix for input data and sources.

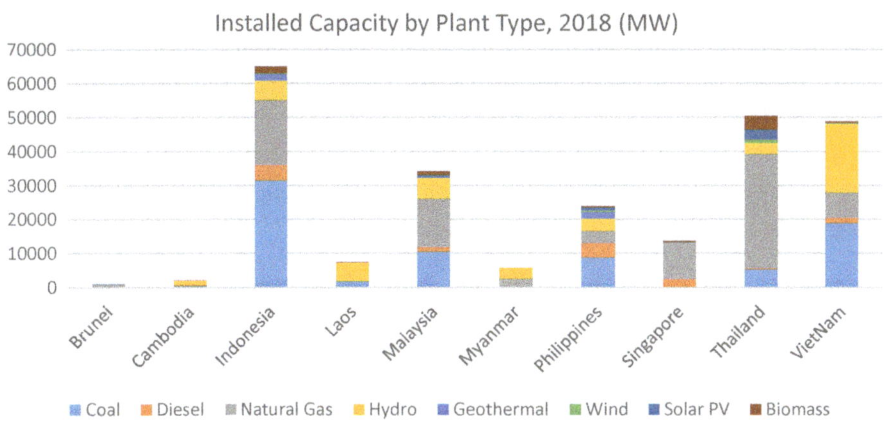

Fig. 1 Installed Capacity by Plant Type in ASEAN, 2018 (MW). *Source* ACE (2022) and IRENA and ACE (2019)

Table 4 Load factors of generation technologies

Types of generation technologies	Load factor
Fossil fuel-fired	0.85
Hydro	0.90
Geothermal	0.95
Wind	0.30
Solar PV	0.11
Biofuel	0.85

Source Chang and Li (2013)

4 Results, Discussions and Policy Implications

This study's key findings are as follows. Table 5 gives a summary of potential and utilized hydropower capacity, i.e., added capacity, by scenario and country.

First, it is anticipated that hydropower potential utilization increases when the minimum electricity import rate increases, suggesting that cross-border power trade appears to promote ASEAN hydropower development. Second, hydropower potential utilization varies by country, with Cambodia, Lao PDR, and Myanmar appearing to fully utilize their potential capacities in the 50% Trade scenario. Third, Malaysia, the Philippines, and Vietnam appear to fully utilize their hydropower potential in either scenario. Fourth, Indonesia and Thailand appear to utilize less of their potential capacities in the of 50% Trade scenario. Fifth, Lao PDR appears not to add any hydropower capacity in the No Trade scenario.

4.1 Hydropower Potential Utilization Increases with Higher Import Percentages

The hydropower potential utilization rate in the No Trade scenario is 53.15%, rising to 80.81% in the 50% Trade scenario. Table 6 shows the hydropower potential utilization rates, and Fig. 2 shows the total added hydropower capacity over the study period.

Table 5 A summary of potential and utilized capacity under scenarios (MW)

Countries	Potential capacity (MW)	Utilized capacity (MW)		Remarks
		No trade (%)	50% trade (%)	
Cambodia	10,300	5468.11 (53.09)	10,300 (100)	Fully utilized
Indonesia	75,459	58,748.11 (77.85)	28,334.15 (37.55)	Less utilized
Lao PDR	18,000	0 (0)	18,000 (100)	Fully utilized
Malaysia	29,000	29,000 (100)	29,000 (100)	No change
Myanmar	108,000	15,528.72 (14.38)	108,000 (100)	Fully utilized
Philippines	13,097	13,097 (100)	13,097 (100)	No Change
Thailand	15,155	13,639.50 (90)	7577.50 (50)	Less utilized
Vietnam	16,000	16,000 (100)	16,000 (100)	No change

Source Author's compilation

Table 6 Hydropower potential utilization rates (%)

Scenarios	Utilization rates (%)	Remarks
No trade	53.15	
50% trade	80.81	

Source Author's calculation

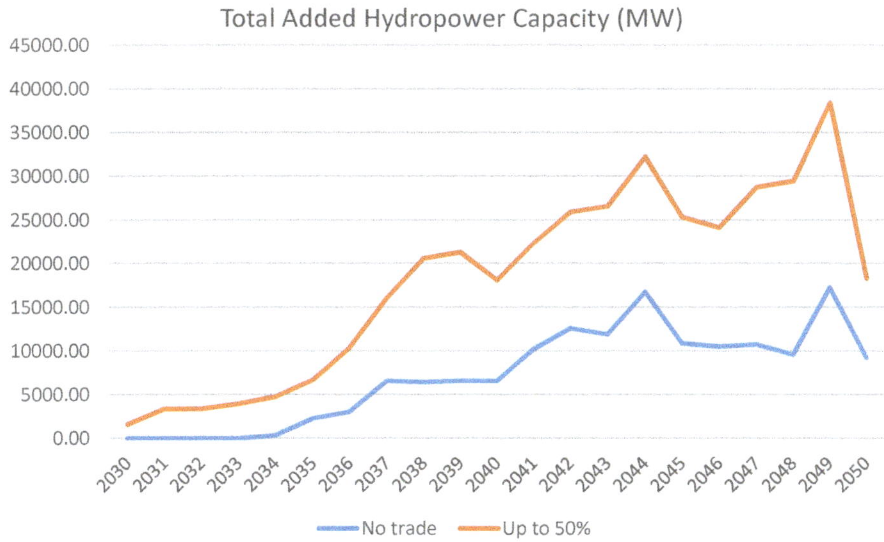

Fig. 2 Total added hydropower capacity (MW). *Source* Author's compilation

As shown in Table 6, more hydropower capacity is added in the 50% Trade scenario over the entire study period than in the No Trade scenario. Vietnam appears to be the first to add hydropower capacity, beginning in 2030, per Fig. 8.

4.2 *Full Potential Hydropower Utilization: Cambodia, Lao PDR, and Myanmar*

Cambodia, Lao PDR, and Myanmar appear to utilize their potential hydropower capacity to the fullest, with Cambodia appearing to start sooner.

4.2.1 Cambodia

In the No Trade scenario, Cambodia adds hydropower to its fuel mix beginning in 2037, achieving some 54% of its potential capacity. In the 50% Trade scenario, it instead adds hydropower beginning in 2033, utilizes more capacity than in the No Trade scenario every year, and fully utilizes its potential by 2044 (Fig. 3).

4.2.2 Lao PDR

While Lao PDR appears not to add any hydropower to its fuel mix in the No Trade scenario, it appears to add its full potential hydropower capacity to its fuel mix in the 50% Trade scenario (Fig. 4).

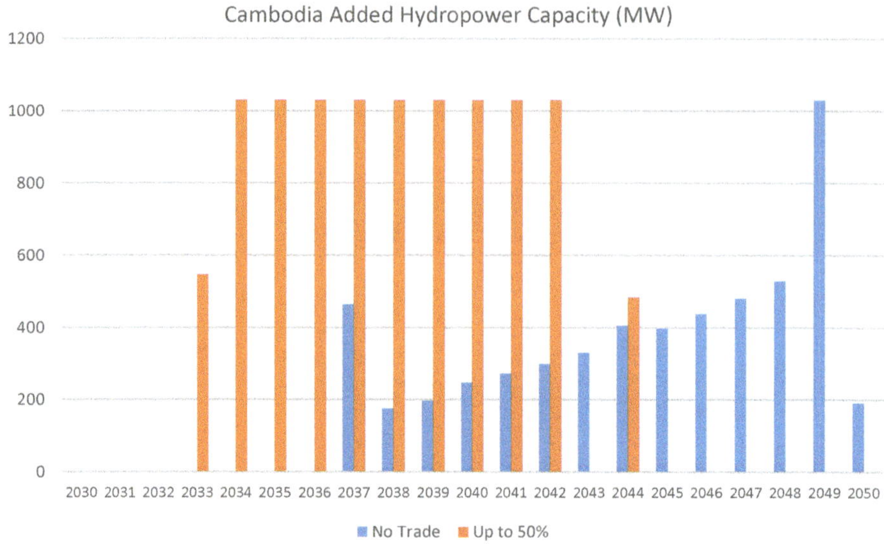

Fig. 3 Cambodia. *Source* Author's compilation

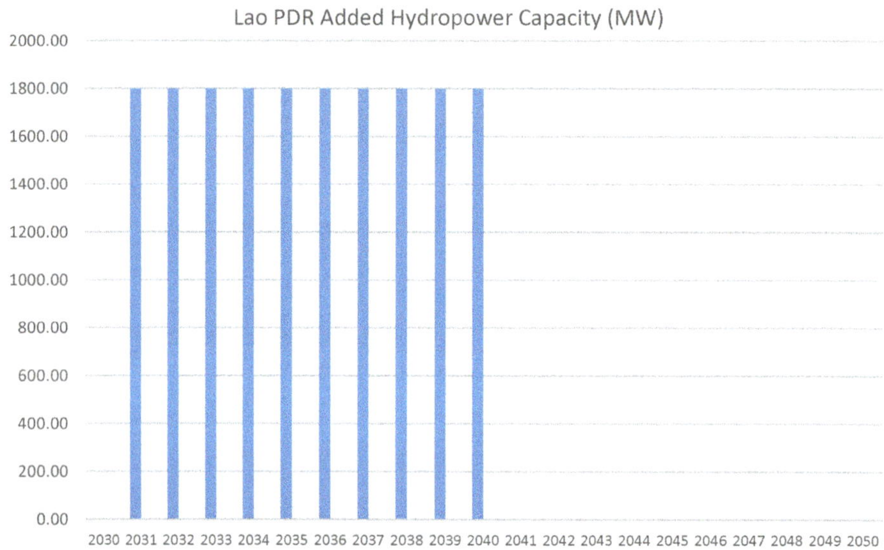

Fig. 4 Lao PDR. *Source* Author's compilation

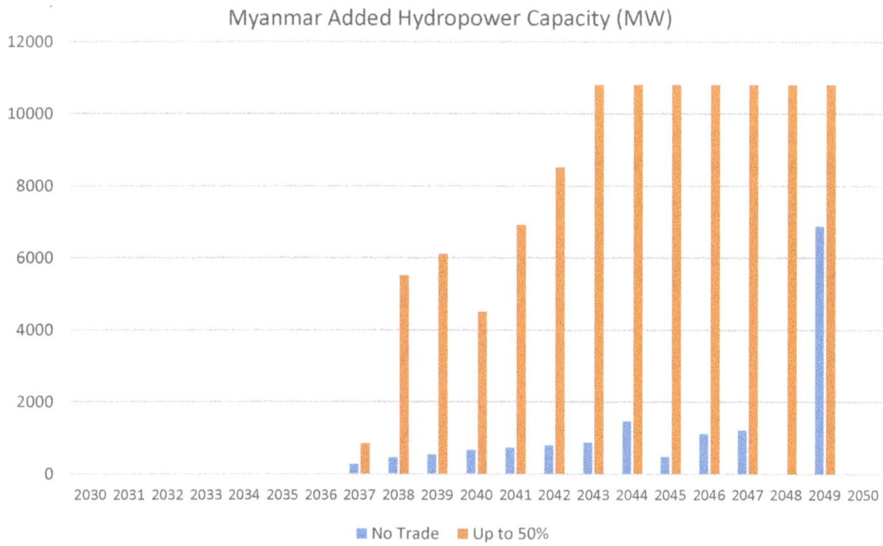

Fig. 5 Myanmar. *Source* Author's compilation

4.2.3 Myanmar

While Myanmar appears to add some 14% of its hydropower potential to its fuel mix from 2037 to 2050 in the No Trade scenario, it appears to add its full hydropower potential capacity to its fuel mix over the same period in the 50% Trade scenario (Fig. 5).

4.3 Malaysia, the Philippines, and Vietnam: Fully Utilized in Either Scenario

Malaysia, the Philippines, and Vietnam appear to fully utilize their potential hydropower capacity in either scenario.

4.3.1 Malaysia

As shown in Fig. 6, Malaysia appears to fully utilize its potential hydropower capacity evenly from 2036 to 2045 in the 50% Trade scenario, while Malaysia's potential hydropower capacity utilization level in the No Trade scenario apparently fluctuates somewhat and it takes slightly longer to achieve such full utilization.

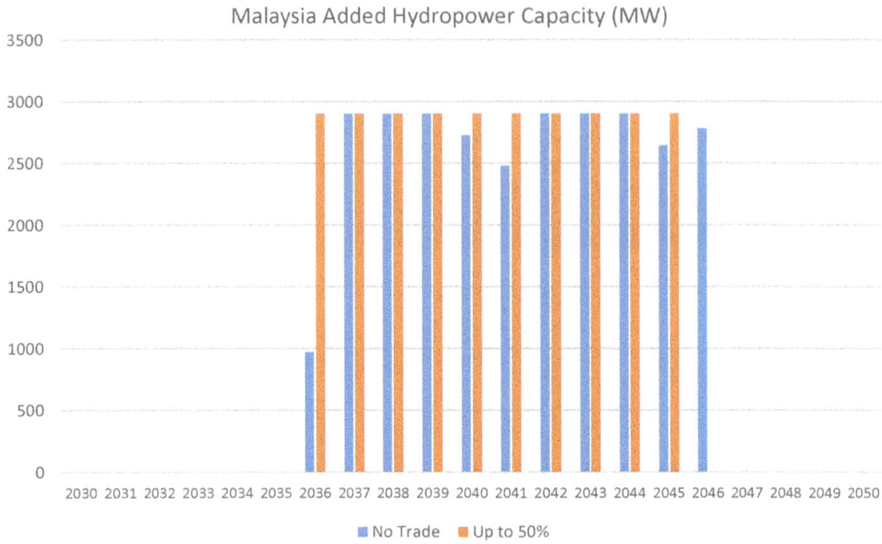

Fig. 6 Malaysia. *Source* Author's compilation

4.3.2 The Philippines

In the 50% Trade scenario, the Philippines appears to fully utilize its hydropower potential later than in the No Trade scenario, while apparently utilizing this potential over a longer period than in the 50% Trade scenario (Fig. 7).

4.3.3 Vietnam

As mentioned above, Vietnam appears to fully utilize its hydropower potential in either scenario. The only difference is that cross-border power trade in the 50% Trade scenario brings the starting of such utilization forward from 2035 to 2030 (Fig. 8).

4.4 Indonesia and Thailand: Under-Utilization of Hydropower Potential

Indonesia and Thailand appear to under-utilize their potential hydropower in the 50% Trade scenario as shown in Figs. 9 and 10.

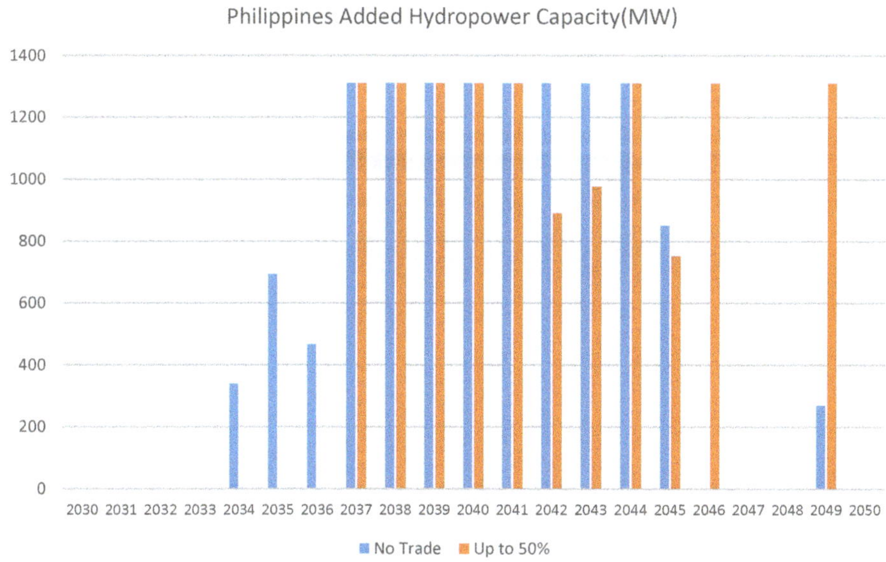

Fig. 7 The Philippines. *Source* Author's compilation

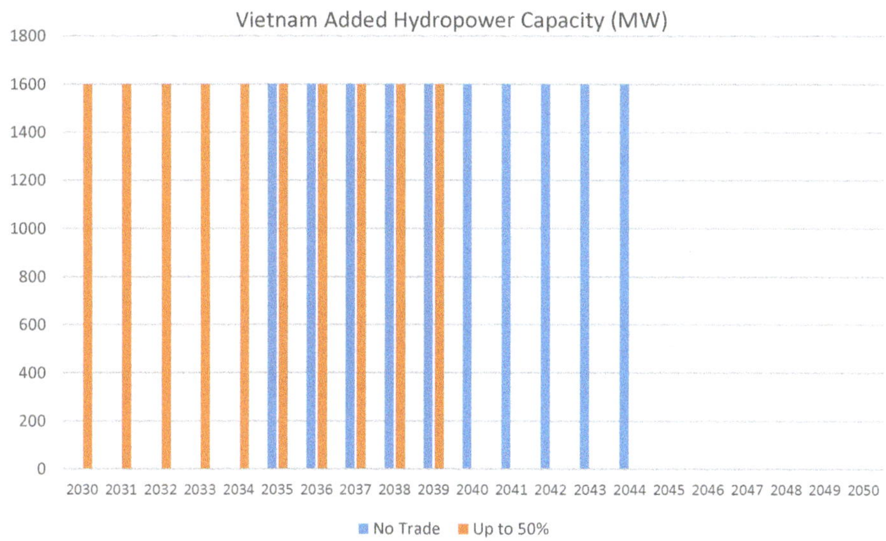

Fig. 8 Vietnam. *Source* Author's compilation

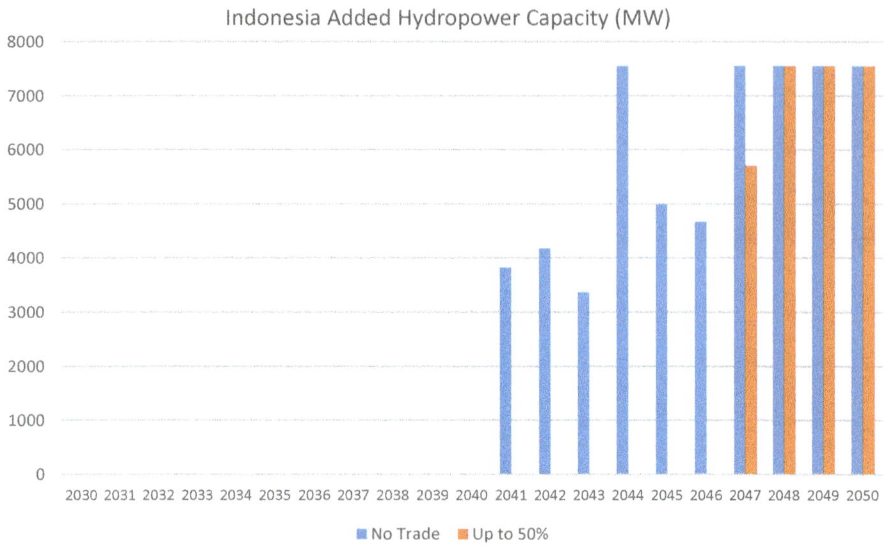

Fig. 9 Indonesia. *Source* Author's compilation

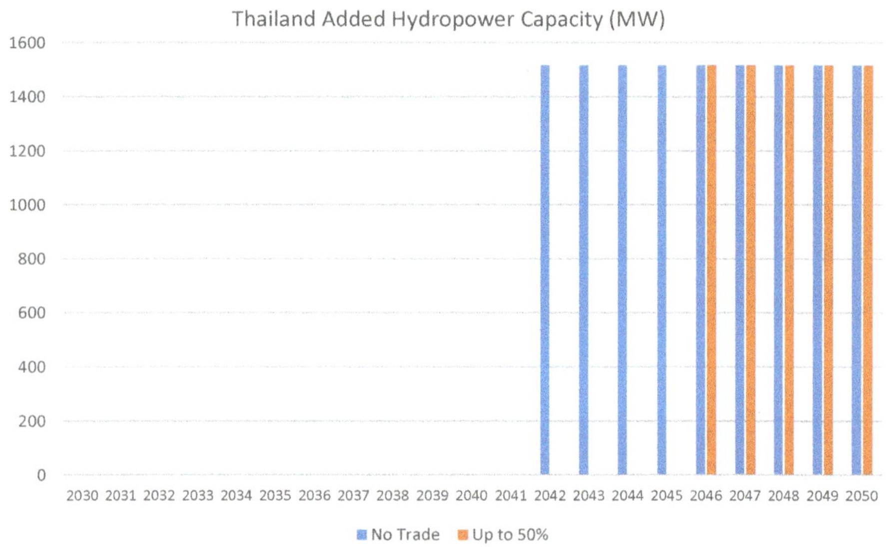

Fig. 10 Thailand. *Source* Author's compilation

4.4.1 Indonesia

Indonesia's potential hydropower utilization rate falls from 77.85% to 37.55% in the 50% Trade scenario. As shown in Fig. 9, Indonesia utilizes more potential hydropower capacity over a longer period in the No Trade scenario than in the 50% Trade scenario, in which such utilization occurs only in the last four years of the study period.

4.4.2 Thailand

Thailand's potential hydropower utilization falls from 90 to 50% in the 50% Trade scenario. As shown in Fig. 10, Thailand utilizes more potential hydropower longer in the No Trade scenario than in the 50% Trade scenario, in which such utilization occurs only in the last five years of the study period, equally distributed.

4.5 No Potential Hydropower Utilization in the No Trade Scenario

While Lao PDR appears not to add any potential hydropower capacity in the No Trade scenario, it apparently adds equal hydropower capacity to its fuel mix from 2031 to 2040 and fully utilize its potential capacity of hydropower, per Fig. 4.

4.6 Policy Implications

The findings of this study offer the following policy implications. First, cross-border power trade appears to accelerate the hydropower development in the GMS as shown in Cambodia, Lao PDR, and Myanmar and the increase in overall potential hydropower utilization in the region. The GMS countries are thus strongly advised to remove barriers to cross-border power trade in the region.

Second, it is anticipated that more hydropower will aid GMS countries' net-zero transition. These states should thus be encouraged to address issues underlying Indonesia and Thailand's potential hydropower under-utilization.

5 Conclusions

The GMS countries have relatively abundant hydropower capacity. Cross-border power trade appears to promote hydropower development therein. The findings of this study suggest that potential hydropower utilization in the GMS appears to increase the greater the permitted electricity imports, exceeding 80% in the 50% Trade scenario versus just over 50% in the No Trade scenario.

Notwithstanding, hydropower potential utilization rates vary by country. Cambodia, Lao PDR, and Myanmar appear to fully utilize their hydropower potential in the 50% Trade scenario. Malaysia, the Philippines, and Vietnam appear to fully utilize their hydropower potential regardless of maximum allowed electricity imports with no net gain, varying only in how much hydropower capacity added to their fuel mixes each year. Interestingly, Indonesia and Thailand appear to utilize less of their hydropower potential in the 50% Trade scenario. Their hydropower potential utilization rates appear to decrease as electricity imports increase. Finally, Lao PDR appears not to add any hydropower capacity at all in the No Trade scenario.

These findings suggest that the GMS countries should strive to remove barriers to cross-border power trade in the region and address issues relating to potential hydropower underutilization to achieve net-zero transitions.

Appendix. Input Data of the Model and Sources of Data

See Tables A1, A2, A3, A4, A5 and A6.

Table A1 CAPEX, OPEX, lifespan, and availability of power generation assets

	Coal*	Diesel	Natural Gas	Hydro**	Geothermal	Wind	Solar PV	Biomass
CAPEX (Million USD/MW)	2.079	1.139	1.054	4.933	6.18	2.187	5.013	4.027
OPEX (USD/MWh)	31.86	229.75	43	4.32	14.23	20.58	19.52	28.87
Lifespan (years)	40	30	30	80	30	25	25	25
Load factor (percentage of year)	0.85	0.85	0.85	0.23–0.64	0.95	0.3	0.11	0.85
Carbon emissions (ton/MWh)	1.0	0.8	0.5	0.001	0.05	0.01	0.05	0.05

Sources IEA (2010) and EUSEC (2008)

*Due to consideration of abundance in coal resources, countries including Indonesia, Malaysia, Thailand, and Vietnam are assumed to have 30% lower CAPEX and OPEX in coal-fired power generation

**Due to consideration of abundance in hydropower resources, countries including Cambodia, Indonesia, Laos, Malaysia, Myanmar, and Philippines are assumed to have 30% lower CAPEX and OPEX in hydropower generation

Table A2 Power generation resources in ASEAN member states (MW)

	Brunei	Cambodia	Indonesia	Laos	Malaysia	Myanmar	Philippines	Singapore	Thailand	Vietnam
Coal	15,000	15,000	50,000	15,000	50,000	30,000	30,000	15,000	50,000	50,000
Diesel	15,000	15,000	15,000	15,000	15,000	15,000	15,000	15,000	15,000	15,000
Natural Gas	15,000	15,000	50,000	15,000	50,000	30,000	30,000	30,000	50,000	50,000
Hydro	0	10,300	75,459	18,000	29,000	0	13,097	0	700	2170
Geothermal	0	0	27,000	0	67	930	2379	0	5.3	270
Wind	0	452	7404	1600	452	1600	7404	0	1600	452
Solar PV	115	3771	37,800	4538	6192	12,967	6336	130.7	300	10,321
Biomass	0	700	49,810	0	29,000	4098	200	50	7000	400

Sources Lidula et al. (2007) and World Energy Council (2010)

Table A3 ASEAN member states' electricity demand and duration

	Brunei	Cambodia	Indonesia	Laos	Malaysia	Myanmar	Philippines	Singapore	Thailand	Vietnam
Peak demand (MW)	454.7	291	23,438	350	12,990	1140	8766	5711	22,586	11,605
Peak duration (hours)	4681.7	4925.2	4681.7	4745	4681.7	2428	4015	5840	4015	2428
Non-peak demand (MW)	257	85	5338	60	8388	162	3394	1324	8692	6862
Non-peak duration (h)	4078.3	3834.8	4078.3	4015	4078.3	6332	4745	2920	4745	6332

Sources HAPUA website; Center for Data and Information on Energy and Mineral Resource (2011); Electricite du Laos (2010); and Zhai (2008, 2009)

Table A4 ASEAN member states' electricity demand growth

	Growth rate (%)
Brunei	1.2
Cambodia	9.9
Indonesia	3.9
Laos	7.7
Malaysia	4.5
Myanmar	9.0
Philippines	4.5
Singapore	4.2
Thailand	4.9
Vietnam	6.7

Sources Institute of Energy Economics, Japan (2011)

Table A5 Change in OPEX and CAPEX

	Rate of change (%)
Coal	2.1
Diesel	1.26
Natural Gas	1.36
Hydro	– 0.5
Geothermal	– 0.5
Wind	– 1.4
Solar PV	– 4.6
Biomass	0.3

Sources EUSEC (2008)

Table A6 ASEAN member states' transmission loss and cost

		Transmission loss (%)	Transmission cost ($/MWh)
Distance*	0–1600 km	0.01	3
	> 1600 km	0.087	5
	> 3200 km	0.174	7.5

Sources Claverton Energy Research Group http://www.claverton-energy.com/
*Estimated distances between member states' capital cities

References

Asian Development Bank (ADB) (2013) Assessment of the Greater Mekong Subregion energy sector development progress, prospects, and regional investment priorities, Philippines
Asia Pacific Energy Research Centre (APERC) (2004) Electric power grid interconnections in the APEC region. Institute of Energy Economics, Japan

ASEAN Centre for Energy (2021) ASEAN Power Grid (APG): Regional integration on electricity interconnectivities, SIEW Thinktank Roundtable, 29 Oct 2021, Singapore
ASEAN Centre for Energy (ACE) (2022) https://aseanenergy.org/work/data-statistic/
British petroleum (BP) (2022) The statistical review of world energy. British Petroleum, London
Bunthoeun P (2012) Country report for the group training course in energy policy (B). Training and Dialogue Programs, presentation slides, Institute of Energy Economics, Japan, Aug 2012. Available at www.eneken.ieej.or.jp/data/4470.pdf. Retrieved 20 Nov 2013
Center for Data and Information on Energy and Mineral Resources (2011) Handbook of energy & economic statistics of Indonesia, Ministry of Energy and Mineral Resources, Indonesia
Chang Y (2021) Integrated regional energy markets: principles and prospects. In: Park SC, Kim CJ, Taghizadeh-Hesary F, Sirivunnabood P (eds) Economic integration in Asia and Europe: lessons and policies. Asia Development Bank Institute, Tokyo, Japan
Chang Y, Tay TH (2006) Efficiency and deregulation of the electricity market in Singapore. Energy Policy 34:2498–2508
Chang Y, Li Y (2013) Power generation and cross-border grid planning for the integrated ASEAN electricity market: a dynamic linear programming model. Energ Strat Rev 2:153–160
Chang Y, Li Y (2015) Renewable energy and policy options in an integrated ASEAN electricity market: quantitative assessments and policy implications. Energy Policy 85:39–49
Chang Y, Phoumin H (2021) Harnessing wind energy potential in ASEAN: modelling and policy implications. Sustainability 13:4279
Chang Y, Phoumin H (2022) Curtailed electricity and hydrogen in Association of Southeast Asian Nations and East Asia Summit: perspectives from an economic and environmental analysis. Int J Hydrogen Energy 47:24548–24557
Chang Y, Fang Z, Li Y (2016) Renewable energy policies in promoting financing and investment among the East Asia Summit countries: quantitative assessment and policy implications. Energy Policy 95:427–436
Chimklai S (2013) Briefing on ASEAN power grid: interconnectivity and cross border trade. Available at http://portal.erc.or.th/aern/images/Panel%201-1%20Briefing%20on%20ASEAN%20Power%20Grid.pdf. Retrieved 20 Nov 2013
Do TN, Burke PJ (2022) Is ASEAN ready to move to multilateral cross-border electricity trade? Asia Pacific Viewpoint. ISSN 1360-7456
Economic Consulting Associates (ECA) (2010) Greater Mekong Subregion (GMS) Transmission & Trading Case Study
Electricite du Laos (2010) Annual report 2010, Vientiane, Laos
EU SEC (2008) Energy sources, production costs and performance of technologies for power generation, heating and transport
Heads of ASEAN Power Utilities/Authorities (HAPUA) (2012) http://www.hapuasecretariat.org/. Assessed in July 2012
International Energy Agency, and Nuclear Energy Agency (2010) Projected cost of generating electricity: 2010 Edition
International Energy Agency (2019). Establishing multilateral power trade in ASEAN (2019) International Energy Agency. France, Paris
International Renewable Energy Agency (IRENA) and ACE (2019) Renewable energy outlook for ASEAN
Institute of Energy Economics, Japan (2011) The third ASEAN energy outlook, paper for the ASEAN SOE leaders and the ministry of the economy, trade and industry, Japan, February 2011
Jamasb T, Pollitt M (2005) Electricity market reform in the European Union: review of progress toward liberalization and integration, Center for Energy and Environmental Policy Research
Li Y, Chang Y (2015) Infrastructure investments for power trade and transmission in ASEAN+2: costs, benefits, long-term contracts and prioritized developments. Energy Econ 51:484–492
Li Y, Rakhmah TF, Wada J (2020) Market design for multilateral trade of electricity in ASEAN: a survey of the key components and feasibility. Asian Econ Papers 19(1):43–60

Lidula NWA, Mithulananthan N, Ongsakul W, Widjaya C, Henson R (2007) ASEAN towards clean and sustainable energy: potentials, utilization, and barriers. Renew Energy 32:1441–1452
Oseni MO, Pollitt MG (2016) The promotion of regional integration of electricity markets: lessons for developing countries. Energy Policy 88:628–638
Otsuki T, Binti A, Isa M, Samuelson RD (2016) Electric power grid interconnections in Northeast Asia: a quantitative analysis of opportunities and challenges. Energy Policy 80:311–329
Qadir S, Dosmagambet Y (2020) CAREC energy corridor: opportunities, challenges, and IMPACT of regional energy trade integration on carbon emissions and energy access. Energy Policy 147:111427
Ricardo Energy and Environment (2019) Greater Mekong Subregion power market development: all business cases including the integrating GMS Case, Didcot, UK
Siddiqui S, Vailancourt K, Bahn O, Victor N, Nichols C, Avraam C, Brown M (2020) Integrated North American energy markets under different futures of cross-border energy infrastructure. Energy Policy 144:111658
Singh A, Jamasb T, Nepal R, Pollitt M (2018) Electricity cooperation in South Asia: barriers to cross-border trade. Energy Policy 120:741–748
Tran TA, Suhardiman D (2022) Laos' hydropower development and cross-border power trade in the Lower Mekong Basin: a discourse analysis. Asia Pac Viewp 61(2):219–235
Turvey R, Anderson D (1977) Electricity economics: essays and case studies. The John Hopkins University Press, Baltimore and London
Trowers and Hamlins LLP (2021) ASEAN's dream for multilateral electricity trade
Watcharejyothin M, Shrestha RM (2009) Effects of cross-border power trade between Laos and Thailand: energy security and environmental implications. Energy Policy 37:1782–1792
World Energy Council (2010) 2010 survey of energy resources
Yu X (2003) Regional cooperation and energy development in the Greater Mekong Sub-region. Energy Policy 31:1221–1234
Yu Z, Bowen BH, Sparrow FT, Siriariyaporn V, Yu L (2005) Integrated energy resources planning for the ASEAN countries and Southern China. Oil Gas Energy Law Intell 3(4):1–12
Yuan M, Tapia-Ahumada K, Reilly J (2021) The role of cross-border electricity trade in transition to a low-carbon economy in the Northeastern U.S. Energy Policy 154:112261
Zhai Y (2008) "ADB's role in supporting power sector restructuring in the Phillipines - electrciity market development project. In: Presentation slides for the orientation program for DMC official, ADB, October 2008
Zhai Y (2009) "Roadmap for energy and power integration in the Greater Mekong Subregion (GMS). In: Presentation slides for the conference on energy development in the Greater Mekong Subregon, Phnom Penh, July 2009
Zhai Y (2010) Energy sector integration for low carbon development in Greater Mekong Subregion: towards a model of south-south cooperation. Available at www.worldenergy.org/documents/congresspapers/52.pdf. Retrieved 20 Nov 2013

Chapter 4
Tradeable Renewable Energy Credit Markets: Lessons from India

Aparna Sawhney

Abstract India has undergone a significant energy mix transformation over the past decade, with renewables accounting for 30% of installed grid capacity and almost 14% of electricity generation today. These achievements, however, fall short of the ambitious targets set for 2030. The policy package for renewables includes a market-based instrument of tradeable renewable energy certificates (RECs), which provide a channel for an alternative valuation of the green attribute of electricity generation in the country. It also provides for spatial flexibility in green power generation in resource-rich areas and compliance with renewable portfolio obligations through REC purchases by states with shortfalls. This paper analyzes the REC market experience over the past decade and examines the implications of changes in trading rules during that time. It highlights that although the renewable certification rate initially rose sharply from 2% in 2011–12 to 15% in 2014–15, it steadily declined to 3–6% during 2017–21 as REC market prices plummeted and unsold RECs accumulated. While the certification rate has picked up following an REC market design overhaul in 2022, problem of unsold RECs inventory persists. The author concludes that the problems of target underachievement and non-compliance with state renewable purchase obligations must be tackled through deep reforms in the functioning of power distribution companies rather than the REC mechanism per se.

Keywords India · Renewable energy certificates · Renewable portfolio obligation

JEL Codes Q 42 · Q 48

An earlier draft was published as ADBI Working Paper 1313 Renewable Energy Certificates Trading in India: A Decade in Review, in 2022.

A. Sawhney (✉)
Centre for International Trade and Development, Jawaharlal Nehru University, New Delhi 110067, India
e-mail: asawhney@jnu.ac.in

H. Phoumin et al. (eds.), *Large-Scale Development of Renewables in the ASEAN*, Economics, Law, and Institutions in Asia Pacific,
https://doi.org/10.1007/978-981-99-8239-4_4

1 Introduction

India, a non-Annex I country, was an early signatory of the 1992 United Nations Framework Convention on Climate Change, the 2005 Kyoto Protocol, and the 2015 Paris Agreement.[1] To mitigate climate change domestically, India implemented a major renewable energy drive as part of the comprehensive National Action Plan on Climate Change (NAPCC) in 2008. The National Solar Mission, one of the eight missions outlined in the NAPCC, was devoted exclusively to solar power as renewable energy. It aimed to increase the share of solar energy in the total energy mix of the country, as well as enhancing the scope of other renewables such as wind and biomass. To ensure sustained demand for grid-connected renewable-based power, the NAPCC proposed a "dynamic minimum" renewable purchase obligation of 5% of the grid's total purchases for 2009–10, increasing by 1% each year for 10 years thereafter (NAPCC 2008: 44). Effectively, 15% of India's electricity was to be produced from renewable resources by 2020.

Under the Paris Agreement, India's Intended Nationally Determined Contribution, submitted to the United Nations Framework Convention on Climate Change (UNFCC) in October 2015, committed to reducing greenhouse gas emissions intensity in its GDP by 33–35% from its 2005 level by 2030, and specified that approximately 40% of the total installed electricity capacity would be non-fossil fuel based by the year 2030 (INDC 2015). At the 2019 United Nations Climate Action Summit, India announced an ambitious goal of 450 GW installed capacity of renewable energy by 2030. At home, an interim goal of 175 GW cumulative installed renewable power capacity was set for 2022, comprising 100 GW solar, 60 GW wind, 10GW biomass, and 5 GW small hydro.

India signified its intent to transform its energy profile with the 2003 Electricity Act, which ushered in the requisite regulatory changes by laying out a framework for greening the country's energy mix (Sawhney 2013). It mandated that State Electricity Regulatory Commissions (SERCs) would promote grid connectivity for electricity generated from renewable sources through tariff regulations, and also specify minimum purchase obligations for renewable power. The 2006 National Tariff Policy stipulated that distribution companies would purchase renewable electricity at preferential feed-in tariffs (FITs)[2] as determined by SERCs. It also provided guidelines for SERCs to use in fixing minimum renewable purchase obligations (RPOs). As electricity is subject to both central and state government regulation in India, SERCs were directed to set RPOs based on knowledge of regional resource availability and retail tariff impacts.

RPOs are a critical policy component in changing energy use profiles by ensuring demand for renewable-based electricity. The National Tariff Policy offers flexibility in SERC determination of RPOs, allowing for regional variations in renewable generation capacity by enabling lower RPOs in renewable resource-poor states. It was

[1] India ratified these agreements in 1993, 2002, and 2016, respectively.

[2] Renewables would have to be cost-competitive with other energy sources in the long run (6.4 (2)).

expected, however, that over time the states would take on their share of the renewable energy mix.[3] In anticipation of possible challenges to states in meeting mandatory RPOs, the NAPCC also offered flexibility in meeting same through tradeable Renewable Energy Credits or Certificates (RECs) (NAPCC 2008:44).

A tradeable REC is a market-based instrument that offers flexibility to obligated entities to meet RPOs in a cost-efficient manner. Under the REC mechanism, a renewable power generator may either sell green electricity to a distribution company, or "discom," or to any other obligated entity,[4] at a prescribed preferential tariff, or sell said green electricity separately from its renewable attributes. Thus, the option to sell unbundled renewable attributes as RECs to obligated entities with locational disadvantages, that is, entities that are unable to buy green electricity directly, means that renewable power generators can earn a green premium through REC pricing. Different tradeable RECs have accordingly been used the world over, including Guarantee of Origin certificates in the European Union, Solar Renewable Energy Certificates in some US states, and others more recently in East Asia.

The 2010 Central Electricity Regulatory Commission (CERC) regulation initially introduced both solar and non-solar RECs, the latter encompassing wind, biomass, small hydro, municipal waste, geothermal, and biofuel cogeneration. Each REC represents 1 MWh of electricity generated from renewable sources, which may be sold to any obligated entity of any Indian state. If an obligated entity fails to comply with its RPOs, it would be subject to penalties pursuant to the 2003 Electricity Act. In states well-endowed with renewable resources, the renewable electricity generation was expected to go beyond the baseline RPOs. It was assumed that RECs generated in renewable-resource-rich states, and associated with excess power sold at non-preferential tariffs, would be available for sale to obligated parties in deficient states, helping them comply with RPOs.

In 2022, a new CERC regulation on Terms and Conditions for Renewable Energy Certificates for Renewable Energy Generation Regulations removed the categorization of RECs by of renewable type, replacing it with a multiplier scheme for fungibility across technologies and common source-neutral RECs. The Ministry of Power issued notices of new RPOs in 2022, specifying wind and hydro RPOs, including that for large hydro, and removing solar RPOs.

Centralized monthly REC trading on the Indian power exchanges, Indian Energy Exchange (IEX) in Delhi and Power Exchange India Limited (PXIL) in Mumbai, which was devised to assist with RPO compliance across states through easy access to certificates and enable price discovery in a national auction market, signals the value of environmental or green attributes in energy generation to potential entrants in renewable power production (Sawhney 2013). Thus, RECs offered the only variable-price policy instrument in the gamut of fiscal policy incentives for RE generation in India, and trading commenced in 2011.

---

[3] Over the years, the tariffs have varied by states across India, since CERC tariffs are guidelines and not binding on states. RPOs are also set at state level to accommodate differences in renewable resource endowment across the large federated nation that is India.

[4] These include open-access consumers and industries consuming captive power.

Trading in RECs was partially suspended briefly in 2017–18 and completely in 2020–21,[5] only resuming in late 2021. Following the REC market redesign, the 2022 CERC regulation removed price controls, such that REC price discovery is achieved through free market bidding between buyers and sellers. REC trading was source-specific until November 2022, with trading in source-neutral RECs commencing December 2022. Market clearing prices and volumes of REC transactions were low during the first eight months of the newly designed market.

This paper reviews the performance of the REC market since its inception in India, government policy controls in the REC mechanism, and lessons learned. The rest of the paper is organized as follows. Section 2 offers a snapshot of India's policy goals for renewable-based power. Section 3 analyzes REC mechanism implementation and trading experiences over the intervening decade. Section 4 summarizes REC market performance assessments. Section 5 concludes with policy recommendations.

2 Enhancing Renewables in the Electricity Energy Mix

As the electricity sector is by far the single largest contributor of carbon emissions, much of India's climate policy has focused on achieving a clean energy mix in its electricity generation. To enhance the share of renewables in its energy mix, India has set targets for installed electricity capacity as well as goals for the share of its electricity generation derived from renewables. The latter are issued in the form of an RPO at the national level, defined by technology or renewable type. While SERCs are responsible under the Electricity Act for specifying minimum purchase renewable power obligations in their respective states, the Ministry of Power sets national RPO targets under the Tariff Policy, in consultation with the Ministry of New and Renewable Energy. The REC market offers all obligated parties a mechanism by which they can achieve cost-efficiency in complying with RPOs.[6]

[5] The Appellate Tribunal for Electricity suspended REC trading in July 2020 following appeals protesting the 2020 CERC abolition of an REC price floor, and resumed trading 24 November 2021. The Terms and Conditions for Renewable Energy Certificates for Renewable Energy Generation Regulations 2022 have removed REC floor and ceiling prices.

[6] The National Load Despatch Centre (NLDC) Power System Operation Corporation Ltd (POSOCO) implement the REC mechanism under the Ministry of Power, including registration of eligible renewable energy generation facilities, issuing RECs, maintenance and settlement of REC accounts, and serving as the repository of REC transactions. This is convenient given that the NLDC, together with Regional Load Despatch Centres, is in charge of integrated operation of regional and national power systems. IEX and PXIL hold closed REC sales auctions at the national level.

2.1 Renewable Energy Mix: Targets Versus Achievements

Under India's National Electricity Plan, the share of renewables in electric power capacity is targeted to reach 54% of installed capacity by 2030, per Table 1. This outlines the optimal technology mix for transition to cleaner power by 2030. It is remarkable that the share of non-fossil fuel-based power for 2030 is set at 65%, which is far more ambitious than the 40% stated in the Intended Nationally Determined Contribution under the Paris Agreement.

By March 2023, power capacity in renewables, excluding large hydro, reached 125GW, accounting for 30% of total installed capacity per Table 1, with solar and wind comprising the lion's share of renewables. While solar has the highest average annual capacity growth, at 46% per annum for the period 2014–23, followed 16.5% for wind per annum for the same period, these achievements remain insufficient to reach the announced targets.

Regional REC capacity registration is concentrated in seven states, which account for more than 75% of total REC registered capacity to-date (Sawhney 2022): Andhra Pradesh, Gujarat, Karnataka, Madhya Pradesh, Maharashtra, Rajasthan, and Tamil Nadu.

Despite being classified as non-fossil fuel technology, large hydropower generation was not initially classified as renewable technology by the Ministry of

Table 1 Energy mix in electricity installed capacity: target and current capacity

Technology	Target, 2029–30		Actual installed, 2023*	
	GW	% share	GW	% share
Thermal capacity	291	35%	237.27	57.03
Of which				
Coal and lignite	266.8		211.85	
Gas	24.3		24.82	
Non-fossil fuels	540	65	178.78	42.9
Of which				
Renewables-based	**450**	**54**	**125.69@**	**30.08**
Solar PV and CSP	300		66.78	
Wind power	140		42.63	
Biopower	10		10.80	
Small hydro			4.94	
Large hydro	73.4		46.85	
Nuclear	16.9		6.78	
Total	**831.5**	**100**	**416.06**	**100**

*As of 31 March 2023

@Excluding large hydro

Source Compiled from CEA installed capacity report, MOP (2023), CEA (2019a, b)

Table 2 RE-based power capacity: total versus REC-registered, 31 March 2023

Renewable type	Capacity (MW)	
	Total RE based power	REC registered
Wind	42,633	2469 (5.8%)
Of which		
Wind (Commissioned before 1 April 2022)		2469
Wind (Commissioned since 1 April 2022)		0
Solar	66,780	1185 (1.8%)
Of which		
Thermal		0
PV		1185
Small hydro (commissioned before 8 March 2019)	4944	210 (4.2%)
Small and large hydro (*commissioned since* 8 March 2019)*	1451	446 (30.7%)
Biomass/co-generation	10,248	733 (7.1%)
Of which		
Biomass		373
Bio fuel/bio-fuel cogeneration		360
Urban or municipal waste	554	0
Others		3
Total	126,610	5046 (4%)

*Includes large hydro projects commissioned since 8 March 2019. Registered REC capacity consists of large hydro projects registered in 2023 until 31 March
Source REC registry and CEA

Power. Following the 2022 REC mechanism redesign, however, large hydro projects commissioned after 8 March 2019 are considered for credit as renewable and are accommodated within newly issued RPOs. Current REC project registration reflects this new definition, and per Table 2, approximately 30% of installed capacity of large hydro projects commissioned since March 2019 are currently registered to issue RECs. This is in sharp contrast to the share of wind-based power capacity registered for RECs, at 5.8%, solar-based power so registered, at 1.8%, or overall RE-based electricity capacity so registered, at 4%.

2.2 *Renewable-Based Electricity Generation and RPO Targets*

As noted earlier, per Ministry of Power directives of 2016 and 2018, target RPOs were differentiated into solar and non-solar technology prior to FY2021. RPOs were defined based on total electricity consumption by obligated entities, and excluded consumption from hydropower. Table 3 summarizes the RPO targets for the period 2016 through FY2021 based on the solar/non-solar distinction, together with a new classification that distinguishes among wind, hydro, and other sources beginning in FY2022. The newly defined hydro RPO can be met with large hydro projects, meaning greater than 25 MW, or small hydro projects commissioned after 8 March 2019. The newly defined wind-based RPOs apply to wind power projects commissioned after 31 March 2022.

RPO increases are intended to drive demand for renewable-based power from obligated entities, particularly distribution companies, and aid the electricity generation transition to a carbon-free energy mix. The share of purchased RE-based power by obligated entities, however, have remained below the RPO targets.

Although the share of renewable-based power in the total electricity generated has steadily increased from 5.56% in 2014–15 to 13.53% in FY2022, per Table 4, contrasted with the NAPCC vision and the path required to meet the 2030 target,

Table 3 RPO Targets, 2016–2030

Year	RPO by RE type			Total (%)
	Non-solar (%)		Solar (%)	
2016	8.75		2.75	11.50
2017	9.50		4.75	14.25
2018	10.25		6.75	17.00
2019	10.25		7.25	17.50
2020	10.25		8.75	19.00
2021	10.50		10.50	21.00
	Wind[a] (%)	Hydro[b] (%)	Other* (%)	
2022	0.81	0.35	23.44	24.61
2023	1.60	0.66	24.81	27.08
2024	2.46	1.08	26.37	29.91
2025	3.36	1.48	28.17	33.01
2026	4.29	1.80	29.86	35.95
2027	5.23	2.15	31.43	38.81
2028	6.16	2.51	32.69	41.36
2029	6.94	2.82	33.57	43.33

Year corresponds to Indian fiscal year, 1 April–31 March

[a]Electricity consumption met from wind power projects commissioned after 31 March 2022, or consumption above 7% from wind projects commissioned earlier

[b]Power consumption met from hydropower projects, large and small alike, commissioned after 8 March 2019

*Refers to renewables categories other than wind and hydro specified in *a* and *b*, including solar, biomass, etc.

Source MOP (2018) and CEA (2023)

Table 4 Total grid-connected re-based electricity generation (MU), 2014–23

Year	Total electricity	RE-based	Annual growth in RE-based generation (%)	Share of RE-based in total generation (%)
2014–15	1,110,392	61,719	–	5.56
2015–16	1,173,603	65,781	6.58	5.61
2016–17	1,241,689	81,548	23.97	6.57
2017–18	1,308,146	101,839	24.88	7.79
2018–19	1,376,096	126,759	24.47	9.21
2019–20	1,389,121	138,337	9.13	9.96
2020–21	1,381,855	147,248	6.44	10.66
2021–22	1,491,859	170,912	16.07	11.46
2022–23	1,504,264	203,552	19.10	13.53

Year corresponds to Indian fiscal year, 1 April–31 March
Source CEA electricity and renewable electricity generation reports; and CEA (2022)

these results fall short of the RPO goal of 24.6% for FY2022. It seems unlikely that the goal of 27% of renewables in total electricity generation by FY2023 will be met.

Per Table 4, the double-digit annual growth rate in RE-based power generation between FY2016 and FY2018 slumped to 9% and 6% respectively during the COVID-19 years of FY2019 and FY2020, to recover beginning in FY2021. Considering the gamut of fiscal incentives built into the policy package to encourage the growth of RE-based electricity, both in terms of installed capacity and actual generation, doubts emerge whether these policy instruments are effective or need overhauling. More pertinent doubts concern the enforcement of the accompanying RPO regulation that was intended to provide the essential institutional framework within which the tradeable REC mechanism works.

3 Key Fiscal Instruments for Renewable-Based Power and REC Trading Schemes

Key fiscal instruments in India's renewable energy policy package include (i) installation incentives in the form of accelerated depreciation, (ii) a generation-based incentive (GBI) or subsidy payout per kWh generated (grid-interactive), (iii) preferential FiTs, and (iv) tradeable RECs. Viability gap funding for solar energy and a long-term interest subsidy for distribution utilities are also included. The intent is for the accelerated depreciation to attract investment and offset the high capital costs in renewables-based power projects, and for GBIs and FiTs to act as incentives for green electricity generation, until economies of scale take hold.

However, sudden changes in conditions, in particular for accelerated depreciation and GBIs, often result in confusion and policy uncertainty.[7] In an analysis of relative cost disadvantages of renewable energy vis-à-vis conventional sources given the variety of fiscal incentives with different timelines, Shrimali et al. (2016) concluded that low-cost long-term debt is the most cost-effective way to make renewable energy cost-competitive, because the high cost of renewable energy is driven by its higher capital costs rather than the higher variable cost component of conventional energy.

To ensure that renewable power providers can cover the steep cost of renewable technology, the CERC was empowered to set FiTs for grid-connected power pursuant to the 2003 Electricity Act and the National Tariff Policy. FiTs are technology-specific, in order to assure providers that they will recover their full costs during the debt repayment period for the useful life of the power-generating station, making them equivalent to levelized tariffs.

If a renewable power-generating company chooses not to sell power at preferential tariffs, it is eligible to obtain an REC. They must also sell electricity to area distribution licensees at a price no higher than the pooled cost of power purchase, i.e., the average power purchase cost (APPC),[8] or to any other licensee or open-access consumer at either a mutually agreed-upon price or at the power exchange market price.[9]

3.1 The Cost of RE Power Generation, RE Power Tariffs, and REC Price

The worldwide average levelized cost of electricity (LCOE) of utility-scale solar photovoltaic (PV) fell by 82%, and that of onshore wind fell by 39%, between 2010 and 2019 (IRENA 2020). The former was due to lower installation costs, driven in turn largely by a decline in module prices. India showed the most spectacular reduction in 2019, with average solar LCOE falling by 85% (ibid.: 70).

Such a decline in solar LCOE will, however, depress the premium for producing renewable over nonrenewable power if the cost of the latter or the APPC does not also decline, which in turn will depress the REC market price. Indeed, the CERC reduced the price band for REC auctions, particularly for solar RECs in 2015, 2017, and 2020, as solar LCOE declined and competitive bidding caused a dramatic fall in solar electricity tariffs. The average price of solar RECs dropped to Rs. 28,750/

[7] For instance, the accelerated depreciation scheme was discontinued in 2012 for wind energy as added capacity was not accompanied by generation. It was reintroduced in 2014. Changes to accelerated depreciation and GBIs affect policy in all states, as these are national assistance instruments, unlike FiTs, which are determined at state level.

[8] Defined as the weighted average pooled power purchase price (APPC) for distribution licensees in states excluding transmission charges.

[9] Eligibility criteria require that providers have no power purchase agreements (PPAs) with obligated parties, whether directly or indirectly through traders to sell power at preferential rates, for the purpose of fulfilling RPOs.

MWh in 2017 and fell further thereafter per Table 7, equivalent to a premium of Rs. 2.8/kWh. As the APPC was more than Rs. 3/kWh, however, one would expect that an obligated entities would have an incentive to purchase solar power directly from providers rather than through RECs, i.e., buying the unbundled power at the APPC and paying separately for the REC. The decline in the solar LCOE thus has a twofold effect: first, a reduction in the price premium of RECs, thereby depressing supply thereof; and second, increasing demand for direct purchase of cheaper solar power, lowering demand for RECs.

Another factor anticipated to affect the REC market is the development of power transmission infrastructure aimed at easing access to reliable grid power supply. With the completion of the Green Energy Corridor,[10] there would be greater access to the renewable power from RE-rich states through the interconnected national grid for the rest of the country. It might be expected that the role of the REC market would be further diminished, as RE-poor states would be able to directly purchase green power from the grid rather than go down the REC route to comply with RPOs. For example, whenever there is excess wind power in Tamil Nadu, it would be accessible through the grid in RE-poor regions thousands of miles away.

Rather, the expectation was that declining competitive RE-based power tariffs and grid improvements would make RE power more attractive to obligated entities than paying the APPC for thermal power. Paradoxically, however, this has not been observed across India. RPO noncompliance is rampant.

3.2 Control Versus Clearing Price in REC Auctions: Signaling Low Green Attribute Value

The REC mechanism was originally conceived as a market-based instrument that would reflect a variable premium above the APPC. Launched in 2010, non-solar REC auctions commenced in March 2011, with solar REC auctions following in May 2012. Trading was conducted on IEX and PXIL with CERC approval.

Initially, the CERC mandated price floors and price ceilings, to provide guarantees for potential participants in REC auctions. Both the APPC and FiTs were incorporated in calculating floors and ceilings of solar and non-solar RECs alike. For example, the highest difference between the APPC and costs of generation, i.e., renewable energy tariff, nationwide is used to determine the forbearance, i.e., ceiling, price for solar and non-solar technologies (CERC 2011). Note that the CERC, recognizing the linkage between ceiling and compliance charges, which deter noncompliance

[10] The Green Energy Corridor, implemented by the RE-rich states of Andhra Pradesh, Gujarat, Himachal Pradesh, Karnataka, Madhya Pradesh, Maharashtra, Rajasthan, and Tamil Nadu, is expected to integrate thermal and renewable energy and aid in transmitting the latter to consumers located thousands of kilometres away. Construction began on the mega intra- and inter-state transmission network system 2017, pursuant to the 12th Plan, 2012–17, and is expected to carry some 20,000 MW large-scale renewable power (MNRE website: http://164.100.94.214/green-energy-corridor).

with RPOs as fixed by SERCs, indicated that lowering the forbearance price could dilute the impact of such deterrence (CERC 2011). Responsibility for this "adequate deterrence" against noncompliance with RPOs was, however, left squarely with the SERCs (CERC 2011: 6).

Price floors ensure cost recovery for meeting target energy generation under the NAPCC. The viability or feasibility requirement implies that RE projects would be able to cover loan repayment and interest charges, operations and maintenance expenses, and, in the case of biomass and cogeneration, fuel expenses (CERC 2011: 8). Recall that these are the same cost components that the CERC considers in determining FiTs. However, while the CERC's preferential FiTs were differentiated by technology, the dispersion of costs across non-solar technologies and states was not considered in arriving at the non-solar REC price floor. Wind, biomass, cogeneration, and small hydro were conflated therein, despite their different viability costs.

Although REC trading took place through double-sided closed-bid auctions on the national exchanges, clearing prices for the respective RECs hugged their respective CERC-imposed price floors, with selling bids consistently outnumbering buying bids. The scheme became highly regulated, with periodic price bracket revisions as summarized in Table 5.

REC durations have also been revised over time. While RECs were valid originally for one year, i.e., 365 days, from date of issue, in view of unsold inventory, the CERC extended the validity to 730 days in 2013 to prevent unsold RECs from expiring. As excess supply persisted and inventories further accumulated, the CERC further extended the validity of RECs to 1095 days in 2014, and also allowed providers to retain certificates to offset RPOs. Under the redesigned REC scheme (CERC 2022), RECs were declared valid in perpetuity or until sold.

Table 5 REC price controls as set by CERC

Year	Price control	Solar	Non-solar
2010–12	Floor	Rs. 12,000/MWh	Rs. 1500/MWh
	Forbearance	Rs. 17,000/MWh	Rs. 3900/MWh
2012–17	Floor	Rs. 9300/MWh	Rs. 1500/MWh
	Forbearance	Rs. 13,400/MWh	Rs. 3300/MWh
2015–17	Floor	Rs. 3500/MWh	
Revised for solar	Forbearance	Rs. 5800/MWh	
2017–20	Floor	Rs. 1000/MWh	Rs. 1000/MWh
	Forbearance	Rs. 2400/MWh	Rs. 3000/MWh
2020–22	Floor	0	0
	Forbearance	Rs. 1000/MWh	Rs. 1000/MWh
2022–present	Floor	0	0
	Forbearance	0	0

Source CERC Orders of 1 June 2010, 23 August 2011, 30 December 2014, 30 March 2017, and 17 June 2020 (CERC 2022)

Court challenges were brought against the drastic CERC-directed reduction in mandated prices of 2017, resulting in REC trading being suspended that May. While trading of non-solar RECs resumed after two months, that of solar RECs remained suspended until April 2018, reflecting market dissatisfaction on the part of RE-based power providers who failed to realize the premium they had expected to earn on the green attribute of electricity generated.

Declining LCOE for RE-based power caused the CERC to further compress price bands over time, especially for solar RECs. In its June 2020 policy revision, the CERC removed the price floor entirely for both solar and non-solar RECs by setting it effectively to zero, which led to petitions in protest and caused the Appellate Tribunal for Electricity (APTEL) to suspend trading until November 2021. The redesigned REC scheme (CERC 2022) took effect as of December 2022.

Table 7 summarizes average clearing prices for solar and non-solar RECs on the IEX. It is evident that these prices closely tracked price floors until 2019. While transactions initially reflected a higher price of solar RECs compared to non-solar RECs, the clearing price of solar RECs declined faster than that of non-solar RECs, reflecting the underlying fall in LCOE for solar power and solar FiTs.

Annual volumes of transacted RECs over the preceding decade, as reported in Table 6, show that REC redemption falls substantially beginning around 2017. As trading in solar RECs was halted between May 2017 and March 2018, the spike in sales thereafter merely reflects inventory clearance, given that RECs were valid for three years at the time. Inventory of unsold RECs peaked in 2018, per Table 6, with the largest category being wind, followed by biomass and solar. While REC auctions resumed in November 2022, following the aforementioned suspension, they have been slow to recover. While a large quantity of RECs has been issued in FY2022–23, many remain unsold.

Persistent low demand, i.e., buy bids, for RECs indicates little regard for RPO compliance. This lack of demand for RECs and the resulting erosion in their market value reduces the incentive for producing RECs as a byproduct of RE-based energy. As mentioned, sell bids plummeted after 2018 for solar RECs, and non-solar REC sell bids fell at the same time to almost one-tenth of 2016 levels. The phenomenon persists in the redesigned REC market, as total buy bids constituted less than 10% of sell bids in 2023, per Table 7.

3.3 Tracking the REC Market and CERC Interventions

Lackluster demand created a large inventory of unsold RECs, especially from 2014 through 2023. While REC redemptions spiked due to exchange in 2017–19, more RECs were issued thereafter. A similar hike in REC issuance in FY2022–23 accompanied another inventory rise, per Table 6.

The expectation of rapid growth of RE-based power generation in resource-rich states, and robust purchase of RECs by resource-poor states to fulfill their RPOs, has not been realized. Discoms purchased just about 61% of all RECs sold between 2011

Table 6 REC Redemption through power exchanges and unsold RECs, 2011–2023

Year	RECs issued	RECs redeemed through exchanges	RECs retained by RE providers	RECs revoked/ deleted	Closing balance (unsold RECs)[@]
2011	1,054,243	1,015,698	0	0	38,545
2012	4,328,198	2,589,814	0	0	1,776,929
2013	6,834,276	2,748,694	0	0	5,862,511
2014	9,624,866	3,061,922	248,232	0	12,177,223
2015	9,733,840	4,955,153	363,942	0	16,591,968
2016	8,195,763	6,487,739	465,313	0	17,834,679
2017*	6,326,816	16,184,151	485,059	0	7,492,285
2018	7,777,341	12,608,795	452,848	0	2,207,983
2019#	12,739,554	8,927,850	286,728	0	5,732,959
2020	5,022,099	920,761	346,967	3,623,895	5,863,435
2021	6,126,631	8,460,403	304,513	0	3,225,150
2022	22,527,251	8,161,604	80,387	432	17,419,929

Year is Indian fiscal year, 1 April–31 March

@Closing balance = RECs issued—RECs redeemed through exchanges—RECs self-retained by generator—RECs revoked

*There was no trade in solar RECs between May 2017 and March 2018, and no trade in non-solar RECs in May–June 2017

#There was no REC trade from July 2020 to 23rd November 2021

Source Monthly REC registry data. https://www.recregistryindia.nic.in/index.php/publics/recs

and 2018, with captive power plants and open-access consumers purchasing the rest (POSCO 2018). Total REC purchases are much less than that required to cover the RPO shortfall at state level, reflecting widespread noncompliance. Little, however, has been done to boost REC demand.

The Indian REC market has overwhelmingly been a story of control of supply, and until recently control of price as well, as the CERC has focused on changing supply characteristcs, including eligibility and validity, and the trading price band, prior to 2021:

(i) In the face of unsold or unredeemed RECs, the CERC periodically extended REC validity, from 365 to 730 days in 2013, 1095 days in 2014, and perpetuity in 2021;
(ii) Mandating lower REC prices did little for demand, instead signaling that there is little premium to be realized for the "green attribute" of electricity for providers; and
(iii) Discoms were made eligible to issue RECs when purchasing renewable energy over and above their RPOs under the Second Amendment to the REC Regulations of 2013. While the CERC considered this an incentive for distribution licensees to go beyond their RPOs, it flew in the face of the substance of REC issuance for premium price discovery by renewable power providers.

Table 7 Annual average REC price, and total REC bids versus clearing on the Indian Energy Exchange (IEX), 2011–2023

Year	Non-solar RECs					Solar RECs				
	Floor price	Clearing price@	Total buy bids	Total sell bids	Cleared volume	Floor price	Clearing price@	Total buy bids	Total sell bids	Cleared volume
2011	1500	2261	1,302,820	658,910	402,862	12,000		30,559		
2012	1500	2127	2,936,049	5,604,813	1,982,614	9300	12,740	36,342	5112	3782
2013	1500	1500	1,200,591	21,903,968	1,200,591	9300	10,600	83,644	259,747	39,173
2014	1500	1500	1,046,397	44,734,718	1,046,397	9300	9300	24,444	1,964,592	24,444
2015	1500	1500	2,693,510	85,636,055	2,693,510	3500	3500	370,574	18,874,807	370,574
2016	1500	1500	2,575,976	91,652,179	2,575,976	3500	3500	398,094	28,123,532	398,094
2017*	1000	1450	9,224,465	85,046,784	9,224,465	1000	2875	286,163	14,232,235	286,163
2018**	1000	1249	8,898,208	9,759,224	5,008,743	1000	1061	6,801,851	14,015,848	3,960,742
2019	1000	1595	9,796,338	6,193,904	4,640,492	1000	2067	6,630,913	2,087,255	1,346,287
2020#	0	1333	2,493,987	9,953,839	1,889,254	0	2100	2,507,770	1,337,629	1,181,468
2021##	0	1000	3,447,241	5,003,364	3,320,132	0	2106	1,817,626	1,494,228	508,098
2022 (Jan–Nov)	0	1000	3,606,676	27,940,690	3,606,676	0	1623	4,382,883	23,030,619	2,483,817
	REC									
	Clearing price@		Total buy bids	Total sell bids	Cleared volume					

(continued)

Table 7 (continued)

Year	Non-solar RECs					Solar RECs				
	Floor price	Clearing price@	Total buy bids	Total sell bids	Cleared volume	Floor price	Clearing price@	Total buy bids	Total sell bids	Cleared volume
2022 (Dec)		1000	688,356	2,123,044	487,356					
2023 (Jan–July)		899	3,429,291	44,325,582	3,136,097					

@ Average clearing price (Rs/REC) is the simple average price for traded months in the calendar year
*There was no trade in solar RECs between May 2017 and March 2018, and no trade in non-solar RECs in May–June 2017
**There was no trade in Solar RECs between January–March 2018
Non-solar RECs issued before April 2017 had an average clearing of Rs. 1501 and Rs. 1771 during 2018 and 2019 (not shown)
#Corresponds to trade until June 2020, as REC trading was suspended between July 2020 and Nov. 2021
##Corresponds to trading in November and December 2021, as trading was suspended between July 2020 and 23 November 2021. Non-solar RECs issued after 1 April 2017 were traded
Source Indian Energy Exchange (IEX) data. https://www.iexindia.com/marketdata/recdata.aspx

3.4 *Persistent Low Demand for RECs, RPO Noncompliance, and Discoms*

The stated objective of the REC mechanism, to wit, the need to enable greater RPOs in states with low renewable generation potential, has not been met, as the shortfall states chose simply neither to raise their RPOs nor enforce those in place. Even renewable resource-rich states have a poor record of RPO compliance. Few states comply with RPOs set by relevant SERCs. Only six states[11] were found to be RPO-compliant between 2010 and 2014. Even in the state of Maharashtra, which has one of the largest registered REC capacities, RE generation has fallen below the RPO for several years, and the RPO backlog has been waived for state-run and private utilities alike (CSE 2019: 120).

Analysts have long drawn attention to the problem of RPO noncompliance and the need for sufficient penalties against obligated entities (Shereef and Khaparde 2013; Shrimali and Tirumalachetty 2013). Interventions pursued so far in the REC market will not revive the scheme without serious nationwide enforcement.

Ongoing RPO noncompliance has caused low demand for RECs year after year, with a few exceptions. As noted above, RPO non-implementation has been overlooked or waived in many states. Thus, obligated entities have had little incentive to purchase RECs despite steady price reductions. Consequently, the REC mechanism has had a negligible role in driving RPO compliance, as seen in audits on renewables. The most recent audit noted that during 2010–2014, only 4.77% of RPO compliance was met with RECs, while the other 95.23% was achieved by direct RE electricity purchases (CAG 2015).

Obligated entities for purchasing RECs, in particular the state Discoms, have long complained of being in dire financial straits, and ill-equipped to implement major reforms. On the one hand, they have outstanding debts to renewable power providers, amounting to Rs32 billion in 2019 (Nirula 2019: 9).

On the other hand, discoms are constrained due to existing long-term PPAs with large thermal plants, making it difficult to switch to renewable power purchases even when RE power is cheaper (Sreenivasan 2019). It is reported that overdue payments to renewable power providers reached almost US $3 billion as of June 2022 (IEA 2023: 59). The discoms' poor financial state and state-level RPO nonenforcement leave little room for REC market participation. Consequently, they are considered the weakest link in the electricity supply chain and to continue to present hurdles to renewables growth.

---

[11] Himachal Pradesh, Karnataka, Meghalaya, Mizoram, Tamil Nadu, and another northeastern state (CAG 2015).

4 Assessing REC Market Performance

In an ideal REC market, the premium value of "green energy" would be realized through a free market price as long as supply and demand reflect true underlying costs and valuation. REC market price would tend toward zero when RE-based power is cost-competitive with thermal energy and the system has moved toward largely RE-based power. That REC prices instead collapse even when the system is predominantly non-RE based, while RE-based power is becoming more cost-competitive, indicates a failure of the market. Such has paradoxically been the case in India, which has experienced one of the lowest generation costs for utility-scale solar-based power, leading to a dramatic reduction in the average levelized cost of electricity of utility-scale solar PV. Notwithstanding, uptake thereof has faltered.

Recall also that REC power projects are not all comparable to utility-scale power projects, as the former are often low-capacity. Thus, the CERC-mandated price calculations based on attracting buyers may have missed otherwise expected premium REC-registered renewable electricity providers. This is especially true of unsold RECs from earlier years, when the cost of RE-based power was higher, but whose bankability was reduced. REC project registrations began declining after the initial growth spurt (CAG 2015).

An important indicator of the REC mechanism's functionality would be its certification rate over time. A robust mechanism would be characterized by increasing certification rates for renewable-based electricity, as this would reflect RE providers expecting good returns on certificate trading, and therefore registering their RE generation with the scheme.

On calibrating the Indian RE-based power certification rate over time, it is evident that there was a steady decline in the rate of certification of RE power for RECs from FY2015–16 through FY2021–22. We find that the RE certification rate initially increased rapidly, from 2.06% in FY2011–12 to 15.6% in FY2014–15, per Table 8, indicating that RE generators were upbeat in the initial years and opted for REC registration and issuance of RECs. However, the certification rate began to decline after 2014–15, and it may be recalled 2014–15 was also the year that witnessed a sharp decline in the auction price of solar RECs. Table 8 shows that the share of REC certification of RE-based power in the total RE-based power generated witnessed a sharp drop from approximately 14.8% in 2015–16 to 10% in 2016–17 and then plummeted to 6% during 2017–20 and further to 3.58% in 2021–22. This is not surprising, as REC trading had been suspended and uncertainty in price realization made RECs an unprofitable option for renewable power generators. The spike in certification rate during 2022–23 to 11% reflects the issuance of newly defined the RECs from large hydro projects recently commissioned. While it shows the creation of potential value for hydro-projects, it remains to be seen whether it helps in moving the REC market toward a robust one.

The record of the past 12 years highlights the fact that policy interventions in the form of revisions in the REC validity period, to deal with REC inventory, have

Table 8 Renewable power certification rate, 2011–23@

Year	RE-based power (BU)	RECs issued (MWh)#	RE power certification rate
2011	51.2	1,054,243	2.06
2012	57.4	4,328,198	7.54
2013	53.1	6,834,276	12.87
2014	61.7	9,624,866	15.60
2015	65.8	9,733,840	14.79
2016	81.5	8,195,763	10.06
2017	101.8	6,326,816	6.21
2018	126.8	7,777,341	6.13
2019	138.3	9,115,659*	6.59*
2020	147.2	5,022,099	3.41
2021	170.9	6,126,631	3.58
2022	203.6	22,526,819^	11.07

@RW power certification rate is calibrated as = 100* (quantum of electricity issued RECs in year t)/ (total renewable-based energy generated in the country in year t)
Year is Indian fiscal year, 1 April–31 March
#One REC equates to 1 MWh of RE-based electricity injected into the grid, where 10^6 Mwh = 1 BU
*Initially 12,739,554 RECs were issued, of which 3,623,895 were later revoked, resulting in 9,115,659 RECs effectively being issued to RE providers, for a 6.59%certification rate
^22,527,251 RECs issued, less 432 revoked, as of 31 March 2023
Source Author's calculations based on data from CEA, MOP (2021b), and REC registry

been ad hoc and ineffective.[12] They extended the RECs' bankable period but not their actual bankability, as failure to enhance REC market transactions and caused prices to continue to decline. Nor has the recent market overhaul enhanced RECs' bankability. The resulting lackluster trade in old RECs is no surprise.

There are also inconsistencies in the policy of awarding RECs to distribution licensees as an incentive for exceeding RPOs. Such an instrument flies in the face of the essence of REC issuance, to wit, engendering premium price discovery by renewable energy providers. It instead increases the supply of RECs in an already over-saturated market.

The major challenge confronting the REC market has been lack of demand by obligated entities for renewable-based power providers. Yet there has been no concerted

[12] The same approach has been adopted in a recent amendment to redesign the REC mechanism, approved by the Ministry of Power in September 2021 (MOP 2021a, b, c), that proposes to further extend REC validity from the current 1095 days to perpetuity until sold. Price floors and ceilings would also be eliminated. *MOP Press release*: https://pib.gov.in/PressReleasePage.aspx?PRID=1759300.

The MOP proposed that the CERC would begin monitoring REC trading to prevent RE providers manipulating prices by hoarding RECs. Doing so, however, would incur further supply-side costs in a languishing system, potentially giving a disincentive to RE providers from participating in REC trading or project registration.

effort to rectify this long-standing problem, emanating as it does from pervasive RPO noncompliance. There is an urgent need to create genuine demand for RECs if the potential benefits of the mechanism are to be realized.

5 Conclusion and Policy Recommendations

Despite promising flexibility and a green premium on renewable-based power generation, the Indian REC mechanism has faltered since its launch in FY2010. Falling prices have not stimulated demand in the face of widespread RPO non-compliance. Moreover, for a country that has significantly reduced RE-based power generation costs, e.g., utility-scale solar-PV, uptake has fallen short due to constraints faced by distribution companies locked in long-term thermal PPAs, among other obligated entity purchasers.

A review of the past decade of the REC mechanism's operation shows that although RE providers were upbeat during the initial years and increasingly chose to register their projects for REC issuance, certification rates dropped sharply after 2015, coinciding with the precipitous fall in the average auction price of solar RECs and the mounting inventory of unsold RECs, solar and non-solar alike. It reflects low underlying REC market sentiments among renewable power providers, who opted to withdraw given that returns were decreasing and uncertain. The increase in certification in FY2022 reflects issuance of RECs to large hydro projects. Whether the redesigned REC mechanism will be more efficient remains to be seen.

REC trading in India has been overly regulated and policy-heavy on the supply side, and thus cannot offer providers the expected premium for the green attribute of RE-based electricity. Abysmally low demand for RECs from obligated entities, particularly state distribution companies, signifies rampant RPO noncompliance nationwide. Insufficient demand led to unsold RECs accumulating, which lower prices fail to clear out.

The CERC responded to this crisis by extending the validity of the certificates and lowering the price band for auction. Neither rectified the problem. The interventions extended the bankable period of the tradeable assets without making them more bankable. As RECs declined in auction value, the certificates' bankability eroded, as their future revenue stream was seen to be dwindling. In the redesigned REC market sans price control, market clearing prices are low, as is demand.

Nationwide RPO enforcement and reform of discoms is required to make the REC market more attractive. Discoms can neither pay for energy purchases nor participate in the REC market. According to the most recent data, in FY2022 the overall share of RE-based electricity was 13.5%, compared to the RPO target of 24.6%. Few states are thus RPO-compliant.

Policy must thus focus on creating stronger demand for renewable energy-based power and enforcing RPO compliance, rather than more supply-side interventions. The redesigned REC market, following the reclassification of renewable technology and extension of the shelf-life of RECs, will otherwise fail. It is thus necessary to systematically appraise distribution companies and address their non-participation in the REC market, so as to create a robust market where these green certificates can then reach their true price in meeting India's climate goals for 2030.

References

CAG (2015) Report of the controller and auditor general of India on renewable energy sector in India, Ministry of New and Renewable Energy (Performance Audit), Report 34 of 2015, Government of India

CEA (2019a) Draft report on optimal generation capacity mix for 2029–30. Central Electricity Authority, Ministry of Power, Government of India

CEA (2019b) Growth of electricity sector in India from 1947–2019. Central Electricity Authority, Ministry of Power, Government of India

CEA (2022) Annual report 2021–22. Central Electricity Authority, Ministry of Power, Government of India

CEA (2023) National electricity plan: generation, volume I, Central Electricity Authority, Ministry of Power, Government of India

CERC (2011) Central electricity regulatory commission petition no. 142/2011, Government of India

CERC (2022) Terms and conditions for renewable energy certificates for renewable energy generation regulations, Central Electricity Regulatory Commission, Government of India

CSE (2019c) The state of renewable energy in India 2019. Centre for Science and Environment, New Delhi

IEA (2023) Renewables 2022: analysis and forecast to 2027, International Energy Agency

INDC (2015) India's intended nationally determined contribution: working towards climate justice, submitted to the United Nations Framework Convention on Climate Change. https://www4.unfccc.int/sites/ndcstaging/PublishedDocuments/India%20First/INDIA%20INDC%20TO%20UNFCCC.pdf

IRENA (2020) Renewable power generation costs in 2019. International Renewable Energy Agency, Abu Dhabi

MOP (2018) Order: long-term growth trajectory of renewable purchase obligations. No. 23/03/2016-R&R, dated 14 June, Ministry of Power, Government of India

MOP (2021) Discussion paper on redesigning the renewable energy certificate (REC) mechanism. No. 23/6/2021-R&R Part-1, 4th June 2021, Ministry of Power, GOI

MOP (2021a) Order: renewable purchase obligation trajectory. No. 23/03/2016-R&R, dated 29 January 2021, Ministry of Power, Government of India

MOP (2021b) Annual report 2020–21. Ministry of Power, Government of India

MOP (2023) Annual Report 2022–2023. Ministry of Power, Government of India

NAPCC (2008) National action plan on climate change, Prime Minister's Council on Climate Change, Government of India

Nirula A (2019) India's power distribution sector: an assessment of financial and operational sustainability. Discussion Note, October, Brookings India, New Delhi

POSOCO (2018) Renewable energy certificate mechanism in India: key learnings, data analysis and way forward. Central Agency, National Load Despatch Centre

Sawhney A (2013) Renewable energy policy in India: addressing energy poverty and climate mitigation. Rev Environ Econ Policy 7(2):296–312

Sawhney A (2022) Renewable energy certificates trading in India: a decade in review. ADBI Working Paper 1313. Asian Development Bank Institute, Tokyo. https://www.adb.org/sites/default/files/publication/794046/adbi-wp-1313.pdf

Shereef RM, Khaparde SA (2013) Current status of REC mechanism in India and possible policy modifications to way forward. Energy Policy 61:1443–1451

Shrimali G, Tirumalachetty S (2013) Renewable energy certificate markets in India: a review. Renew Sustain Energy Rev 26:702–716

Shrimali G, Trivedi S, Srinivasan S, Goel S, Nelson D (2016) Cost-effective policies for reaching India's 2022 renewable targets. Renew Energy 93:255–268

Sreenivasan P (2019) Powerless firms. Down to Earth, Centre for Science and Environment, New Delhi

Chapter 5
Rooftop PV with Batteries for Improving Self-consumption in Vietnam: A Cost–Benefit Analysis

Linh Dan Nguyen and Han Phoumin

Abstract Vietnam's energy sector has become one of Southeast Asia's most vibrant in recent times. Since the adoption of feed-in-tariffs (FiTs) in 2017, the national electricity system's installed capacity rose from 47GW to 78GW in 2021, 68% of which are contributed by variable renewable energy growth. Market design and transmission capacity deficiencies complicated extending or reforming FiTs for wind and solar after 2020. Vietnam must expand the use of renewables to achieve net zero emissions by 2050 while meeting growing economic demand, necessitating initiatives including energy storage. This study examines the costs and benefits of rooftop solar plus battery in a sample factory in Ha Tinh province, using roughly 115 MWh of grid-connected electricity annually in manufacturing building materials, and installing 137 kWp solar with battery to be self-sufficient. Calculated by PVsyst as a stand-alone system, based on the current policy scheme and the average battery cost, the company can hardly recover its investment. Therefore, the study considers other assumptions such as subsidies, electricity sales together with social and intangible impacts of corporate social responsibility, improved branding, potential CO_2 credit trading, and reduced curtailment risk. It concludes with policy recommendations towards sustainable development target for Vietnam.

L. D. Nguyen (✉)
Hanoi University of Science and Technology, Hanoi, Vietnam
e-mail: dan.nguyenlinh@hust.edu.vn

H. Phoumin
Economic Research Institute for ASEAN and East Asia, Jakarta, Indonesia

H. Phoumin et al. (eds.), *Large-Scale Development of Renewables in the ASEAN*, Economics, Law, and Institutions in Asia Pacific,
https://doi.org/10.1007/978-981-99-8239-4_5

Fig. 1 Vietnam GDP growth rate, 2011–2020. *Source* GSO

1 Introduction

1.1 General Developments

Vietnam is one of the fastest growing economies in Southeast Asia, with an average annual growth rate of approximately 6–7% over the past decade (Fig. 1). It has transformed significantly from a centrally-planned economy to a socialist market-oriented economy, emphasizing export-oriented manufacturing.

While the COVID-19 pandemic caused a slowdown in Vietnam's economic activity and a decline in demand for its exports, Vietnam's successful containment of the virus enabled it to rebound quickly, making it one of three ASEAN member states with a positive growth rate in 2020, rising to more than 6% thereafter. Its economy grew to over USD343 billion, surpassing Singapore, at USD337.5 billion, and Malaysia, at USD336.3 billion, to become the region's fourth largest economy (GSO 2020). Vietnam has made progress in improving access to education and healthcare, reducing child mortality rates, and increasing life expectancy. Its poverty rate has also decreased, from 58% in the early 1990s to 2% in 2020.

1.2 Energy Status and Issues

The World Economic Forum reported recently that, according to Ember's Global Electricity Review 2023, all nations are approaching "the beginning of the end of the fossil age," when power-generation emissions begin to fall and energy from solar and wind reaches 12% of the total (Thomson 2023). While European nations may achieve decarbonization first, Asian nations, despite getting off to a later start, are catching up rapidly thanks to significant contributions from Vietnam and other emerging economies.

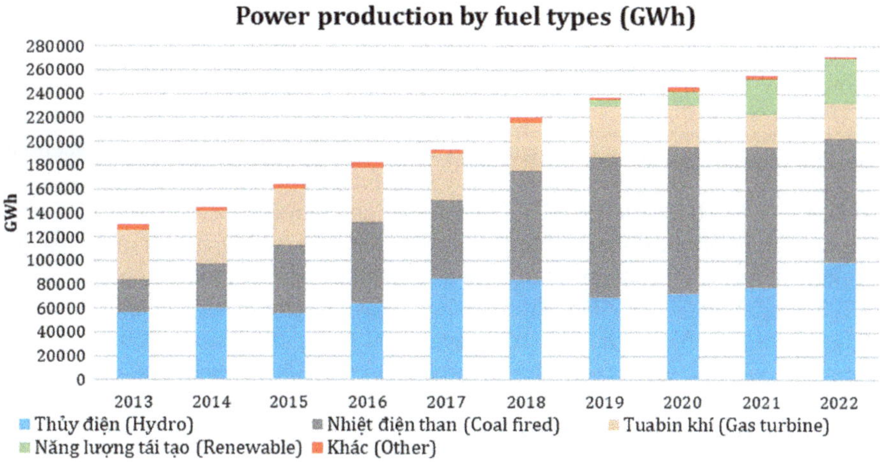

Fig. 2 Vietnamese power production by fuel type, 2013–2022. *Source* NLDC 2022

Vietnam has diverse energy sources, including coal, oil, and natural gas, as well as hydropower and other renewables. The country's total installed capacity as of 2021 was 76.6 GW, an increase of 60% from 2018s 47.8 GW. While having traditionally relied heavily on coal for power generation, Vietnam has significant potential for hydropower, wind, and solar (Fig. 2). It has a long coastline and high average wind speeds, making it ideal for wind development, especially offshore, which is estimated at over 470 GW technical potential within 200 km of the coast. The World Bank (2021) suggested that a target of 10 GW by 2030 and 25 GW by 2035 would likely drive Vietnam's industrial development and help the country meet its emissions targets. Abundant sunshine makes it an attractive location for solar, particularly in the south, with potential estimated at 12–15 GW. The average annual solar energy received on a horizontal surface in Vietnam varies between approximately 1200 and 2000 kWh/m^2.

Renewable energy has developed strongly in Vietnam over the past five years, with total power from such sources rising from practically nothing before 2018 to 21 GW by the end of 2021, accounting for one-third of the country's national power capacity. The country has achieved 100% electrification for communes and more than 99% for rural households at relatively lower cost than its neighbors (EVN 2021).

1.3 Energy Policy

At COP26, governments demonstrated a strong commitment to addressing climate change, with 197 countries signing the Glasgow Climate Pact. Vietnam also formally pledged itself to net-zero targets.

The Vietnamese government has taken steps to reduce emissions, including introducing the first-ever national development strategy for renewable energy (2015), followed by feed-in tariffs (FiTs) and other renewable energy development incentives. Some key policies for solar power include the government's Decision No. 11/2017/QD-TTg (2017), its amendment No. 02/2019/QD-TTg (2019) and replacement No. 13/2020/QD-TTg (2020), and MOIT's[1] Circular No. 16/2017/TT-BCT (2017) and Circular No. 05/2019/TT-BCT (2019). The government's Resolution 55 (2020) assigns high priority to sustainable energy development, setting a goal of increasing the share of renewable energy sources in total primary energy production to 15–20% by 2030 and 25–30% by 2045. These policies have promoted significant renewable energy growth in Vietnam, with solar and wind being the fastest-growing energy sources. Regarding fossil fuels, the country will not build any new coal-fired power plants after 2030 and will eliminate coal-fired thermal power plants entirely by 2050. Liquid natural gas (LNG) is set as an alternative to heavily polluting coal, although the feasibility of gas-fired power projects is subject to further review.

While some favorable renewable energy policies have expired, the government has not enacted further supporting mechanisms for solar or wind, leaving many renewable energy projects in "transition" (Nguyen 2022). It is reported that about 4.6 GW of variable renewable energy have completed construction but not yet been put into operation, or for which tariffs have not yet been agreed on, due to missing deadlines for decisions.

1.4 Challenges to Renewable Energy Development

Vietnam faces significant challenges in achieving its renewable energy targets to transition to a low-carbon energy system, including grid infrastructure, project financing, and land acquisition. Domestic supply does not meet demand, leading to increasing energy imports. Also, many power projects are behind schedule, and some energy security indicators are fluctuating negatively. Resource management and extraction remains limited, and energy exploitation and use is still not very efficient (Politburo 2020). The country's infrastructure and grid system still rely heavily on coal, and there are concerns about grid stability and reliability as more intermittent renewable energy sources come online.

According to extensive assessments by EREA and DEA (2021), photovoltaic (PV) and wind systems have relatively high initial costs, making financing more decisive in their adoption, and lower capacity factors than other generation technologies. They depend on the weather and the time of day, their output can vary greatly over short periods, and some only produce power in direct sunlight. PV output can only be adjusted negatively according to demand, i.e., reduced feed-in, as production essentially follows daily and annual solar irradiation variations, as production capacity is not held back during generation. They also vary by region. Large-scale projects

[1] Ministry of Industry and Trade.

may require significant amounts of land. They cannot be installed where there is limited space or competing land use, causing transmission challenges in Vietnam or other developing countries as areas of demand are usually far from renewables plants. Large-scale projects may also disrupt natural habitats and environs with a corresponding possible negative impact on biodiversity and food security. Storage is thus necessary to support power regulation and to reduce load on the national grid owing to supply uncertainty. A range of energy storage technologies have been identified for long-term policy planning (EREA and DEA 2023), including hydro-pumped storage, lithium-ion batteries, flywheels, compressed air energy storage, vanadium redox flow batteries, and hydrogen storage.

Batteries can play an important role in solar systems by storing excess energy generated during the day for use at night or other times when the sun is not shining. They store excess electricity using electrochemical storage batteries such as lithium-ion, redox flow, lead-acid, high temperature sodium sulphide (NaS), or sodium nickel chloride (Na $NICl_2$). Batteries have many potential applications in electricity systems, ranging from supporting weak distribution grids to providing bulk energy services or off-grid solutions (EREA and DEA 2023).

As the grid cannot handle more variable capacity in the short-term, energy storage by batteries is one of the most feasible solutions to promote self-consumption solar rooftops in industry. Battery energy storage systems (BESS) have a wide range of applications, from residential systems to large-scale utility projects that help with peak shaving, frequency regulation, and backup power. In areas where the grid is unreliable or inaccessible, batteries can provide backup power in case of outage or other emergency. Using saved energy during times when electricity prices are higher allows a solar power system to help reduce electricity bills and save money over the long term, easing greenhouse gas (GHG) emissions and contributing to a more sustainable, self-sufficient energy future.

Some projects have been carried out to embed energy storage in large-scale PV systems in Vietnam. A remarkable example is a US-sponsored project on the order of USD3 million awarded to the solar power plant of AMI AC Renewables Company in Khanh Hoa province, demonstrating the growing interest in battery storage as a means of integrating renewable energy into the grid and improving system reliability and efficiency. That was a proposal to install a lithium-ion BESS with an initial design capacity of 15 MWh/7.5 MW in a 50 MWp under-operation power plant in central Vietnam, to provide grid stability and reliability by mitigating the variability and intermittency of solar power generation (US Embassy 2021). Once the system comes online, other parties can benefit from its experience as it is reported that the operator will publish all BESS-related data.

The above mentioned project originated in a study that Electricity Vietnam (EVN) conducted in 2018, funded by a grant from the U.S. Trade and Development Agency (USTDA), to examine the feasibility of deploying advanced energy storage technologies in Vietnam. GE Power, a US energy company owned by General Electric, providing equipment, solutions, and services across the energy value chain from generation to consumption, participated in evaluating the potential benefits of BESS for EVN's grid, including improving reliability, making the system more efficient, and reducing GHG emissions (GE 2019). It was intended to identify potential locations for BESS deployment, assess technical and economic feasibility of different storage technologies, and provide recommendations for developing a BESS roadmap for Vietnam.

In 2021–2022, Shizen Energy, a Japan-based international renewable energy company with a track record of 21 MW wind and 35 MW solar in Vietnam, conducted a similar study to test their innovative digital micro-grid controlling service for expanding adequate renewable energy usage in the country (Shizen Energy 2022). They proposed studying the feasibility of introducing solar power generation with storage batteries and Shizen's energy management system (SDS). They have won several Joint Crediting Mechanism projects (GEC 2020) to reduce CO_2 emissions in developing countries and are going to apply for the same grant for this BESS project. However, detailed outcome of such project is not generally available as the company is only required to report to the primary beneficiary, the Japanese Ministry of Economy, Trade and Industry (METI).

Understanding the importance of promoting PV self-consumption, domestic companies have taken action on their own to do so in Vietnam. Provincial EVN in Ho Chi Minh City has initiated various renewable energy and smart grid projects intended to effectively integrate distributed power sources, including rooftop solar installations on public buildings, developing wind power plants, and implementing smart grid technologies to improve efficiency and reliability (EVNHCMC 2021). The company ran a pilot microgrid project that integrates renewable energy and battery storage systems with a capacity of some 350 kWh at its Data Center, which is expected to come online by mid-2022 (EVNHCMC 2022).

2 Research Objectives

In this study, we focus on systems of smaller, more practical scale that might better suit Vietnam's current requirements. We analyze the costs and benefits of deploying rooftop solar plus battery at a factory in an industrial zone, and the potential of such a system for wider application. While the self-consumption market offers much potential for investors and consumers alike, it remains immature due to lack of policy support. We hope to provide some socioeconomic and technical information on policymakers' decision-making processes, as well as ideas for investors or prospective prosumers, that is, electricity consumers who produce electricity for their own consumption, to fulfill their renewable promotion initiatives. More importantly, we examine all perspectives from a holistic viewpoint, economic as well as social, to ensure this business model's sustainability.

3 Literature Review

Several studies indicate the feasibility of attaching battery systems to renewables to promote self-consumption instead of grid connections. The African Technology Policy Studies Network (ATPS 2013) focuses on developing standard procedures for designing large-scale institutional grid-connected PV systems on rooftops and parking garages. The paper presents a pre-feasibility study using RETScreen software and a literature review of solar PV systems. It simulates designing a 1 MW grid-connected solar PV system for a university in Ghana (KNUST), with results showing that with an annual energy yield of about 1159 MWh, or 12% of KNUST's annual electricity consumption, the institution would have CO_2 emissions savings of some 792 tons. Net present value (NPV) and a payback period on the order of 50 years improve when higher FiTs, grants, and capital subsidies are introduced into the simulation.

Luthander and his colleagues (2015) summarize past researches in the field of self-consumption of electricity from residential PV systems. At the time of research, most of the papers studied PV-battery systems with storage capacities of 0.5–1 kWh times the installed PV capacity in kW, due to the high cost of such systems, meaning that batteries were used for short-term storage, normally less than one day. Without incentives, profits derived from electricity buying and selling price differentials, necessitating a balance between consumption, PV production, and storage capacity. Self-consumption increased in relation to storage capacity and rated PV power between 13 and 24 percentage points. Although storage battery cost does not significantly decline overtime, real price per lithium-ion technologies capacity declined 13% annually (Ziegler and Trancik 2021) between 1992 and 2016, comparatively altering the profitability of such projects over time.

Nyholm et al. (2016) investigate the potential benefits of using domestic energy storage in the form of batteries to increase self-consumption of electricity generated by PV installations in households. They examine 122 different combinations of PV installation sizes and battery capacities for different categories of single-family dwellings in Sweden. Using relative battery capacity, i.e., battery energy storage capacity in kWh divided by expected annual PV panel electricity output in MWh, they show that at 2.5–4.0, a battery can increase self-consumption by 18–48 percentage points. The ability of the battery to increase self-sufficiency increases with PV capacity, with the highest observed on the other 30 percentage points for an installation with an array-to-load ratio of 6. This work does not include an economic assessment, however.

Julakarn et al. (2019) examine the benefit of "prosumers", electricity consumers who produce electricity for their own consumption, using distributed energy technologies including distributed solar PV in Thailand. For all the schemes assessed, i.e., no compensation for excess electricity, net metering, and net billing, all four customer groups surveyed can make more profit than interest on savings accounts or government bonds. Even though there is no compensation for excess generation, investing in PV was economically viable for these groups, with net metering offering the most benefits. Measuring indicators are IRRs, NPV, and PB.

In an Australian case study, Roberts et al. (2019) examined energy and financial flows in five apartment buildings with PV and BESS using real apartment interval-metered load profiles and simulated PV generation profiles. They claim that there are clear financial benefits to combined PV, BESS, and an embedded network (EN) or microgrid system for many sites in Australia.

Other background studies, such as Braun et al. (2009), Castillo-Cagigal et al. (2011), Merei et al. (2016), McKenna et al. (2018), and Keiner et al. (2019), suggest that Vietnam could benefit from applying this model, given the abundant radiation and favorable energy policies mentioned above.

4 Case Study and Methodology

Cong Khanh is one of the four industrial clusters in Hong Linh, a town in Ha Tinh, a province in central Vietnam. It is established within an area of some 45 ha in Dau Lieu ward, where the main businesses are building materials manufacturing, supporting industries, mechanical processing, agricultural and forestry products, civil products, and packaging. Total investment is up to VND255 billion (HTPC 2016). The power supply is provided by the 35 kV line running along the National Highway 1A bypass section of Hong Linh. One 35/0.4 kV substation has been built, including three divisions of 5000, 450, and 560 kVA for production, lighting, administration, and services. The medium voltage grid uses underground cables to supply power from the 35 kV line outside the site along the sidewalks of the industrial cluster's main roads to the substations. The low voltage grid and lighting grid use underground cables to transmit electricity from substations to factories.

We take a factory in Cong Khanh specializing in production and trading of cinderblock building materials with a core product of aggregate cement cinderblocks as a case study. Products are diverse in size and standards, ranging from solid brick, used to build foundations and load-bearing walls, to brick with porosity exceeding 40%, used to build lightweight partitions. As this is one of the most energy-intensive sectors, after aviation, shipping, and the chemical industry, energy security and conservation for such units are important tasks in meeting the industry sector's overall emissions targets.

To facilitate the study, we investigated the recommended PVsyst, HOMER, and RETScreen software packages used for designing, analyzing, and optimizing renewable energy systems. PVsyst is popular and constantly updated for designing and optimizing PV systems, while HOMER is said to be an option for a more comprehensive energy system design and optimization tool. RETScreen is good for small to medium-scale renewables projects with simple configurations. For this paper, we chose PVsyst because of its user-friendly interface to help us decide on a proper system for the site and calculate the cost–benefit ratio. The simulation method is based on hourly energy balances over the course of a year, tracking the behavior of the system to calculate the appropriate combination that would obtain a system with the maximum amount of energy, in functions of such climatic variables as global radiation, wind speed, and temperature, and taking into account the PV system's installed capacity. PVsyst's results will be combined with further calculations of CO_2 emissions savings, other avoided costs, and evidence of intangible impacts, for a better evaluation of the model.

5 Results

5.1 *Technical Aspects*

With the location determined, PVsyst provides global horizontal irradiation (kWh/m^2), horizontal diffuse irradiation, local temperature, wind velocity, and other indicators on a monthly basis. We design the system with plane tilt of 18° to optimize losses for best annual irradiation yield (Table 1).

After we input the factory's load profile, based on 3000 m^2 maximum available rooftop area, following PV module and storage battery design is suggested (Table 2).

Battery voltage of 256 V suffices for industrial purposes. The storage battery market in Vietnam is not yet mature, limiting choices. For first-time users, BYD batteries may be a popular brand at reasonable prices, compared to Panasonic or Tesla. The system is set to accept 5% of the time during the day that the operator does not meet the load needs, i.e., empty tank, and one backup day for the battery to be on stand-by. At 80% depth of discharge (DOD), referring to the percentage of a battery's total capacity that has been discharged, stored energy from the battery can reach 575 kWh. It is important to monitor DOD, because deep discharges can shorten battery life and reduce overall capacity over time. In current temperature mode with air conditioning at 20 °C, total stored energy over the life of the battery is estimated at 3886 mWh.

Regarding PV modules, Longi solar is one of the available sellers in the Vietnamese market, and their model LR5-66HIH with capacity of 490 Wp and 32 V is appropriate. The performance ratio is an indicator of the availability of solar energy for final uses, shown in Fig. 3. A portion of the energy used internally is included in this evaluation. For these configurations, the total available roof space is not used, allowing the factory to expand the installation if load demand increases in the future.

Table 1 Main PV system set-up indicators

Plane tilt	18°
Transposition factor FT	1.00
Loss with respect with optimum	– 0.6%
Global on collector plane	1316 kWh/m^2

Table 2 Battery and PV technical design for selected case

Storage	
Battery model	BYD B-box pro 7.5, Lithium-ion LFP
Battery voltage	256 V
Nominal capacity	2496 Ah
Stored energy (80% DOD)	575.1 kWh
Discharging min. SOC	10%
No. of cycles	7500
Loss-of-load probability	5%
Requested autonomy	1 day
Number of batteries	80 (5 series and 16 in parallel)
Temperature	Fixed 20 °C
PV module	
PV model	Longi solar LR5-66HIH
Technical information	490 Wp 32v
Nominal STC	137 kWp
Operating mode of the controller	MPPT converter
Modules	279 (31 strings × 9 in series)
MPPT converter	
Module area	655 m^2

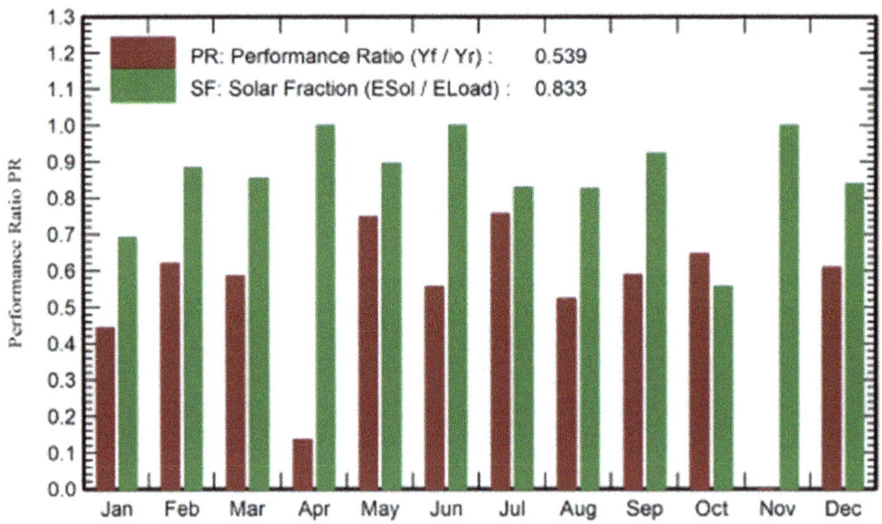

Fig. 3 Performance ratio and solar fraction of the system

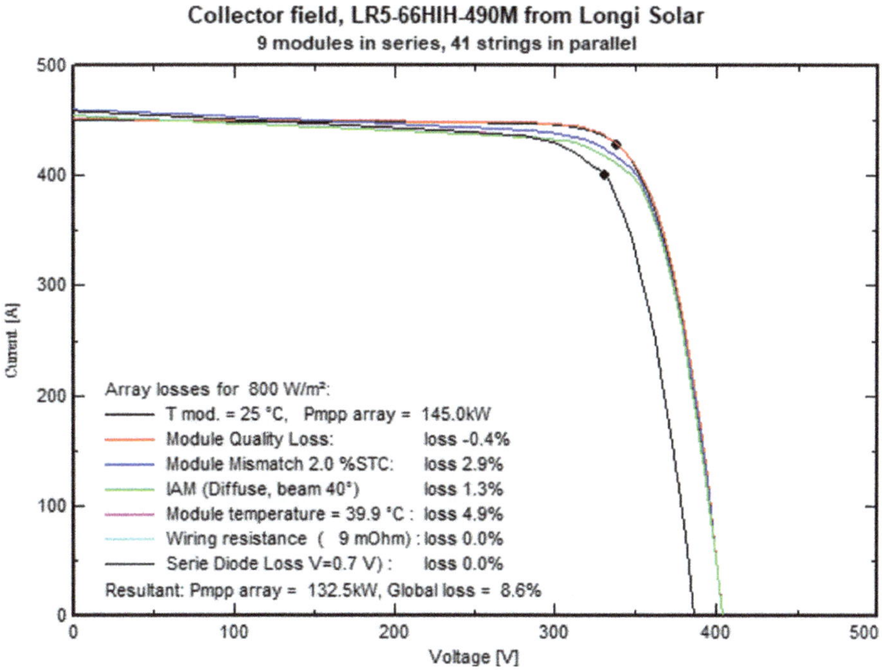

Fig. 4 System losses

Various loss types were addressed, including incidence angle, soiling, irradiance, and thermal or module quality losses, per Fig. 4. Setting free mounted modules with air circulation, we have thermal loss factor U = Uc + Uv* Wind velocity, a constant loss factor (Uc) of 29 W/m^{2}k and wind loss factor (Uv) of almost 0, assuming that module quality is satisfactory.

The PV panels can produce 148,050 kWh annually, of which 48,582 kWh goes unused. Cycles state of wear is 98.6% and static state of wear is 90% (Fig. 5). Battery lifetime is set at 10 years although the current warranty for BYD series is 5 years on average.

5.2 *Financial Aspects*

We estimate PV module unit price at USD122.5, and battery at USD1600 including tax. Assuming other fees such as other equipment or land cost are minimal, total installation cost is on the order of USD162,753 (Table 3).

Solar PV maintenance costs are relatively low compared with conventional energy systems, potentially ranging from $100 to $500 annually. Possible costs are for cleaning solar panels, inspecting wiring, connections, and inverter, and monitoring and labor. These costs, however, increase with off-grid systems where batteries are

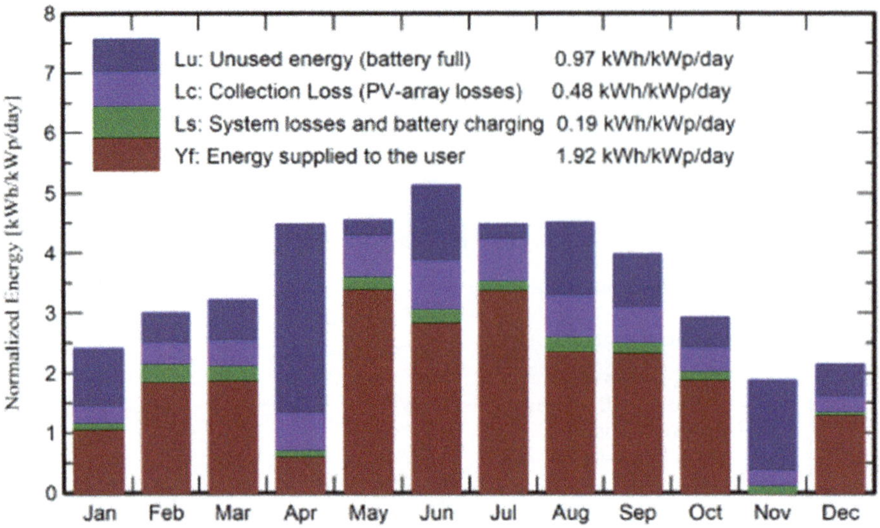

Fig. 5 Normalized productions per installed kWp

Table 3 Total installation and maintenance costs

Item	Quantity units	Cost in USD	Total USD
PV modules	279	122.5	34,177
Batteries	80	1600	128,000
Controllers			576
		Total depreciable assets	162,753
O&M for PV			1100
Provision for battery replacement			8533
Inflation 2.5%			
		Total OPEX	13,162

used. Effective regular monitoring by data loggers should help keep maintenance costs down, making 5–6% of the initial system cost a reasonable threshold for this design.

Project lifetime is set at 25 years under optimal conditions. According to the World Bank, inflation in Vietnam was at 2.7% in 2020, a slight increase from 2.2% in 2019. As the Vietnamese government has implemented various measures to help control inflation, we can expect it not to exceed 3% annually on average for the foreseeable future. Discount rate is on the order 6% per year.

In the most moderate scenario, investors use their own funds without any government subsidies. Excess energy cannot be sold to the grid or any third party, as there is no pertinent regulation as of this writing (Table 4).

Table 4 Financial analysis results

Project lifetime	25 years
Start year	2024
Inflation	2.5%/year
Production variation	– 0.5%/year
Discount rate	6%/year
Financing	Own funds
Electricity sale	0
Net present value	– 319,098
Internal rate of return	– 100%
Return on investment	– 196%

It is clearly that the project is unprofitable. Even though economic indicators are set at minimal levels, the project investors would have no incentive in investing in such system on their own.

5.3 *Electricity Bill and Emissions Savings*

The government regulates electricity pricing in Vietnam through the Ministry of Industry and Trade (MOIT) and the Electricity Regulatory Authority of Vietnam (ERAV). Different schemes are used for three types of customers. For households, a six-tiered pricing system was implemented beginning in 2019, ranging from VND1678, or USD0.071 per kWh for the first 50 kWh consumed, to VND2927 (USD 0.125) per kWh for usage exceeding 400 kWh monthly. Industrial and commercial customers are charged based on their contracted power capacity and consumption volume. The factory in this case is assumed to fall into the scheme shown in Table 5.

As per the load profile, the factory needs to pay about VND186 million, or USD7965 in annual electricity bills. This amount is saved through using self-consumed electricity instead of connecting to the grid.

Based on the emission factor of coal and gas for electricity and their share in national power generation (IEVN 2021), CO_2 emissions are estimated on the order

Table 5 Billing scheme for manufacturing sector

	Time frame	VND/kwh	USD/kwh
Standard time	4.00 a.m.–9.30 a.m. (5 h 30 min) 11.30 a.m.–5.00 p.m. (5 h 30 min) 8.00 p.m.–10.00 p.m. (2 h)	1,555	0.067
Off-peak time	10 p.m.–4 a.m. (6 h)	1,007	0.043
Peak load time	9.30 a.m.–11.30 a.m. (2 h) 5.00 p.m.–8 p.m. (3 h)	2,871	0.123

of 22 tons. International carbon prices have fluctuated widely, from as low as €2 per ton in 2013 to over €50 per ton in 2020. The European Union's (EU) market carbon price has doubled since the start of 2021, due to factors including soaring gas prices that have also prompted some power generators to switch to coal, resulting in higher emissions and demand for permits, according to Chestney et al. (2022). The price of permits on the EU carbon market is roughly €100 a ton, indicating that the model can save one factory an additional USD2410 annually.

6 Discussion

6.1 Alternative Scenarios

As mentioned above, the project is not profitable under this scheme. Lacking regulation, the PV owner is unable to sell the excess power to other loads, whether the national grid or neighboring systems. Nor are there any subsidies or other incentives promoting battery storage, leaving current battery prices high. Investors must provide funds entirely out of pocket, without any support from banks or other sources of funding.

Battery storage costs have been declining in recent years due to advancements in technology and increased production volumes. According to the International Renewable Energy Agency (IRENA), the cost of lithium-ion batteries has declined by approximately 89% since before 2010 and will fall further over the next decade. In 2021, battery pack prices were cheapest in China, at USD111/kWh (BNEF 2021). Using the lower battery cost assumption changes the financial results significantly.

As mentioned above, CO_2 prices are likely to change in time to come, as some 140 countries have committed to net zero, accounting for some 90% of global emissions. While Vietnam has yet to implement a national carbon pricing policy, the country has taken steps toward implementing carbon pricing, including conducting pilot programs and developing a draft decree on carbon pricing that is currently under review (No. 06/2022/ND-CP). By 2027, Vietnam should develop regulations on managing carbon credits and exchanging GHG emissions quotas for carbon credits, develop regulations on operating a carbon credit trading floor, and implement a pilot mechanism and related guidelines for carbon credit exchange and offset in potential fields. Vietnam is expected to implement and regulate an official carbon credit exchange beginning in 2028, and connections with regional and global carbon markets and exchanges of domestic carbon credits therewith should be available. Apart from its net-zero commitment, in 2022 Vietnam released a new updated Nationally Determined Contribution (NDC) under the Paris Agreement, wherein the country increases its unconditional GHG emissions reduction target from 9% in the previous 2020 version to 15.8% by 2030 relative to a business-as-usual scenario from the reference year of 2010 and including land use, land-use change, and forestry (LULUCF). With bilateral and multilateral cooperation alike, NDC 2022 raises its emissions reduction target from 27 to 43.5%, contingent on international support and financing.

While carbon trading scheme would be one of the most effective mechanisms enabling Vietnam to meet this target, other means are available. JCM, as mentioned above in the case of Shizen Energy as well as many other Japanese companies in the Vietnamese market, is an ongoing carbon trading framework that can be taken advantage of at this time. Projects in Vietnam can undertake GHG reduction measures with Japanese funding, and the resulting emission reductions can be jointly credited toward both countries' mitigation targets. Domestic companies have opportunities to access favorable loans from international banks and environmentally-oriented funding institutions if there is an open and appropriate scheme for them to do so.

It is also possible to make extra income from excess power that the system produces. One option is to participate in net metering, which transfers the excess power to the grid in exchange for credits. Selling to neighboring factories or households is feasible if transmission charges are sufficiently low. Under the current pricing scheme, electricity tariffs for businesses, such as service or commercial buildings, are highest during peak load time. There would be consumers willing to buy such clean energy as long as the selling price is less than USD0.18/kwh, which is set to further increase (GOVVN 2023). Still another option is to work with a third-party energy provider or energy aggregator who can help sell excess power on the open market. Taking these ideas into consideration, we developed another scenario that incorporates possible support, including:

- Cutting battery costs in half;
- Selling excess power at USD0.124/kWh, only slightly above LCOE in the previous scenario; and
- Investors can get loans at 1% interest instead of paying out of pocket.

Under these assumptions, the payback period falls to 24.8 years and NPV turns positive, with CO_2 emissions reductions and electricity bill savings treated as added income.

6.2 *Intangible Benefits*

Intangible positive effects can increase a project's total benefits more than with other conventional energy projects. Leaving aside the established benefits of renewable energy, installing storage batteries with solar panels can help stabilize grids, including preventing blackouts. The National Load Dispatch Center (NLDC) reports that there were 29 blackouts in the Vietnamese electrical system in 2022. Excess solar energy generated by day can be stored for use at night or during cloudy weather, reducing dependence on the grid and increasing energy independence.

While we conducted our case study for a factory in an industrial cluster in central Vietnam, it is also applicable to far-flung towns with limited access to reliable electricity. In some mountainous areas, self-consumption microgrids are the only option for meeting the government's 100% full electrification target. Solar panels with storage can improve quality of life for individuals and communities by providing access to such basic amenities as lighting, refrigeration, and communications. Replacing traditional fossil fuel-based electricity generation with solar panels and storage batteries in communities also has such environmental benefits as reduced air pollution and residents' associated health improvements.

This model can also bring social responsibility and business opportunities. Big companies are requiring their supply chains to use at least some renewable energy. Intel has declared that it will run on 100% renewables by 2030, by investing in environmental projects, setting companywide environmental targets, and exercising direct control over manufacturing processes. Unilever reports that it is using 100% renewable grid electricity across all of its factories, offices, R&D facilities, data centers, warehouses, and distribution centers. Apple has announced that its global suppliers generate more 13 GW of renewable electricity around the world in 2022, an increase of approximately 30% over the previous year. Microsoft, Google, and Amazon have similar targets that will necessitate further support for renewable energy utilization if Vietnam intends to work with them. This may also provide economic benefits to the surrounding community through green job creation or attracting foreign direct investment (FDI). Installing storage batteries with solar panels may also create jobs in such areas as design, installation, maintenance and operation, sales and marketing, and other support services, potentially having a positive impact on the labor market by providing new employment opportunities.

Avoiding noise pollution and visual disruptions are secondary benefits from not using electricity from the grid, for which fossil fuels currently account for more than 50%. While renewable energy alone cannot replace the stabilization and security assurance role that fossil fuels currently play, renewable energy with storage can perform the task far better.

7 Conclusion

The world is collectively attempting many possible solutions, especially regarding green transitions, to achieve the emissions reduction goals outlined in the Paris Agreement and limit global warming to well below 2 °C above pre-industrial levels. Vietnam's energy system is in a state of transition too, with the government seeking to balance the need for economic growth with the need to reduce GHG emissions and increase renewables.

Under the current scheme, the only options for further renewables development involve additional solutions such as storage. Overall, as mentioned above, installing solar panels with storage batteries can have a positive impact on both individuals and society as a whole by increasing energy independence, reducing GHG emissions, improving energy access, and increasing grid stability. Nevertheless, battery life being shorter than the lifetime for PV modules, large investments, and no economic incentives make storage a burden for investors despite its tangible and intangible benefits alike.

The government's efforts to increase the share of renewables in Vietnam's energy mix suggest a growing recognition of the need to transition to a more sustainable and environmentally friendly energy system, although it must address regulatory and financing challenges to encourage more investment in such initiatives. Several policies should be strengthened and guided in more detail, such as those for energy storage. Demand response can be an effective strategy for utilities use to manage electricity consumption during peak load times, chiefly by offering incentives to encourage customers to reduce their energy usage during such periods. Investors also need to have access to domestic and international financial assistance more easily and cheaper compared to the market rate.

A liberalized electricity market would enable domestic and international trading, increasing energy sector efficiency by giving producers incentives to produce power at the possible lowest cost, making energy production more cost-effective. Electricity retailers and users negotiate and agree on electricity prices based on Power Purchase Agreements (PPAs), necessitating guidance for such practices to enable such a market seamlessly.

Generation cost, transmission charge, distribution rate, and ancillary service fee constitute electricity prices. To boost the competitiveness, transmission/distribution and generation should be separated, and all be independent from the governmental administration. Otherwise, the government should seriously consider repricing electricity tariffs if they remain regulated for energy security reasons. Electricity prices in Vietnam are subject to periodic adjustments based on changes in fuel prices, exchange rates, and inflation. While the government implemented various subsidies and incentives to promote energy efficiency and renewables deployment, those favorable regulations have since expired due to uncontrollable development. If such variable sources are discontinued, it is likely that there will be more coal-fired power generation and EVN will suffer heavier losses due to high prices for imported coal, not to mention massive losses of transitional projects. The minimum average electricity price was set to increase by approximately 13.7% to VND1,826.22, or USD0.78/kWh beginning February 3, 2023 (Decision 02/2023/QD-TTg, GOVVN 2023). This replaces Decision 34/2017/QD-TTg, which was issued six years prior. As the same tariff has been in effect since 2019, meaning that annual changes were suspended during the COVID-19 pandemic, Vietnam's electricity price is 50% lower than that of the Philippines, and is also lower than in other ASEAN member states such as Indonesia and Thailand. Pricing for manufacturing in particular is so much lower than other sectors that factories have no motivation to switch to green electricity. We thus recommend raising the tariff to cover the costs of investing in more expensive systems, such as battery storage.

Increasing tariffs and subsidies for promoting renewable energy may drive up inflation otherwise effect businesses in ways that many will oppose. The government will need to give the public transparency and clarity about production costs and pricing if it is to succeed.

References

ATPS (2013) Design and analysis of a 1MW grid-connected solar PV system in Ghana. ATPS research paper No. 27

BNEF (2021) Battery pack prices fall to an average of $132/kWh, but rising commodity prices start to bite. November. https://about.bnef.com/blog/battery-pack-prices-fall-to-an-average-of-132-kwh-but-rising-commodity-prices-start-to-bite/#_ftn1

Braun M, Büdenbender K, Magnor D, Jossen A (2009) Photovoltaic self-consumption in Germany—using lithium-ion storage to increase self-consumed photovoltaic energy

Castillo-Cagigal M, Caamanõ-Martín E, Matallanas E, Masa-Bote D, Gutiérrez A, Monasterio-Huelin F, Jiménez-Leube J (2011) PV self-consumption optimization with storage and active DSM for the residential sector. Solar Energy 2338–2348

Chestney N, Abnett K, Twidate S (2022) Europe's carbon price nears the 100 euro milestone. February. https://www.reuters.com/business/energy/europes-carbon-price-nears-100-euro-milestone-2022-02-04/

EREA, and DEA (2021) Vietnamese technology catalogue

EREA, and DEA (2023) Vietnamese technology catalogue for energy storage technologies

EVN (2021) EVN annual report

EVNHCMC (2021) en.evn.com. August. https://en.evn.com.vn/d6/news/Piloting-a-Renewable-Energy-and-BESS-Integrated-Microgrid-Project-66-163-2459.aspx

EVNHCMC (2022) smartgrid.evnhcmc. June. https://smartgrid.evnhcmc.vn/bai-viet/microgrid-project-at-evnhcmc-data-center

GE (2019) GE.com. April. https://www.ge.com/news/press-releases/vietnam-electricity-awards-ge-battery-energy-storage-feasibility-study-funded-us

GEC (2020) Introduction of 32MW rooftop and floating solar power system to factories. https://gec.jp/jcm/projects/20pro_tha_07/

GOVVN (2023) Decision 02/2023/QD-TTg dated February 3, 2023 on the price bracket of the average retail price of electricity. Prime Minister of Vietnam

GSO (2020) Viet Nam economy in 2020 the growth of a year with full of bravery

HTPC (2016) Ha Tinh People's committee decision on detailed planning for Cong Khanh industrial cluster, No. 585/QĐ-UBND. March. https://hatinh.dcs.vn/kinh-te/news/thanh-lap-cum-cong-nghiep-cong-khanh-1-tai-thi-xa-hong-linh.html

IEVN (2021) Vietnam Energy Statistics 2021. Institute of Energy

Keiner D, Ram M, De Souza Noel Simas Barbosa L, Bogdanov D, Breyer C (2019) Cost optimal self-consumption of PV prosumers with stationary batteries, heat pumps, thermal energy storage and electric vehicles across the world up to 2050. Solar Energy 406–423

Luthander R, Widén J, Nilsson D, Palm J (2015) Photovoltaic self consumption in buildings: a review. Appl Energy 80–94

McKenna E, Pless J, Darby S (2018) Solar photovoltaic self-consumption in the UK residential sector: new estimates from a smart grid demonstration project. Energy Policy 482–491

Merei G, Moshövel J, Magnor D, Uwe Sauer D (2016) Optimization of self-consumption and techno-economic analysis of PV-battery systems in commercial applications. Appl Energy 171–178

Nguyen L-D (2022) Vietnam's renewable energy policies and opportunities for the private sector. Clean EDGE Asia of the NBR

NLDC (2022) Vietnam power system and power market operation. EVN-NLDC

Nyholm E, Goop J, Odenberger M, Johnsson F (2016) Solar photovoltaic-battery systems in Swedish households—self-consumption and self-sufficiency. Appl Energy 148–159

Politburo (2020) Resolution on orientations of the Viet Nam's national energy development strategy to 2030 and outlook to 2045. Central Committee of the Communist Party of Vietnam

Roberts MB, Bruce A, MacGill I (2019) Impact of shared battery energy storage systems on photovoltaic self-consumption and electricity bills in apartment buildings. Appl Energy 78–95

Shizen-Energy (2022) Feasibility study project on microgrid control solutions utilizing digital technology for the expansion of introduction of renewable energy in Vietnam. https://www.shizenenergy.net/pt-br/2021/09/21/shizen-energys-proposal-chosen-for-metis-feasibility-study-in-vietnam/

Thomson E (2023) Shaping the future of energy, materials and infrastructure. World Economic Forum

Tongsopit S, Julakarn S, Wibulpholprasert W, Chaianong A, Kokchang P, Vu Hoang N (2019) The economics of solar PV self-consumption in Thailand. Renew Energy

USEmbassy (2021) USEmbassy.gov. October. https://vn.usembassy.gov/wp-content/uploads/sites/40/KH-BESS-Fact-Sheet-20211007-Bilingual.pdf

World-Bank (2021) Offshore wind roadmap for Vietnam

Ziegler MS, Trancik JE (2021) Re-examining rates of lithium-ion battery technology improvement and cost decline. Energy Environ Sci 1635–1651

Chapter 6
The Role of Battery Energy Storage Systems and Market Integration in Indonesia's Zero Emission Vision

Pramudya, Muhammad Indra al Irsyad, Han Phoumin, and Rabindra Nepal

Abstract Indonesia has committed to achieving net zero emissions by 2060, with emphasis on the electricity sector eliminating harmful gas emissions by that year. Using the Balmorel energy model, this study simulated the impact of the target on optimal capacity expansion, electricity production mix, emissions, and electricity supply costs across 230 grid systems. The results indicate the substantial benefits of integrating solar photovoltaics (PV) and Battery Energy Storage Systems (BESS). Solar energy sees a remarkable capacity increase, reaching 288.7 GWp by 2060. Other renewable sources, including hydro and wind energies, also exhibited significant growth, increasing from 6.2 GW and 130 MW in 2030 to 29.4 GW and 22.5 GW, respectively, by 2060. Intermittent renewables' growth necessitates a rise in BESS capacity from 1 MW in 2022 to 73.4 GW by 2060. The study also underscores to replace phased-out coal-fired power plants with nuclear power by 2060. The study concludes with policy implications arising from these findings.

Keywords Balmorel energy model · Regional electricity systems · Power plant expansion · Electricity production cost · Super grids

Pramudya
Directorate General of Electricity, Ministry of Energy and Mineral Resources, Jakarta, Indonesia

M. Indra al Irsyad (✉)
Faculty of Technology, Policy and Management, Delft University of Technology, NL-2600 GA Delft, The Netherlands
e-mail: mind001@brin.go.id

Research Center for Behavioral and Circular Economics, National Research and Innovation Agency, Jakarta, Indonesia

H. Phoumin
Economic Research Institute for ASEAN and East Asia (ERIA), Jakarta, Indonesia

R. Nepal
School of Business, University of Wollongong, Wollongong, NSW, Australia

H. Phoumin et al. (eds.), *Large-Scale Development of Renewables in the ASEAN*, Economics, Law, and Institutions in Asia Pacific,
https://doi.org/10.1007/978-981-99-8239-4_6

Abbreviations

ABB e7	The ASEA Brown Boveri (ABB) Ability e7 platform modeling software
ABM	Agent-Based Modelling
AIM	Asia-Pacific Integrated Model
BESS	Battery Energy Storage System
CCS	Carbon Capture Storage
CF	Capacity Factor
CFPP	Coal-Fired Power Plants
CGE	Computable General Equilibrium
CO_2e	Carbon dioxide equivalent
EV	Electric Vehicles
ExSS	Extended Snapshot Tool
HSD	High-Speed Diesel
IAM	Integrated Assessment Model
IESR	Institute for Essential Services Reform
IPP	Independent Power Producers
JAMALI	Java-Madura-Bali
LCOE	Levelized Cost of Electricity
LCOS	Levelized Cost of Storage
LEAP	Long-range Energy Alternatives Planning system/Low Emissions Analysis Platform
LPG	Liquefied Petroleum Gas
MEF	Ministry of Environment and Forestry
MEMR	Ministry of Energy and Mineral Resources
NPP	Nuclear Power Plants
NZE	Net Zero Emission
OSS	Online Single Submission
PLN	State-owned Electric Company
PPA	Power Purchase Agreement
PPU	Private Power Utility
PtX	Power to Hydrogen
PV	Photovoltaic
REBED	Renewable Energy-Based Economic Development
REBID	Renewable Energy-Based Industrial Development
ROR	Run-Off-River
RUKN	National Electricity General Plan
RUPTL	Electricity Supply Business Plan
Simple-E	Simple Econometric Simulation System
TIMES	Integrated MARKAL-EFOM1 System
VRE	Variable Renewable Energy
WASP	Wien Automatic System Planning
WH	Wellhead
ZE	Zero Emissions

1 Introduction

The threat of climate change has led to a global call for action to reduce emissions in all economic sectors, including energy. East Asian countries, including Indonesia, face similar concerns, with a projected increase in emissions from two million tons CO_2e in 2018 to 25 million tons in 2050 due to energy consumption and the absence of effective intervention (Kimura and Phoumin 2021). Indonesia, the world's largest coal exporter, confronts unique challenges in providing clean energy to its 272 million population. Coal remains the primary source of power in the country, accounting for 62% of electricity generation in 2020, causing emissions levels of 273 million tons CO_2e in 2019 (MEF 2021b).

Indonesia is currently committed to ensuring zero emissions in its electricity sector by 2060, with one proposed solution being to phase out coal-fired power plants and increase renewable energy utilization. Whileeveral studies have explored optimal low-carbon energy mixes for Indonesia's power plants, only a few have analyzed optimal generation expansion plans for regional electricity systems (Al Irsyad et al. 2019, 2020; IESR et al. 2021; PLN 2021). The PLN (2021) study was the most comprehensive, as it analyzed isolated small systems, although it focused only on PLN's electricity supply without giving due consideration to CCS.

This study aims to address gaps in previous research by asking the following questions about Indonesia's goal of achieving net zero emissions in the electricity sector by 2060: What the optimal generation expansion plan under the NZE target would be, how much BESS capacity said plan would require, and what impact would CCS have on these. The hypothesis is that VRE capacity will increase significantly. The rests of this study are as follows: Literature review in Sect. 2, data and methodology in Sect. 3, findings in Sect. 4, policy implication discussions in Sect. 5, and conclusions in Sect. 6.

2 Literature Review

Indonesia has set an ambitious target of achieving NZE in all economic sectors by 2060, as shown in Table 1. Food and land use sectors are expected to play a significant role in reaching this target, reaching negative emissions by 2030. The energy sector, including electricity, industry, transportation, and buildings, is expected to follow, reaching peak emissions by 2030 before gradually declining to 153 million tons CO_2e by 2060. The electricity sector alone is expected to reach zero emissions by 2060 after peaking at 1022 million tons CO_2e by 2030, resulting in − 6 million tons CO_2e net emissions by 2060.

Several studies have explored low-carbon generation expansion plans in Indonesia using different energy models, as outlined in Table 2. Siagian et al. (2017) used the AIM/CGE global energy model and recommended geothermal and hydropower development to reduce emissions. Van Soest et al. (2021) analyzed the possibility of

Table 1 Indonesia's proposed NZE roadmap by economic sector

Sector	2010	2020	2030	2040	2050	2060
Energy	453	688	1022	978	684	153
• Electricity	140	198	421	342	140	0
• Industry	145	208	241	345	312	62
• Transportation	96	151	191	102	94	65
• Buildings	73	132	169	189	138	26
Agriculture	84	88	94	98	102	101
Food and land use	470	98	− 140	− 246	− 304	− 326
Industrial processes and product use	35	55	62	55	50	45
Waste	89	139	198	170	120	87
Net emissions (million tons of CO_2)	**1131**	**1068**	**1244**	**1038**	**540**	**− 6**

Source MEF (2021a)

achieving NZE in the energy sector by 2070, sooner than the 2080 global average forecast, using six IAM models to evaluate carbon neutrality targets for 10 major emitting countries. Fragkos et al. (2021) applied the AIM/ExSS to predicted that the NZE vision would drive renewable energy share to at least 30% of primary energy consumption by 2050. Reyseliani and Purwanto (2021) used the TIMES model and reported that the including nuclear power in the 100% renewable energy 2050 vision would potentially reduce electricity production costs by 9.7% over the same vision without nuclear.

Studies shown in Table 2 using bottom-up energy models provide a more detailed analysis of Indonesia's electricity systems. The energy model commonly used in developing countries is LEAP (Al Irsyad et al. 2017), as applied by Kumar (2016) to estimate the impact of renewable energy development on emissions reductions in Indonesia and Thailand. Phoumin et al. (2020) used it to appraise the potential hydrogen production from renewable energy development in the Association of Southeast Asia Nations (ASEAN) region. Kimura and Phoumin (2021) used LEAP to update the long-term energy outlook for the East Asia Summit plus the United States (US) (EAS17). Handayani et al. (2022) applied it to assess ASEAN member states' roadmaps to NZE in electricity sector, projecting that solar capacity and storage would reach 78% of total installed power by 2050.

Other bottom-up energy models may provide more robust, detailed analyses. Al Irsyad et al. (2019, 2020) developed PowerGen-ABM, a hybrid energy model, to optimize power plants owned by PLN, IPP/PPU, and rental services in 15 primary electricity systems. Their studies projected high electricity shares from solar energy in North Sulawesi, Southeast Sulawesi, East Nusa Tenggara, Maluku, and North Maluku. IESR et al. (2021) applied the LUT Energy System Transition Model to analyze seven main electricity systems in eight regions; it was the only study to consider rooftop solar PV in Indonesia's optimal generation expansion plan.

Table 2 Studies of power plant expansions in Indonesia

Energy model	Study	NZE	Multi-country analysis	Regional electricity system	Energy storage	Rooftop solar PV	Nuclear power plant	Electricity grid integration	CCS
ABM	Al Irsyad et al. (2019, 2020)	×	×	✓	×	×	×	×	×
AIM/CGE	Siagian et al. (2017)	×	×	×	×	×	✓	×	✓
AIM/ExSS	Fragkos et al. (2021)	✓	✓	×	×	×	✓	×	✓
Balmorel	MEMR (2019)	×	×	✓	✓	×	×	✓	×
Balmorel	Prasodjo et al. (2016)	×	×	×	×	×	×	×	×
IAM	Van Soest et al. (2021)	✓	✓	×	×	×	×	×	✓
LEAP	Handayani et al. (2022)	✓	✓	×	✓	×	✓	×	✓
LEAP	Kimura and Phoumin (2021)	✓	✓	×	×	×	✓	×	×
LEAP	Kumar (2016)	×	✓	×	×	×	✓	×	×
LEAP	Phoumin et al. (2020)	×	✓	×	✓	×	✓	✓	✓
LUT energy system transition model	IESR et al. (2021)	✓	×	✓	✓	✓	✓	✓	×
TIMES	Reyseliani and Purwanto (2021)	✓	×	×	✓	✓	✓	×	✓
WASP	PLN (2021)	×	×	✓	×	×	✓	×	×

The official bottom-up energy models for the generation expansion plan in Indonesia are WASP and Balmorel. PLN (2021) used WASP together with ABB e7 and Energy Exemplar Plexos to prepare RUPTL, with due consideration for programs related to electric vehicles (EV), rooftop solar PV, pumped storage, BESS, and electricity systems in each province. Meanwhile, MEMR officially used Balmorel for Energy Outlook Indonesia (Prasodjo et al. 2016) and RUKN (MEMR 2019). While Prasodjo et al. (2016) integrated Balmorel and LEAP, their analysis neglected regional electricity systems, energy storage, rooftop solar PV, and system integration. MEMR (2019) projected power plants operated by PLN and PPU in every province, albeit similarly overlooking nuclear. This study aims to extend MEMR (2019), which was conducted to analyze power plant expansions, to meet the NZE vision by duly encompassing nuclear power plants, CCS, green hydrogen, and power plants owned by PLN and PPU in its analysis.

3 Methodology and Data

3.1 Methodology

Figure 1 shows that there are two stages in this analysis, electricity demand projections and the optimal generation expansion plan. Electricity demand projections combine the results of Simple-E, LEAP, and additional exogenous electricity demand from priority programs such as smelter projects, new industrial clusters, special economic zones, priority tourism locales, and integrated fishery and marine centers. First, Simple-E was used to estimate electricity demand models on residential, commercial, public, and industrial sectors using 20 years of provincial data on consumption, numbers of customers, gross domestic product (GDP), inflation, population, and average electricity prices.

Second, econometric regression analysis was applied to Simple-E to estimate the sectoral electricity demands for 230 electricity grid systems in 34 provinces. The electricity demand projections were later aggregated into total national electricity demand projections and used as an input in LEAP. This was followed by the re-estimating total electricity demand projection in LEAP by considering the impact of energy switching programs, including replacing LPG stoves with induction cookers, EV, new energy development including green hydrogen and green fuel, and energy conservation programs. Additional exogenous electricity demand was added from various prioritized development programs to the 230 grid systems. The total electricity demand projection from LEAP was migrated again to Simple-E and disaggregated to provide projections for the aforementioned 230 electricity grid systems. Transmission and distribution losses are forecast to decline from 9% in 2021 to 4.5% by 2060.

The Balmorel model (Wiese et al. 2018) was later used to simulate the optimal generation expansion plan from 2022 to 2060 for the 230 grid systems. comprising 39

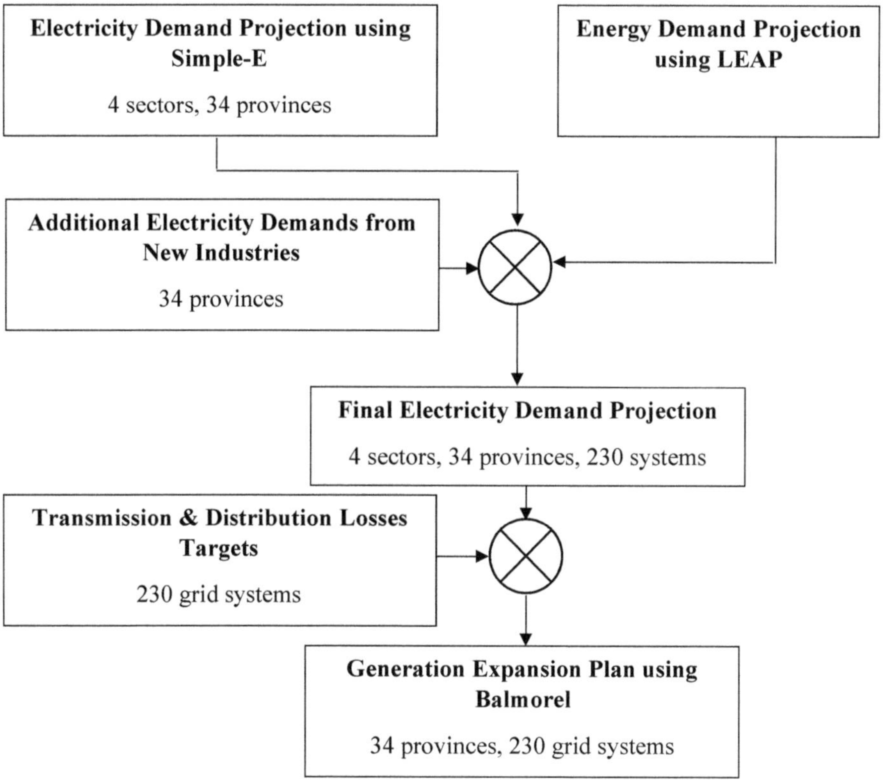

Fig. 1 Analysis flowchart

national PLN, 90 remote PLN, and 101 PPU grid systems. The simulations used more than 1000 power plants, 208 time slices, and 8736 hourly dispatches, annually. The objective function of the model was to minimize the Z costs of capacity expansion costs, unit commitment, and economic dispatch on system y in year t:

$$\begin{aligned} Min\ Z_y = &\ electricity\ production\ cost + hydrogen\ production\ cost + fuel\ cost \\ &+ new\ power\ plant\ investment\ cost \\ &+ new\ transmission\ investment\ costs + Unit\ starting\ cost \\ &+ Online\ O\&M\ cost \end{aligned}$$

$$\begin{aligned} Min\ Z_y = &\sum_{g,t} c^e_{g,t} \cdot G^e_{g,t} + \sum_{g,t} c^h_{g,t} \cdot G^h_{g,t} + \sum_{g,f,t} c^f_{g,t} \cdot F^f_{g,t} + \sum_{g} \left(a \cdot c^I_g + c^{fix}_g \right) I_g \\ &+ \sum_{g} a \cdot c^I_x \cdot I_x + \sum_{g,t} c^s_{g,t} \cdot S_{g,t} + \sum_{g,t} c^o_{g,t} \cdot O_{g,t} \end{aligned}$$

Key:

Indexes					
g:	Technology	h:	Hydrogen	x:	Transmission line
c:	Cost	f:	Fuel	a:	Areas
e:	Electricity	t:	Time	w:	Emissions
Coefficients/relationships					
a:	Annual capacity recovery	κ:	Nominal unit size	Loss:	Loss factor
η:	Marginal efficiency	r:	Variable resource	A:	Annual resource
c:	Extraction coefficient	K:	Capacity	T:	Target
c^e:	Back pressure coefficient	m:	Minimum unit load	W:	Emission factor
k:	Idle fuel consumption				
Variables (endogenous)					
G:	Generation (MW)	I:	Investment (MW)	O:	Units online (units)
D:	Demand (MW)	S:	Start units (units)	L:	Storage level (MWh)
X:	Transmission (MW)	Dn:	Shutdown (units)	Z:	System costs

Subject to:

(a) Balance of electricity supply, i.e., electricity production and imported electricity, and demand, i.e., exported electricity + local electricity demand:

$$\sum_g G^e_{g,t} + (1 - loss_x)X^{Import}_{x,t} = \sum_x X^{Export}_{x,t} + D^e_t$$

(b) Balance of hydrogen supply and demand:

$$\sum G^h_{g,t} = D^h_t$$

(c) Fuel costs for generating electricity, hydrogen, and idle fuel consumption:

$$F^f_{g,t} = G^e_{g,t}\Big/\eta^e_g + G^h_{g,t}\Big/\eta^h_g + k^f_g.\kappa^f_g.O^f_{g,t}$$

(d) Fuel input of power plant g at hour t should be adequate for the minimum electricity production, i.e., the product of minimum unit load, nominal unit size, and the number of online units):

$$F^{f}_{g,t} \geq m_g \cdot \kappa^{f}_{g} \cdot O_{g,t}$$

(e) Total availability of fuel f cannot exceed the annual resource of fuel f:

$$\sum_{g\,f,t} F^{f}_{g,t} \leq A_f$$

(f) For the power plant, g,, electricity production and its hydrogen production at hour t cannot exceed the power plant capacity K:

$$G^{e}_{g,t} - c^{v}_{g}.G^{h}_{g,t} \leq K^{e}_{g}$$

The electricity production at hour t is greater than or equal to its hydrogen production:

$$G^{e}_{g,t} \geq c^{b}_{g}.G^{h}_{g,t}$$

(g) The capacity of the hydrogen generator is equal to electricity demand divided by generator efficiency:

$$G^{h}_{g,t} = \frac{D^{e}_{g,t}}{\eta_h}$$

(h) Total capacity of new and existing power plants cannot exceed the annual resource of fuel f:

$$\sum_{g\,f} (K_g + I_g) \leq A_f$$

(i) Tal capacity of new and existing power plants should be greater than or equal to the capacity target of power plant g:

$$\sum_{g\,f} (K_g + I_g) \geq T^{K}_{f}$$

(j) Total electricity production in each hour t should be greater than or equal to the full load hour requirement:

$$\sum_{t} G^{e}_{g,t} \geq FLH_g \cdot (K_g + I_g)$$

(k) Electricity production of VRE g at hour t cannot exceed variable resources f multiplied by the sum of power plant capacity and investment:

$$G_{g,t} \leq r_t^f \cdot (K_g + I_g)$$

(l) Energy storage level L of hydropower plant g in the following year (t + 1) is the sum of the energy storage level in year t and hydro energy production minus electricity production from hydro:

$$L_{g,t+1} = L_{g,t} + r_t^{HY}.(K_g + I_g) - G_{g,t}^e$$

(m) Transmission capacity X is less than or equal to existing transmission capacity K plus new transmission line capacity:

$$X_{x,t} \leq K^x + I^x$$

(n) Total emissions, i.e., the product of emission factor W and fuel consumption f, cannot exceed the emission target:

$$\sum_{g \sim f} W_w^f \cdot F_{g,t}^f \leq T_w$$

The Balmorel model was used to simulate the NZE scenarios defined in Table 3. The BaU scenario was not focused on achieving NZE, which is why it allows new CFPP construction, whereas the ZE scenario prohibits new CFPP construction beyond the commitment made and under construction as stated in the RUPTL PLN 2021–2030. The phasing-out of fossil-fueled power plants was based on a lifespan of 30 years for coal-fired and 25 years for gas- and oil-fueled. The NZE scenario does not strictly aim for zero electricity emissions, and offers possible reduction of residual emissions in other sectors. This scenario thus allows construction of new CFPPs equipped with CCS. Both ZE and NZE scenarios allow only renewable power plant construction after 2030.

Sensitivity analysis was conducted by changing electricity demand projections, solar capacity growth limit, and demand flexibility. Assumptions low and high electricity demands in 2060 were 1942 TWh and 2366 TWh, respectively. The low electricity demand scenario considered increased energy efficiency in all sectors, whereas the high scenario anticipated a massive shift in industrial energy demand from gas and coal to electricity causing an increase in the electricity share to 80% of total industrial energy demand by 2060 compared with 51% in the BaU scenario. As recorded solar PV capacity in 2021 stood at only 190 MWp and significant growth in the near future was deemed unrealistic, maximum solar PV growth in 2060 was limited to 200 GWp (low), 400 GWp (medium), and 600 GWp (high).

Last, the electricity load pattern was changed by shifting portions of evening peak loads to daytime. This scenario was used to anticipate naturally flexible electricity

Table 3 Scenario definitions

Scenario	BaU	ZE	NZE
NZE target	No emission reduction target	Zero carbon by 2060 or earlier	Residual carbon sink by other sectors
New CFPP	Yes	No, except as stipulated in RUPTL	Only CFPP with CCS
CCS	Yes	No	Yes
NPP	Yes	Yes	Yes
Initial capacity	Existing, ongoing, committed, and planned power plants in RUPTL	Existing, ongoing, committed, and planned power plants in RUPTL	Existing, ongoing, committed, and planned power plants in RUPTL
Investments	No constraints	• No new investments in coal and diesel power plants • Investments for other fossil energy power plants allowed until 2030 • Investments beyond 2030 only for renewables and NPP	• No new investments in CFPP without CCS and diesel power plants • Investments for other fossil energy power plants allowed until 2030 • Investments beyond 2030 only for renewables, NPP, and CCS
Flexible electricity demand	None	• EV smart charging • Green hydrogen plants • Super grid infrastructure	• EV smart charging • Green hydrogen plants • Super grid infrastructure

demand and EVsmart charging, PtX, flexible demand response, and super grid, which is the interconnection of electricity grid systems in 51 regions to transmit renewable energy production among same. It was hypothesized that the flexible demand could reduce power plant peak load and energy storage requirements.

3.2 Data

Data for the simulation were obtained from sources including retrieval of technology and cost data from DEA et al. (2021), which provided a power plant technology and cost database for Indonesia. In Fig. 2, the LCOE of intermittent renewables was projected to decline over time between 2020 and 2060. Solar power plants incur added technology cost when equipped with BESS. Energy storage LCOS was also projected to show a decline from US $0.127/kWh in 2020 to US $0.086/kWh in 2030, US $0.069/kWh in 2040, and US $0.052/kWh in 2050–2060. Figure 2 also

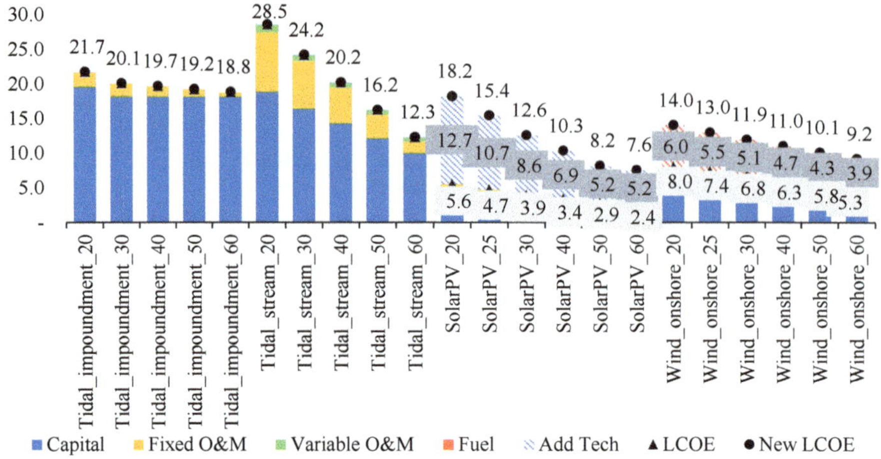

Fig. 2 LCOE (¢US$/kWh) for tidal, solar energy, and wind turbine. *Source* DEA et al. (2021)

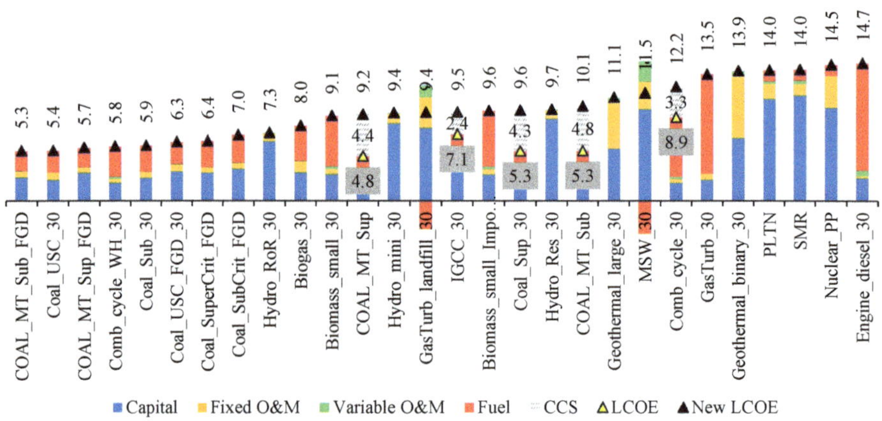

Fig. 3 LCOE (¢US$/kWh) for dispatchable power plants. *Source* DEA et al. (2021)

shows added technology cost for offshore wind turbines, which are considered more expensive than onshore. Figure 3 shows the possibility of the assumed LCOE of dispatchable power plants increasing due to the higher fuel costs. This study further assumed that the CFs are 80% for most power plants, except gas engine/turbine at 40%, diesel engine at 50%, reservoir type hydro at 42%, mini- and ROR- hydro at 50%, geothermal at 95%, nuclear at 90%, tidal at 35%, solar PV at 18%, and onshore wind turbines at 31%.

DEA et al. (2021) also provided assumptions for energy prices from 2022 to 2060, in which real prices for imported biomass, local biomass, mine-mouth coal, gas, well-head gas, biogas, and municipal solid waste (MSW) were relatively stable at US $96/ton, US $80/ton, US $32/ton, US $12/MMBTU, US $6/MMBTU, US

Table 4 Renewable energy potentials

Renewable	Potential (GW)	Utilization in 2021 (MW)
Solar	3295	203.7
Hydro	95	6601.9
Bioenergy	57	1920.4
Wind	155	154.3
Geothermal	24	2276.9
Ocean	60	0
Total	**3686**	**11,157**

$2/MMBTU, and US $-32/ton, respectively. The negative MSW price indicates its application as energy to generate income of US $32/ton processed. Average real coal price was projected to decline from US $130/ton in 2022 to US $74/ton in 2025 and 2060. Average real prices for fuel oil and gasoil were assumed to fluctuate, with the former falling from US $88/barrel in 2022 to US $81/barrel in 2025, then rising to US $98/barrel in 2030 before gradually falling again to US $95/barrel by 2060. A similar trend was assumed for the gasoil price, which was projected to decline from US $60/barrel in 2022 to US $53/barrel in 2025, rising to US $70/barrel in 2030, and falling again to US $67/barrel by 2060. The assumed price for uranium was US $1540/kg.

Table 4 shows renewable potential data provided by the Survey and Testing Agency for Electricity, New-Renewable Energy, and Energy Conservation. The largest renewable potentials were recorded for solar energy at 3295 GWp, with the highest solar potential observed in East Nusa Tenggara, West Kalimantan, and Riau. The second largest potential was wind energy at 155 GW, and East Nusa Tenggara, South Kalimantan, West Java, South Sulawesi, Aceh, and Papua were observed to have the highest such values. Hydro energy potential was recorded at 95 GW, mainly in North Kalimantan, Aceh, West Sumatera, North Sumatera, and Papua. Tidal is potentially available in all regions, especially Maluku, East Nusa Tenggara, West Nusa Tenggara, and Bali, with a total of 60 GW. Bioenergy and geothermal potentials were estimated at 57 and 24 GW, respectively, with the latter scattered along the ring of fire in Sumatera, Java, East Nusa Tenggara, and Maluku. Only 0.3% of this potential has been utilized, making increases in massive renewable energy exploration technically feasible. Indonesia also has uranium and thorium resources estimated at 89,483 tons and 143,234 tons, respectively.

4 Results

4.1 National Aggregated Results

Total electricity demand under BaU was projected to increase from 322 TWh in 2021 to 578 TWh in 2030, 1050 TWh in 2040, 1588 TWh in 2050, and 1942 TWh in 2060, as shown in Fig. 4. Demand growth arises from implementation of such policies as increase in electricity share of total industrial energy demand, power to green hydrogen, 100% EV sales by 2040, and the program to substitute LPG cookers with induction cookers, which are projected to increase the electricity share to 51% of total energy demand by 2060. Electricity demand per capita will increase from 1.2 MWh per capita in 2021 to 2 MWh in 2030, 3.4 MWh in 2040, and 5.9 MWh by 2060. Total electricity demand in the high-demand scenarios were estimated to be 2366 TWh or 7.1 MWh per capita by 2060.

Coal-fired power plant capacity was projected to increase continuously from 43.3 GW in 2022 to 103 GW by 2060 in the BaU scenario, as shown in Fig. 5. Other power plant technology capacities also increased, except for gas- and HSD-fueled. Total power plant capacity in 2060 was estimated at 456.6 GW, with 76% sourced from renewable sources. Solar energy had the most remarkable capacity increase, from 490 MWp in 2022 to 17.3 GWp in 2030, 66.9 GWp in 2040, 161.6 GWp in 2050, and 288.7 GWp by 2060. Others showing significant increases include hydro and wind energies, from 6.2 GW and 130 MW in 2030 to 29.4 GW and 22.5 GW by 2060, respectively. Such increases in intermittent renewables' capacities will lead to an increase in BESS capacity from 1 MW in 2022 to 5.6 GW in 2030 and 73.4 GW by 2060. Figure 5 also shows that coal-fired power plants had the highest electricity production share, contributing 51% to the 1456 TWh total electricity production in

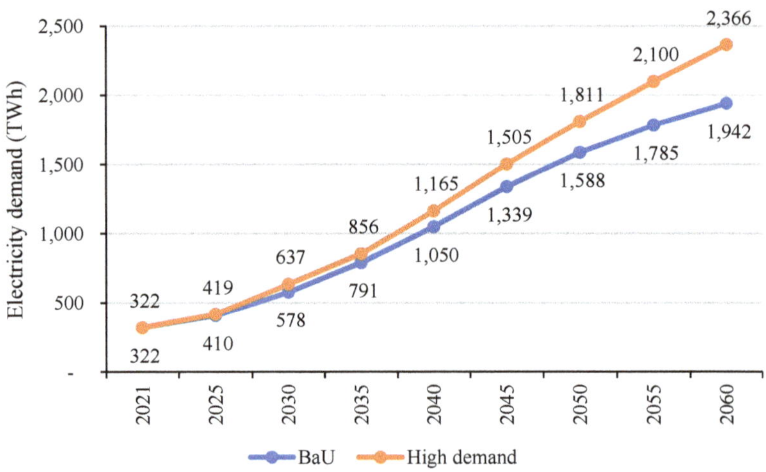

Fig. 4 Electricity demand projection

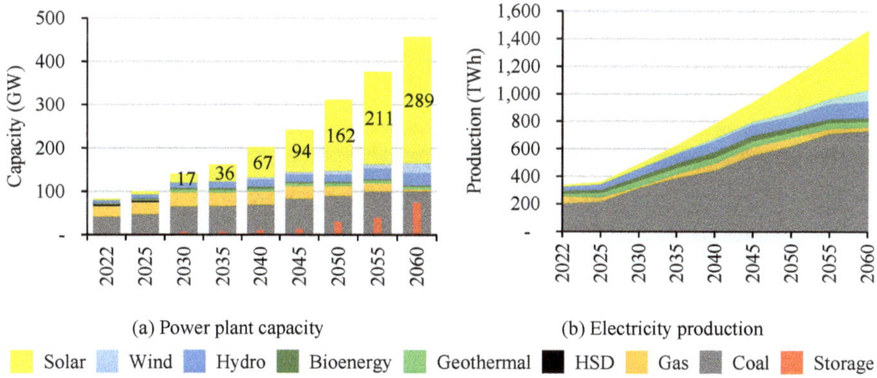

Fig. 5 Power plant capacity expansion and electricity production—BaU scenario

2060, while solar energy, hydro, and wind accounted 29%, 9%, and 6% respectively. Other renewable energy shares will be less than 3% each.

Phasing-out of coal-fired power plants in the ZE scenario requires constructing massive new capacities for renewable sources, specifically VRE, as shown in Fig. 6. By 2060, all power plants will be operating on new and renewable energy, with a total capacity of 708 GW. The capacity for solar, wind, hydro, bioenergy, nuclear, geothermal, and ocean energy will be 421 GW, 94 GW, 72 GW, 60 GW, 31 GW, 22 GW, and 8 GW, respectively. Electricity production in 2060 was projected at 2080 TWh, the highest share going to solar at 29%, followed by bioenergy at 22%, wind and hydro at 14% each, nuclear at 12%, geothermal at 8%, and tidal at 1%. Storage capacity required in the ZE scenario was projected at 61 GW. Conversely, peak coal-fired electricity will be 350 TWh in 2025, gradually declining to zero in 2060, also as shown in Fig. 6. Gas-based electricity generation was forecast to peak at 191 TWh by 2045 before eventually declining to zero by 2060. All oil-fueled power plants were forecast to shut down by 2030.

In the NZE scenario, CCS was found to be more competitive than nuclear and tidal. Figure 7 shows that the simulation conducted with due consideration for CCS technology recommended excluding nuclear and tidal in achieving the NZE target. CCS reduced the emission factor of a coal-fired power plant, and therefore, the capacity of coal-fired power plants was forecast to increase to 88 GW by 2060. Electricity generated from coal-fired power plants was thus forecast to increase from 205 TWh in 2022 to 229 TWh in 2030 and 654 TWh by 2060. CCS also allows low-emission electricity from gas-fueled power plants, leading to a projection of 168 TWh by 2040 before an eventual decrease to 5 TWh by 2060. The coal and gas electricity share in 2060 was forecast to be 13% of the 2088 TWh total, while required energy storage was found to be 54 GW, which was lower than the ZE scenario.

The results also showed that emissions from electricity without the reduction target increased from 226 million tons CO_2e in 2022 to 674 million tons by 2060, as shown in Fig. 8. The ZE scenario was forecast to produce zero emissions by

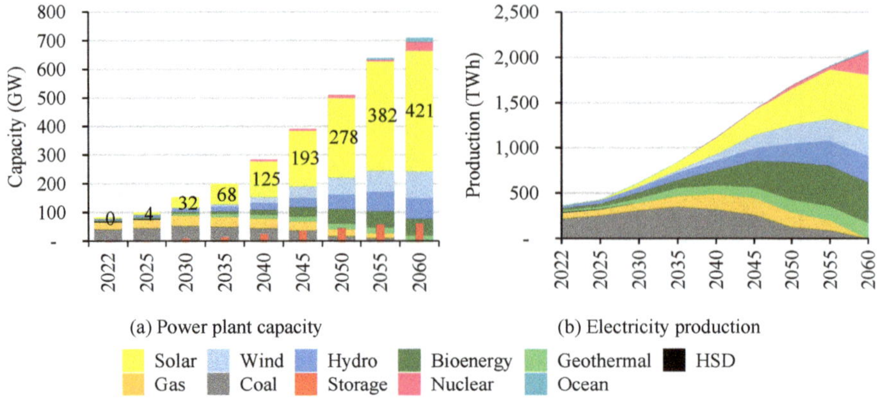

Fig. 6 Power plant capacity expansion and electricity production—ZE scenario

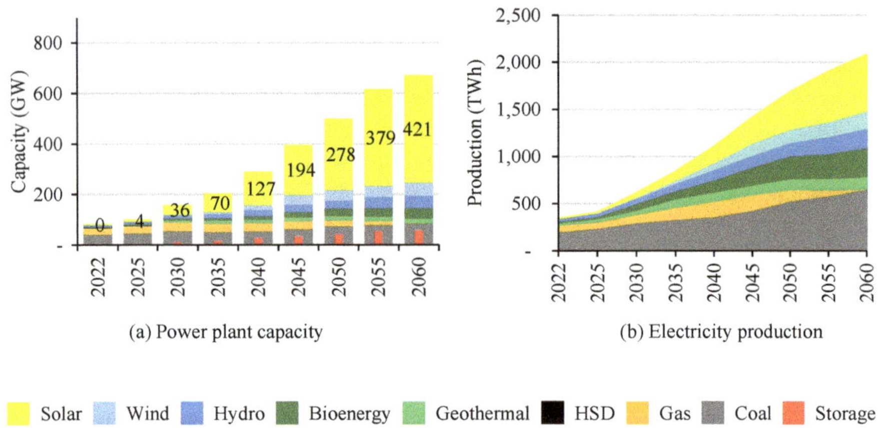

Fig. 7 Power plant capacity expansion and electricity generation—NZE scenario

2060 with a projected peak recorded in 2035 at 395 million tons CO_2e. Meanwhile, emission peak in the NZE scenario was projected to occur in 2040 at 385 million tons of CO_2e, with further emissions in 2060 of 108 million tons CO_2e due to coal and gas production. It was forecast that the forestry sector would compensate for these residual emissions.

Figure 8 compares the electricity supply costs across the scenarios. In the BaU scenario, costs would decline significantly, from US \$0.065/kWh in 2022 to US \$0.048/kWh in 2040, due to more low-cost electricity generated by coal-fired power plants, increasing slightly thereafter to US \$0.051/kWh between 2050 and 2060, due to rising coal prices. While the ZE scenario had the highest electricity supply cost due to having the highest capacities of renewables, energy storage, and nuclear. Initially, the cost would decline to US \$0.052/kWh by 2030 due to an increased share of

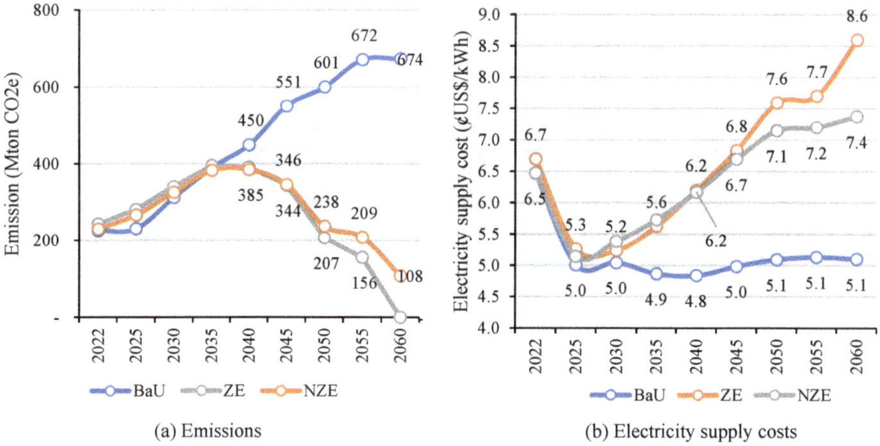

Fig. 8 Emissions and electricity supply costs

coal-based electricity supply, but thereafter, the cost gradually would increase to US \$0.086/kWh by 2060. The findings showed that the NZE scenario had a relatively lower electricity supply cost by 2060 compared to ZE of US \$0.074/kWh due to CCS-equipped coal-fired power plants generating 13% of the total electricity supply as described above.

4.2 *Regional Results*

Figure 9 compares the regional energy mix in 2020 and 2060 for each scenario. The BaU scenario shows a lower coal share in almost all regions except Sumatera, where the coal share increases from 38% in 2022 to 47% by 2060 as indicated in Fig. 9b versus the data for 2020 as shown in Fig. 9a. Solar energy was projected to grow significantly in Nusa Tenggara to a 57% share by 2060, while hydro energy was forecast to increase tremendously in Kalimantan, from 3% in 2022 to 39% by 2060. Another renewable source with a significant share increase will be wind, especially in Java and Bali, where it was projected to contribute 9% to the regional energy mix, followed by Sulawesi with 12% and Nusa Tenggara with 5%.

The ZE scenario will generate zero coal share in all regions by 2060, as shown in Fig. 9c. The scenario calls for solar energy to have the largest regional energy mix shares in Nusa Tenggara at 50%, Maluku and Papua at 46%, Sumatera at 37%, and Java and Bali at 31%. The second largest renewable sources by share will be wind, with Nusa Tenggara at 33% and Java and Bali at 27%, and bioenergy, with Maluku and Papua at 31% and Sumatera at 27%. Sulawesi is also forecast to rely on bioenergy at 33%, wind at 22%, and solar at 21%. Kalimantan was found to have the largest hydropower share at 33%, with nuclear at 33%, solar at 15%, and bioenergy

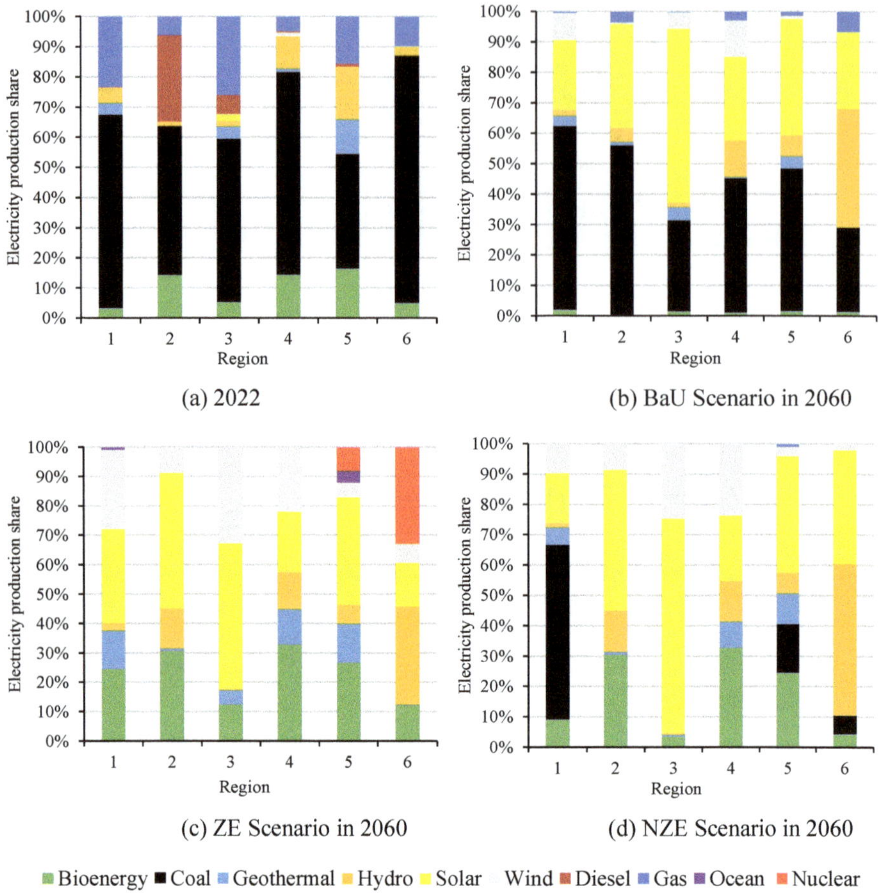

Fig. 9 Regional electricity production mix. *Legend* (1) Java and Bali; (2) Maluku and Papua; (3) Nusa Tenggara; (4) Sulawesi (5) Sumatera; (6) Kalimantan

at 12%. The ZE scenario calls for a nuclear plant to be constructed in Sumatera that would generate 8% of its energy mix by 2060.

The NZE scenario suggested coal share to reach 58% in Java and Bali, 16% in Sumatera, and 6% in Kalimantan by 2060 as shown in Fig. 9d, with no coal-fired power plants operating elsewhere. Renewables increased significantly in all regions, with Nusa Tenggara having the largest portions, solar at 71% followed by wind at 24%. Sulawesi is projected to have bioenergy at 33% followed by the second largest wind share at 24%. Hydropower is expected to contribute 50% of Kalimantan regional energy mix, followed by solar at 38%.

Table 5 shows that higher VRE capacity does not always require higher BESS capacity. The flexible electricity demands in the ZE and NZE scenarios may potentially reduce the BESS required to 56.3 GW and 50.2 GW respectively, while the BESS capacity in BaU is 69.4 GW in 2060. However, BESS capacity in the ZE and

Table 5 Energy storage by type in 2060

Region	BESS (GW)			PS (GW)		
	BaU	ZE	NZE	BaU	ZE	NZE
Java and Bali	21.8	3.5	2.2	3.7	3.7	3.7
Maluku and Papua	4.3	8.4	8.5	–	–	–
Nusa Tenggara	2.9	17.6	14.5	–	–	–
Sulawesi	8.0	4.5	5.3	–	–	–
Sumatera	24.7	20.4	14.9	0.5	0.5	0.5
Kalimantan	7.9	1.8	4.7	–	–	–
Total	**69.4**	**56.3**	**50.2**	**4.2**	**4.2**	**4.2**

NZE scenarios is expected to be higher than the value for BaU in Maluku, Papua, and Nusa Tenggara, due to their relatively low electricity demand and lack of connection to larger grid systems, signifying that they are forecast to have low grid flexibility in 2060.

5 Sensitivity Analysis

The simulation results showed that only the ZE scenario reaches zero emissions due to substantial increases in solar and energy storage capacities. The sensitivity analysis was thus limited to these together with nuclear and electricity supply cost in this scenario. Figure 10a shows that while the increase in solar PV capacity was sensitive to solar PV growth limit assumption, it was less sensitive to changes in electricity demand growth and flexibility. Figure 10b shows that CCS substitutes perfectly for nuclear, as shown in the NZE scenario analysis. Another alternative is solar PV, such that increasing its capacity was discovered to reduce nuclear capacity and vice versa. Energy storage capacity was most sensitive to demand flexibility as shown in Fig. 10c, while also highly sensitive to solar PV capacity growth. Demand flexibility thus significantly influences electricity supply cost, as shown in Fig. 10d. Supply cost was also forecast to increase with higher electricity demand and lower solar PV capacity growth limits, driving the simulation to select other plants with higher LCOE.

6 Policy Implications

The ZE scenario's flexible electricity demands require super grid infrastructure to transmit electricity from sources to regions, as shown in Fig. 11. PLN (2021) includes a 500 kV interconnection grid project for Sumatera-Malaysia and 150 kV for

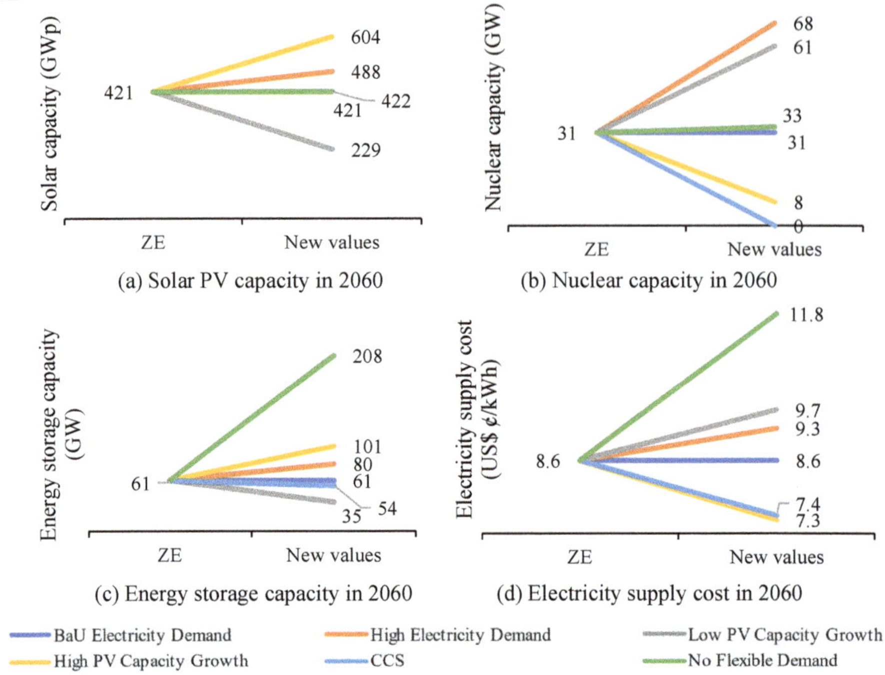

Fig. 10 Sensitivity analysis results for ZE scenario

Sumatera-Bangka, Kalimantan, and North Sulawesi-South Sulawesi. Other potential grid projects requiring further analysis include interconnections for Sumatera-Singapore, Sumatera-Java, Bali-Lombok, Bangka-Belitung, Belitung-Kalimantan, and Bau-Bau-South Sulawesi. Beyond these, based on the ZE scenario as shown in Fig. 11, this study proposes super grid projects connecting Kalimantan–Java, South Kalimantan–South Sulawesi, Bali–West Nusa Tenggara–East Nusa Tenggara, Maluku, North Maluku, and West Papua–Papua.

Investments required for the ZE scenario 2022–2060 were estimated at US $1.14 trillion, an annual average of US $29 billion as distributed in Fig. 12. Approximately 86% of the total would be for new power plants, specifically nuclear at 9%, hydro at 15%, solar PV and wind at 14%, and bioenergy at 11%. BESS and pumped storage were estimated to require US $37 billion and US $3 billion, respectively, and new transmission grids approximately US $116 billion, or 10% of the total. This last could be reduced by implementing REBID and REBED policies to foster industry and other economic activity close to renewable power plants.

Last but not least, phasing out coal-fired power plants requires a roadmap, government regulations, and presidential decrees to be obeyed by PLN, IPP, and PPU. Regulations should clearly state that IPP- owned coal-fired power plant contracts cannot be extended beyond existing PPAs, and that granting of new operational permits for those owned by PPU is prohibited. The Ministry of Investment's OSS system must also block all new permit applications related to coal-fired power plants.

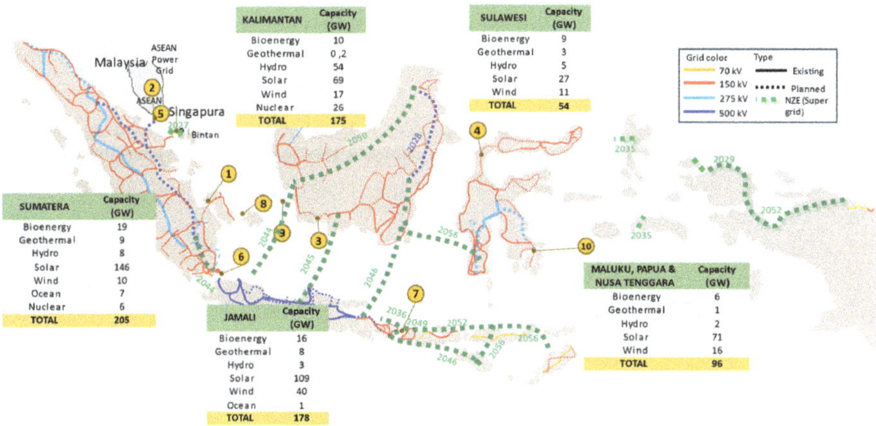

Fig. 11 Proposed super grids for implementing the ZE scenario

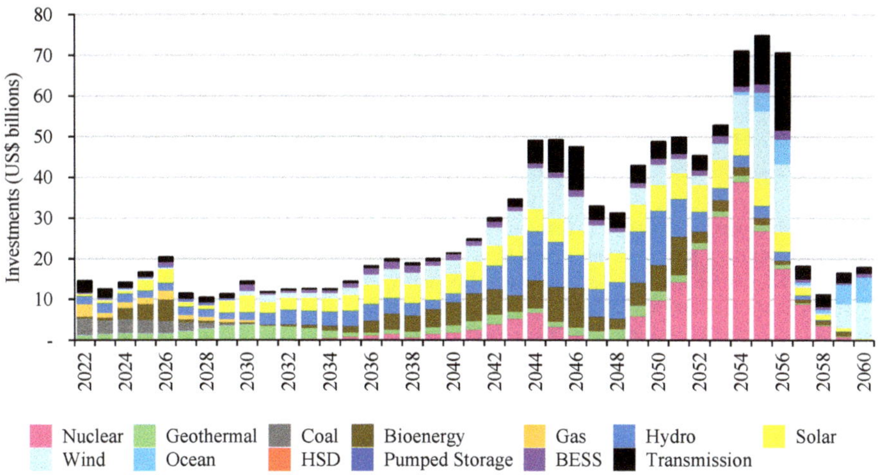

Fig. 12 Estimated investment requirements for ZE scenario electricity generation expansion 2022–2060

7 Conclusions

This study used the Balmorel model to estimate the impact of Indonesia's ZE vision on electricity generation expansion between 2022–2060. The most comprehensive analysis was provided with due consideration for all power plant owners, i.e., PLN, IPP, and PPU, nuclear power, CCS, and green hydrogen as an energy storage option. The simulation was conducted using BaU, ZE, and NZE scenarios, followed by a sensitivity analysis based on electricity demand growth, solar PV growth limits, CCS, and demand flexibility for ZE.

The results showed that the BaU and NZE scenarios generated emissions totaling 674 million tons and 108 million tons CO_2e, respectively. The remaining NZE emissions should be compensated by reductions in other sectors. While the ZE scenario generates zero emissions, it incurs the highest electricity supply cost, as indicated by the projections of US $0.086/kWh for 2060 versus US $0.051/kWh and US $0.074/kWh recorded for the BaU and NZE scenarios, respectively. The ZE scenario forecast constructing renewable power plants beginning with solar PV, followed by onshore and offshore wind turbines. Green hydrogen plants and BESS systems are to be deployed extensively in 2031 and 2034 respectively, to support intermittent renewables plants. Geothermal sources are to be gradually exploited and hydropower potential should be also exploited. Electricity generated thereby should be transmitted to other islands in order to balance intermittent renewables supply. The simulation also recommended constructing hydro-pumped storage beginning in 2025 and continuous nuclear development beginning in 2039 to achieve total capacity of 31 GW by 2060.

This study has two shortcomings that are associated with the Balmorel model. First, it does not have a feature for modeling BESS capability to smoothen and balance the frequency of electricity grids. In this light, BESS was treated as a power plant technology with larger required capacity than needed for frequency balancing alone. Future studies should consider this shortcoming and revise the Balmorel algorithms to take this into account. Second, the model was unable to simulate annual dynamic load demand profile. While this study applied different profiles for each electricity system grid, profiles were fixed during the analysis periods, i.e., 2022 and 2060.

Acknowledgements The authors thank the Economic Research Institute for ASEAN and East Asia (ERIA) for funding provided through Research Project 2022 (Ref. No: ERIA-RD/RA-1-1-2212/04/FY22). Pramudya also thanks the Danish Energy Agency for its contributions in improving the Balmorel model's robustness. The views expressed in this study are those of the authors and do not represent the official positions of ERIA, the Ministry of Energy and Mineral Resources, the TU Delft, the National Research and Innovation Agency, or the University of Wollongong.

References

Al Irsyad MI, Halog AB, Nepal R, Koesrindartoto DP (2017) Selecting tools for renewable energy analysis in developing countries: an expanded review. Front Energy Res 5

Al Irsyad MI, Halog A, Nepal R, Koesrindartoto DP (2019) The impacts of emission reduction targets in Indonesia electricity systems. Indonesian J Energy 2:118–130

Al Irsyad MI, Halog A, Nepal R, Koesrindartoto DP (2020) Economical and environmental impacts of decarbonisation of Indonesian power sector. J Environ Manage 259:109669

DEA, MEMR, ED, EA (2021) Technology data for the Indonesian power sector—catalogue for generation and storage of electricity. Danish Energy Agency (DEA), Ministry of Energy and Mineral Resources (MEMR), Embassy of Denmark (ED), and Ea Energy Analyses (EA), Jakarta, Indonesia, p 216

Fragkos P, Laura van Soest H, Schaeffer R, Reedman L, Köberle AC, Macaluso N, Evangelopoulou S, De Vita A, Sha F, Qimin C, Kejun J, Mathur R, Shekhar S, Dewi RG, Diego SH, Oshiro K, Fujimori S, Park C, Safonov G, Iyer G (2021) Energy system transitions and low-carbon

pathways in Australia, Brazil, Canada, China, EU-28, India, Indonesia, Japan, Republic of Korea, Russia and the United States. Energy 216:119385
Handayani K, Anugrah P, Goembira F, Overland I, Suryadi B, Swandaru A (2022) Moving beyond the NDCs: ASEAN pathways to a net-zero emissions power sector in 2050. Appl Energy 311:118580
IESR, Agora Energiewende, LUT University (2021) Deep decarbonization of Indonesia's energy system: a pathway to zero emissions by 2050. Institute for Essential Services Reform (IESR), Jakarta, Indonesia
Kimura S, Phoumin H (2021) Energy outlook and energy saving potential in East Asia 2020. Economic Research Institute for ASEAN and East Asia (ERIA), Jakarta, Indonesia
Kumar S (2016) Assessment of renewables for energy security and carbon mitigation in Southeast Asia: the case of Indonesia and Thailand. Appl Energy 163:63–70
MEF (2021a) Indonesia long-term strategy for low carbon and climate resilience 2050 (Indonesia LTS-LCCR 2050). Ministry of Environment and Forestry (MEF), Jakarta, Indonesia, p 156
MEF (2021b) Laporan inventarisasi gas rumah kaca (GRK) dan monitoring, pelaporan, verifikasi (MPV). Ministry of Environment and Forestry (MEF), Jakarta, Indonesia, p 167
MEMR (2019) Ministerial regulation of energy and mineral resources number 143 K/20/MEM/2019 on the national electricity general plan. Ministry of Energy and Mineral Resources (MEMR), Jakarta, Indonesia, p 441
Phoumin H, Kimura F, Arima J (2020) Potential renewable hydrogen from curtailed electricity to decarbonize ASEAN's emissions: policy implications. Sustainability 12:10560
PLN (2021) Rencana Usaha Penyediaan Tenaga Listrik PT PLN (Persero) 2021–2030. PT Perusahaan Listrik Negara (PLN), Jakarta
Prasodjo E, Nurzaman H, Walujanto, Rosdiana D, Ismutadi P, Malik C, Santosa J, Nurrohim A, Widiastuti KD, Pambudi SH, Wibowo JL, Sauqi A (2016) Outlook Energi Indonesia 2016. In: Abdurrahman S (ed) Secretary of National Energy Council, Ministry of Energy and Mineral Resources, Jakarta, Indonesia, p 140
Reyseliani N, Purwanto WW (2021) Pathway towards 100% renewable energy in Indonesia power system by 2050. Renew Energy 176:305–321
Siagian U, Yuwono B, Fujimori S, Masui T (2017) Low-carbon energy development in Indonesia in alignment with intended nationally determined contribution (INDC) by 2030. Energies 10:52
Van Soest HL, den Elzen MG, van Vuuren DP (2021) Net-zero emission targets for major emitting countries consistent with the Paris agreement. Nat Commun 12:1–9
Wiese F, Bramstoft R, Koduvere H, Alonso AP, Balyk O, Kirkerud JG, Tveten ÅG, Bolkesjø TF, Münster M, Ravn H (2018) Balmorel open source energy system model. Energ Strat Rev 20:26–34

Chapter 7
Deployment of Renewable Energy and Utility-Scale Batteries in Australia: Lessons Learned and Policy Implications for Other Countries

George Grozev, Ty Christopher, and Pascal Perez

Abstract The huge potential of renewable energy to reduce greenhouse gas emissions has already been demonstrated in Australia, which is positioned well at the forefront of the renewable energy transition despite often changing energy policy. This chapter reviews the most recent trends and outcomes of renewable energy utilization in Australia's National Electricity Market (NEM). The purpose of this review is (1) to update the most recent renewable energy and battery developments in the NEM, (2) to describe the energy dynamics in South Australia, the most advanced Australian state in terms of penetration of wind and solar PV generation, (3) to summarize current and future cost projections of renewable generation technologies in Australia, and (4) to summarize the main policy support schemes used in Australia to facilitate renewable energy investments. This chapter could help inform energy and climate policy decision making in Australia and other countries, including in Southeast Asia.

Abbreviations

ABC	Australian Broadcasting Corporation
ACT	Australian Capital Territory
AEMO	Australian Energy Market Operator
AER	Australian Energy Regulator
ARENA	Australian Renewable Energy Agency
ASEAN	Association of Southeast Asian Nations
AU¢	Australian cent
AUD	Australian Dollar

G. Grozev (✉) · T. Christopher
University of Wollongong, Wollongong, NSW 2522, Australia

P. Perez
Australian Urban Research Infrastructure Network (AURIN), Victoria 3010, Australia

H. Phoumin et al. (eds.), *Large-Scale Development of Renewables in the ASEAN*, Economics, Law, and Institutions in Asia Pacific,
https://doi.org/10.1007/978-981-99-8239-4_7

BESS	Battery Energy Storage System
CSIRO	Commonwealth Scientific and Industrial Research Organisation
CO_2-e	Carbon Dioxide Equivalent
CER	Clean Energy Regulator
DCCEEW	Department of Climate Change, Energy, the Environment and Water (Australian Government)
DELWP	Department of Environment, Land, Water and Planning (Victoria)
DEBS	Distributed Energy Buyback Scheme (Western Australia)
ERIA	Economic Research Institute for ASEAN and East Asia
ESCSA	Essential Services Commission of South Australia
ESS	Energy Storage System
EBITDA	Earnings Before Interest, Taxes, Depreciation and Amortization
FCAS	Frequency Control Ancillary Services
FiT	Feed-in tariff
FTE	Full-Time Employment
FY	Financial Year
GHG	Greenhouse Gas
GST	Goods and Services Tax
GW	Gigawatt
GWh	Gigawatt-hour
HVDC	High-Voltage Direct Current
IEA	International Energy Agency
ISO	Independent System Operator
IPART	Independent Pricing and Regulatory Tribunal (NSW)
ISP	Integrated System Plan
kV	Kilovolt
LCOE	Levelized cost of electricity
LGC	Large-scale Generation Certificate
LRET	Large-scale Renewable Energy Target
MRET	Mandatory Renewable Energy Target
Mt	Megatonne
MW	Megawatt
MWh	Megawatt-hour
MWp	Megawatt peak
NEM	National Electricity Market
NSW	New South Wales
NT	Northern Territory
NZE	Net Zero Emissions
PV	Photovoltaic
QLD	Queensland
RE	Renewable Energy
REBS	Renewable Energy Buyback Scheme (Western Australia)
RERT	Reliability and Emergency Reserve Trader
RET	Renewable Energy Target
SA	South Australia

SBS	Solar Bonus Scheme
SIPS	System Integrity Protection Scheme
SRES	Small-scale Renewable Energy Scheme
STC	Small-scale Technology Certificates
TWh	Terrawatt-hour
VIC	Victoria
VRE	Variable Renewable Energy

1 Introduction

Approximately 78% of human-caused global greenhouse gas (GHG) emissions between 1970 and 2021 were due to fossil fuel combustion (Bergero et al. 2021). In Australia, GHG emissions in 2021 are estimated at 488 Mt CO_2-e, down 23% (145.7 Mt CO_2-e) from 1990, but up 0.8% (4.1 Mt CO_2-e) from 2020 (DCCEEW 2022). Electricity generation produced 32.9% of the total. Emissions per capita were 18.95 t CO_2-e, 48.9% lower than 1990 levels. While only a few countries in the world have higher per capita emissions than Australia, the country has such distinctive characteristics as vast, sparsely populated territory, with several big urban settlements far from each other and predominantly close to the coastline. Australia also has a big, energy-hungry, export-oriented resource-extraction industry.

Renewable energy is one of the best and cheapest candidates to replace fossil fuel-based energy sources and thus reduce anthropogenic GHG emissions. It can be derived from various sources such as solar, wind, hydro, geothermal, or tidal. In Australia the main sources of renewable energy are hydro, wind, and solar. Hydropower has been used for a long time and, except for hydro-based pumping storage, there are not many acceptable locations for new reservoirs or hydropower stations. Utility-scale wind and solar PV generation capacities have grown significantly over the last decade, however.

Many authors (Jones 2010; Lang and Miller 2011; Nelson et al. 2021; Simpson and Clifton 2014; Simshauser and Gilmore 2022) have studied the role and characteristics of renewable energy policies in Australia. The aim of these policies is to promote renewable energy investment, increase the share of renewable generation in the electricity generation mix, and ultimately reduce GHG emissions from fossil fuel-based generation. The main questions for these policies are what framework to implement and how much government support is required to develop a renewable energy industry (Simpson and Clifton 2014). An example Australian policy framework is the renewable energy target legislation, which the federal government enacted in 2000 as the Renewable Energy (Electricity) Act 2000 (Australian Government 2022). The mechanism of this policy obliges electricity retailers to source some of the electricity delivered to their customers from "clean" generation technologies by purchasing renewable energy certificates, which are created when renewable energy is generated by a specified type of generator, such as wind or solar farms. The level of support in this policy is defined by the specific proportion of renewable energy that retailers must achieve.

Bergero et al. (2021), Pablo-Romero et al. (2021), and Praveen et al. (2020) provide an international perspective of renewable energy target policies for different groups of countries. Bergero et al. (2021) use qualitative comparative analysis to investigate the policy diffusion in 187 countries between 1974 and 2017. Their analysis demonstrates that there are multiple paths for renewable energy target adoption.

Byrnes et al. (2013) provide a good overview of Australian renewable energy policy. The authors briefly describe the Australian governmental and energy systems, and offer a comprehensive diagram of the energy regulatory environment that includes all important categories of market players and stakeholders. The paper identifies the following barriers to deployment of renewable energy in Australia: administrative hurdles, costly procedures for grid connection, policy instability, lack of social acceptance, cost competitiveness, and government support for existing electricity generators. Martin and Rice (2015) discuss administrative hurdles in detail, providing information about planning and approval of renewable energy projects in Australia. The paper presents a block diagram for a renewable energy project planning and permitting framework associated with the roles that federal, state, and local governments play in these processes. The authors also list policies, legislation, and regulations that a company trying to develop a renewable energy project has to comply with.

Feed-in tariffs (FiTs) is a very popular policy tool used in Australia and other countries to stimulate uptake of residential renewable energy, especially solar PV. Poruschi et al. (2018) provide a good relatively recent review of FiTs in all states and territories in Australia, analyzing the number of small solar generation units installed by state/territory, average cost of residential solar PV systems, and historical information about the rate of FiTs in each jurisdiction, and investigates the link between FiTs and disconnections from the grid. Other relevant papers discussing FiTs in Australia are Chapman et al. (2016), Li et al. (2021), and Martin and Rice (2013).

Martin and Rice (2021) provide a comprehensive literature review on energy storage systems (ESS). The authors explain the importance of ESS services across the energy supply chain and for future renewable energy growth. Four main groups of literature related to ESS are identified and reviewed: benefits, technical applications, technology cost and economics, and policy support. This research suggests that ESS-related policies have received less attention than renewable energy policies. The authors discuss ESS supporting policies and regulatory options in the Australian context.

McGreevy et al. (2021) comprehensively describe the renewable energy transition in South Australia between 2004 and 2018. This state has demonstrated a highly successful, sustainable transition to a low-GHG emission energy system. The paper claims that when renewable energy achieves a critical uptake, it produces a path-dependent trajectory, which is difficult to change even by governments with different ideologies. As South Australia's renewable energy transition has been prominent overseas as well as in Australia with its challenges and achievements, we dedicate a whole section to it in this paper.

Crowley and Jayawardena (2017) discuss energy disadvantages in Australia, linking energy pricing, energy policy, climate change impact, and disadvantage in the country. As of early 2022, the topic is even more important in the context of high energy prices related to Russia's invasion of Ukraine, high inflation, and post-pandemic supply chain disruptions. The authors point out that energy poverty and disadvantage are not only third-world problems, but also impact poor, elederly, indigenous, remote, and other disadvantaged citizens of Australia. This paper claimed that renewable energy has a role to play in alleviating energy poverty, listing a number of policy recommendations to this end.

The purpose of this paper is to review the most recent trends and outcomes of Australia's renewable energy policies, with a focus on the country's National Electricity Market (NEM). Section 2 presents recent updates on renewable energy and battery developments. Section 3 describes in more detail the energy dynamics in South Australia, the most advanced Australian state in terms of penetration of wind and solar PV generation. Section 4 summarizes current and future cost projections of renewable generation technologies in Australia. Section 5 discusses the main policy support schemes used in Australia to facilitate renewable energy investment. Some policy recommendations are discussed in the last Section. When discussing issues herein, we apply the complex system science approach where possible (Batten and Grozev 2006). When applying this approach to electricity markets, we treat them as complex interactions between physical infrastructure, i.e., electricity grids with supply and demand, economics, i.e., price, cost, and market players, and environment, i.e., greenhouse gas emissions, resource use, etc., including policy and regulation.

As Australia is in an advanced stage of renewable energy uptake and related policies have faced many challenges and zigzags, this experience presents useful lessons for policy makers not only in Australia but in other countries as well, specifically in Southeast Asia.

2 Renewable Energy Uptake in the NEM

2.1 Energy and Generation Capacity

Since commencing operations on December 13, 1998, Australia's National Electricity Market (NEM) has grown to encompass the five states of Queensland, New South Wales (NSW), Victoria, South Australia (SA), and Tasmania, together with the Australian Capital Territory (ACT) (AEMO 2022b; Hu et al. 2005; Nepal and Foster 2016). While the NEM operates on one of the largest electricity grids in terms of geographical area coverage and distance, however in terms of connectivity, it is sparsely connected by transmission lines, usually having only one or two transmission interconnections between any two adjacent market regions, which conform to

Table 1 Generation capacity (2022) and energy (2021) in the NEM

Region	Fuel source	Fuel type	Registered capacity			Generation—2021	
			(MW)	(%)	Number of facilities	GWh	(%)
NEM	Coal	Fossil	23,049	39.04	16	128,008	68.18
NEM	Gas	Fossil	10,444	17.69	51	11,713	6.24
NEM	Distillate	Fossil	1436	2.43	20	101	0.05
NEM	Hydro	Renewable	9285	15.73	59	15,811	8.42
NEM	Wind	Renewable	8385	14.20	80	22,968	12.23
NEM	Solar (Utility)	Renewable	5346	9.05	67	8824	4.70
NEM	Battery	Renewable	657	1.11	9	129	0.07
NEM	Bioenergy	Renewable	440	0.75	35	188	0.10
NEM	Total		59,042	100.00	337	187,741	100.00

the states. The NEM is a real-time, energy-only, gross pool market, operating on 5-min settlement periods since October 1, 2021, versus 30-min settlement periods prior to this date. Simshauser (2022) provides a detailed analysis of microeconomic reform of electricity utilities in Australia and an excellent overview of NEM performance, achievements, and challenges.

Black and brown coal generators have long dominated NEM generation capacity. In the NEM's early years, coal-fired generation contributed more than 90% of total electricity generated. In recent years this contribution has dropped to approximately 60%. Current generation capacities in the NEM, more than 59 GW, and electricity produced in 2021, are presented in Table 1 and Fig. 1 (Open NEM 2022). The following dispatchable firm capacities are currently available, based on Australian Energy Market Operator (AEMO) considerations (AEMO 2022a): 23 GW from coal-fired generation, 11 GW from gas and liquid fuels, 7 GW from hydropower, excluding some pump hydro, and 1.5 GW from dispatchable energy storage, i.e., battery storage and pump hydro.

Shi et al. (2022) studies the role of gas-powered generation in the NEM and claims that it is negatively related to generation from VREs and positively related to electricity demand gap and electricity prices.

2.2 Distributed Solar PV and Decreasing Daily Electricity Demand

Australia has one of the highest uptakes of residential solar PV installations in the world, and more deployment is expected in the near future (Young et al. 2019).

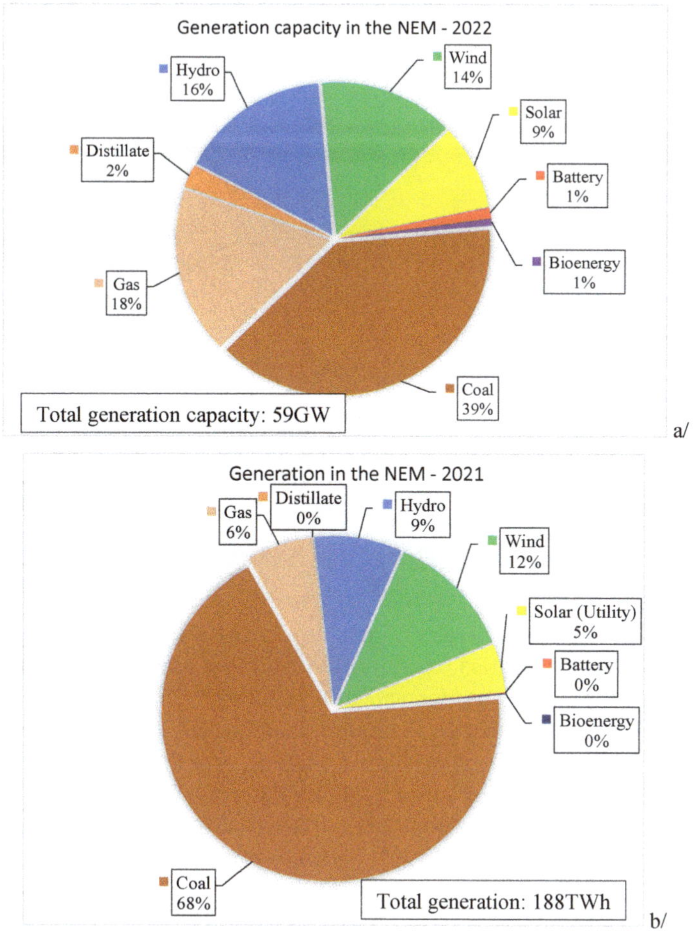

Fig. 1 Proportions of generation capacities (**a**) and generation (**b**) in the NEM[1]

Approximately 30% of detached homes in the NEM regions have solar PV panels with 15 GW aggregate capacity (AEMO 2022a).

Two policies have substantially supported residential solar PV uptake, the Renewable Energy Target at the federal level, and FiTs at state level; see Sect. 5 for particulars. Uptake is expected to continue to grow in the coming years.

The increase of distributed and utility-scale solar PV changes the operational electricity demand profile significantly. With growing distributed solar PV generation during the day when the sun shines, operational electricity demand decreases significantly during the same period, with the biggest reductions occurring around midday, when solar irradiation is strongest. Operational demand must be balanced

[1] Battery and bioenergy-based generation are less than or equal to 0.1% of the total, and are thus displayed as 0% on the pie chart.

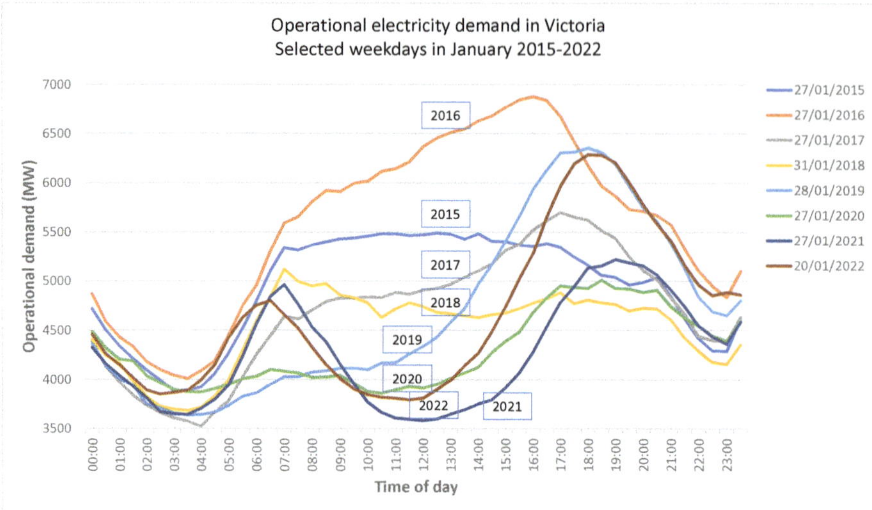

Fig. 2 Operational electricity demand in Victoria on selected weekdays in January 2015–2022

by supply from all other generators, including utility-scale solar PV. In 2012, the California Independent System Operator (CAISO) first used the term "duck curve" to label changes in electricity demand due to solar PV contributions (California ISO 2016). Figure 2 provides an example related to this phenomenon, presenting several daily electricity demand profiles in Victoria over the preceding seven years. Days selected are weekdays in January, when solar irradiation is usually high. From Fig. 2, the trend of declining operational demand close to midday is clear as the valley of the curve, or the "duck belly", becomes bigger and lower. The operational demand curves show several other changes as well. The minimum electricity demand declines, rapidly, potentially occurring during midday instead of early in the morning, i.e., nearer to 4:00 am, as in prior times. The usual afternoon peak demand related to air conditioning use in summer is moving to early evening. This summer peak electricity demand used to be a key driver for network investment, so its reduction benefits network utilities. Close to sunset, when solar irradiation disappears and solar PV generation becomes null, electricity demand sharply shifts, increasing quickly toward the evening peak. All these changes require new approaches to electricity supply management.

2.3 *Variable Renewable Energy*

The proportion of renewable energy in the NEM has increased steadily in the past several years. Simshauser and Gilmore (2022) define the period from 2016 to 2021 as an investment super-cycle for the NEM, in which AUD26.5 billion was invested across 135 projects, mostly for wind and utility-scale solar PV with 16 GW aggregate generation capacity. Another 6 GW is expected to be operational in the next several years in either committed or anticipated projects (AEMO 2022a).

According to AEMO data (AEMO 2022e), instantaneous renewable generation reached 64.1% of total generation in the 30-min interval ending at 11:30 am on September 18, 2022. Total NEM generation includes generation from all big generators plus distributed, i.e., residential, solar PV. Renewable generation includes output from all renewable generators, battery generation, and distributed solar PV. Figure 3 presents the trends in minimum, average, and maximum instantaneous renewable generation in the NEM for the preceding four years (minus one quarter).

Variable renewable energy (VRE) is a term adopted by industry specialists to classify the fastest growing component of renewable energy. While it includes wind and solar PV generation, for example, it excludes hydropower. VRE produced in the NEM has also increased significantly over the preceding several years. Generation from utility-based wind and solar set multiple records over this period, with the latest records from the third quarter of 2022 listed in Table 2. During the 30-min interval on September 18, 2022, with the highest instantaneous renewable generation share of 64.1% of total generation, the distributed solar PV contribution was 32% of total generation and that of VRE was 29%. The average VRE generation in this quarter reached 4465 MW, which was 483 MW higher than the corresponding generation in Q3 2021 (AEMO 2022e).

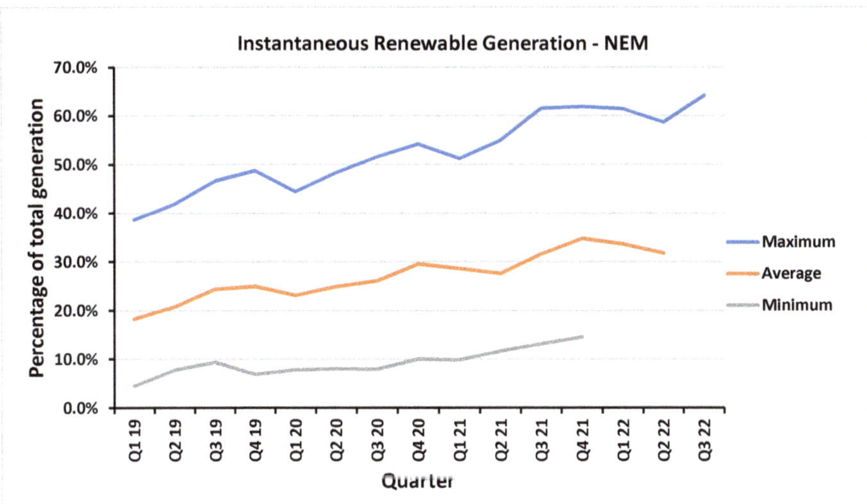

Fig. 3 Instantaneous renewable generation in the NEM (based on AEMO 2022e)

Table 2 Records of renewable generation in the NEM (based on AEMO 2022e)

VRE type	Generation	Date	Time	Comment
NEM highest instantaneous renewable generation share of total generation	64.10%	18-Sep-22	11:30	Instantaneous renewable generation = Grid-scale wind + grid-scale solar + hydro + biomass + battery generation + distributed solar PV
NEM highest wind output	7271 MW	04-Aug-22	21:00	8% higher than the previous record from Q2 2022
NEM highest grid-scale solar output	4628 MW	04-Sep-22	10:00	
NEM highest VRE output	9112 MW	22-Aug-22	09:30	Wind and grid-scale solar

2.4 Batteries

Martin and Rice (2021) and Arraño-Vargas et al. (2022) provide a comprehensive review of the Energy Storage Systems (ESS) literature. They describe benefit realization, technical applications, technical performance, technology cost, and popular policy support for ESS applications. Battery technologies are versatile and need to be adapted to many different technical applications based on such characteristics as type, capacity, response time, and discharge duration. Common technical applications of ESS are for energy storage, peak shaving, emergency backup power, renewable energy integration, i.e., intermittency mitigation, power quality maintenance, grid stability, spinning reserves, transmission and distribution grid deferral, and end user applications and services.

A list of grid-scale energy battery systems in Victoria, Queensland and NSW in given in Table 3, and for SA in Table 5. This list is extracted from AEMO's Registration and Exemption List Excel spreadsheet (AEMO 2022c). The total current NEM battery capacity is 841 MW. More information about battery ESS grid services and existing and proposed battery ESS is provided by Arraño-Vargas et al. (2022).

The Victorian Big Battery is the largest lithium-ion battery in the Australia and one of the largest in the world. Commissioned in 2021, its maximum capacity is 300 MW/450 MWh, although its registered capacity is 360 MW (DELWP 2022). During the summer months of November to March, 250 MW of its capacity is reserved for the System Integrity Protection Scheme (SIPS), with the remaining 50 MW available for commercial NEM participation. At other times, the whole capacity of the battery can be operated on a commercial basis. During the summer months the battery stabilizes the grid in case of unscheduled power outages, allowing AEMO more time to resolve the impact of such outages and potentially avoiding widespread blackouts. The battery thus helps increase the import power flow limit of Victoria to NSW interconnectors by up to 250 MW.

Table 3 Registered energy battery systems in Victoria, Queensland, and NSW—June 2022

Market participant	Station name	Region	Dispatch type	Category	Classification	Registered capacity (MW)
Energy Australia Pty Ltd.	Ballarat Battery Energy Storage System	VIC1	Generator	Market	Scheduled	30
Bulgana Wind Farm Pty Ltd.	Bulgana Green Power Hub - Battery Units 1–40	VIC1	Generator	Market	Scheduled	24
Energy Australia Pty Ltd.	Gannawarra Energy Storage System	VIC1	Generator	Market	Scheduled	31
Victorian Big Battery Pty Ltd.	Victorian Big Battery	VIC1	Generator	Market	Scheduled	360
Kennedy Energy Park Pty Ltd.	Kennedy Energy Park Battery Units 1–4	QLD1	Generator	Market	Non-scheduled	2
AGL Sales (Queensland Electricity) Pty Ltd.	Wandoan Battery Energy Storage	QLD1	Generator	Market	Scheduled	123
Iberdrola Australia Wallgrove Pty Ltd.	Wallgrove BESS 1	NSW1	Generator	Market	Scheduled	50

Snowy 2.0 is the largest multi-billion dollar renewable energy project currently in construction in Australia with government support. It is a pump-based hydro extension of the existing Snowy scheme, which consists of nine power stations and 16 major dams, located between Melbourne and Sydney. Snowy 2.0 will use existing dams and its estimated capacity will be 2 GW/350 GWh, or 175 h of operation (Snowy Hydro 2022). It will have six generating units, the first of which is expected to provide power in 2025. Snowy 2.0 will provide firm, dispatchable generation capacity and bulk, long-term storage, which could utilize excess renewable energy and provide electricity on demand.

3 Renewable Energy Dynamics in South Australia

Australia's NEM is experiencing one of the fastest-growing VRE transitions in the world, raising new challenges to system security and reliability.

South Australia has demonstrated a highly successful, sustainable transition to a low-GHG emission energy system (McGreevy et al. 2021). It leads Australia in this transformation with significant wind and solar PV generation capacity, installing the first utility-scale lithium-ion battery in 2017, and more recently, commissioning four synchronous condensers in November 2021. One of the latest policy decisions underpinning these developments was the Government of SA's enactment of a new energy policy in 2017 (Government of SA 2017).

South Australia is a state in the southern, central mainland Australia with territory of 983,482 km^2 and a population of 1.8 million according to the 2021 Census (Australian Bureau of Statistics 2021). 80% of the population live in the capital Adelaide and its surrounding metropolitan areas. There are four other large population settlements in this vast territory, with landscapes including deserts, mountain ranges, and agricultural land, as well as a coastline of more than 3700 km.

The South Australia electricity system was privatized in 1999 and the state-owned monopoly vertically disaggregated into separate businesses. The state has transformed its energy system, increasing its renewable energy share from 1% to more than 68% over the preceding 15 years (Government of SA 2022). The state has a goal of 100% net renewable energy by 2030. In 2021 the daily electricity generated by renewable sources exceeded the daily demand on 180 days, almost 50% of the time. Registered and maximum generation capacities, as well as generation in South Australia in 2022, are presented in Table 4 and Fig. 4. With its consistent energy policy, South Australia has attracted more than AUD6 billion investment in large-scale renewable and storage projects over this period and has a pipeline of projects surpassing three times this historical investment.

This recent energy development in South Australia has involved challenges, bold policy decisions, innovations, and some unexpected market phenomena, some of which are briefly described hereinafter.

Table 4 Registered and maximum generation capacity in South Australia in 2022

Region	Indicator	Registered capacity		Maximum capacity	
	Fuel source—primary	MW	%	MW	%
SA	Fossil	3244.46	51.28	3462.65	52.82
SA	Wind	2351.41	37.16	2454.00	37.43
SA	Solar	490.22	7.75	401.22	6.12
SA	Battery—Generator	221.36	3.50	217.00	3.31
SA	Renewable/Biomass/Waste	18.14	0.29	20.00	0.31
SA	Hydro	1.44	0.02	1.00	0.02

Table 5 Registered energy battery systems in South Australia—June 2022

Market participant	Station name	Region	Dispatch type	Category	Classification	Registered capacity (MW)
Accel Energy Retail Pty Ltd.	Dalrymple North Battery Energy Storage System	SA	Generator	Market	Scheduled	30.00
Lake Bonney Wind Power Pty Ltd.	Lake Bonney Battery Energy Storage System	SA	Generator	Market	Scheduled	25.00
South Australian Water Corporation	Adelaide Desalination Plant	SA	Generator	Market	Scheduled	7.76
South Australian Water Corporation	Bolivar Waste Water Treatment Plant	SA	Generator	Market	Scheduled	3.08
South Australian Water Corporation	Happy Valley Water Treatment Plant	SA	Generator	Market	Scheduled	5.52
Hornsdale Power Reserve Pty Ltd.	Hornsdale Power Reserve	SA	Generator	Market	Scheduled	150.00

3.1 2016 Black System Event

On September 28, 2016, South Australia experienced a so called "black system event", a sequence of cascading events resulting in loss of electricity supply to all customers in the state (AEMO 2017). First, several tornados with wind speed up to 260 km/h damaged three transmission lines. After that nine wind farms reduced their output or disconnected from the grid due to grid instability, reducing generation by 456 MW in less than seven seconds. The Victoria-SA Heywood interconnector tripped due to a sudden increase in imported power and the SA power system separated from the rest of the NEM. All supply to SA (except Kangaroo Island) was lost at 4:18 pm with 850,000 customers losing power for several hours, some of them for several days. AEMO suspended the market in SA for twelve days.

The main question this event raises is how to adapt and make resilient the aging electricity infrastructure, grid and transmission towers and lines alike, against increasingly frequent climate change-influenced extreme weather events.

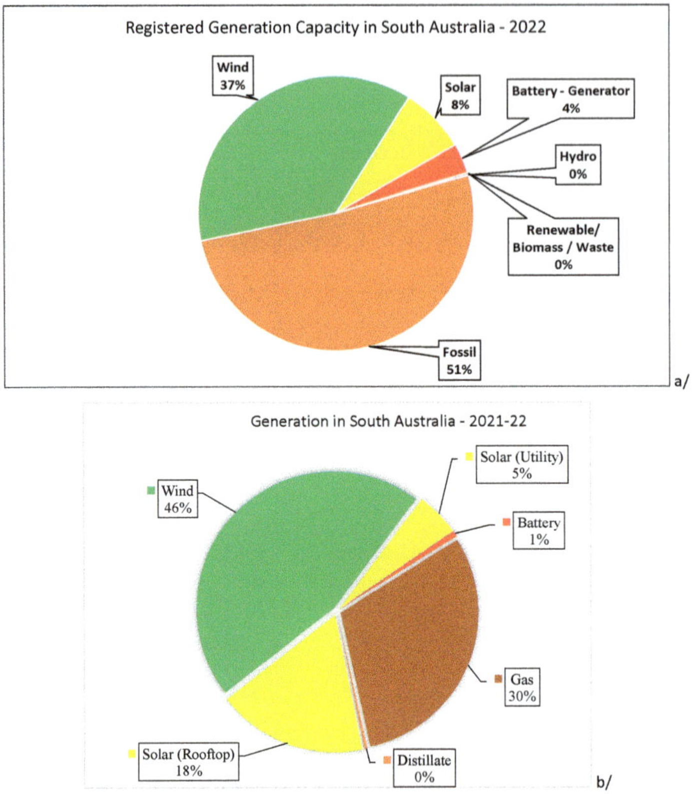

Fig. 4 Proportions of registered generation capacity (June 2022) (**a**) and generation (FY2021–22) (**b**) in South Australia

A secondary question is how to better integrate renewable generation into the electricity grid. Wind farms, like other renewable generators, are asynchronous and use different control systems to ride out disturbances. In the black system event, several wind farms had the same default settings for riding out disturbances, causing simultaneous disconnections that exacerbated the problem. The Australian Energy Regulator (AER) subsequently sued four wind farm operators for not complying with generator performance standards for riding out grid disturbances (Australian Energy Regulator 2019).

3.2 *100 MW Battery in 100 days*

Approximately six months after the black system event in 2016, Elon Musk, boss of Tesla and Space X, announced that Tesla could install a big battery in South Australia to fix its power system problems (ABC 2021). Interestingly, he offered to build a

100 MW battery in less than 100 days or deliver it for free. Four months later, the SA government announced an agreement with Tesla to build a 100 MW battery near Jamestown.

The Hornsdale Power Reserve, the world's first large lithium-ion battery, at 100 MW/129 MWh—was completed on schedule on December 1, 2017 (Hornsdale Power Reserve 2022). In 2020, its capacity was expanded by 50% and a functional was implemented allowing inertia support services to the grid.

During the two South Australian power system separation events, as described hereinafter, the three grid-scale batteries installed there provided a high degree of Frequency Control Ancillary Services (FCAS) support and generated approximately AUD50 million in spot FCAS revenue (AEMO 2020). It was also reported that the Hornsdale Power Reserve had delivered AUD88 million earning (EBITDA) in the first two and half years of operation, making it practically pay for itself (Renew Economy 2020).

A list of grid-scale energy battery systems in South Australia in given in Table 5, extracted from AEMO's Registration and Exemption List Excel spreadsheet (AEMO 2022c). Each battery is registered twice, once as "generator" and once as "load", and the respective registered capacities may vary. Total registered battery capacity as generators in SA stands at 221.36 MW.

3.3 *Grid Separation Events*

Two high-voltage transmission lines link South Australia with Victoria: the 275 kV AC Heywood interconnector with 650 MW bidirectional capacity and the Murraylink 220 MW High-Voltage Direct Current (HVDC) link. These transmission lines are the only links between SA and the NEM. A full outage of the Heywood interconnector would lead to system separation of SA from the NEM as Murraylink, being an HVDC interconnector, does not provide system strength or inertia support.

On January 31, 2020, a severe storm brought down the 500 kV transmission line in Western Victoria, leading to an 18-day separation of the South Australian and Victorian power systems (AEMO 2020). There was another separation event on March 2, 2020, lasting approximately 8 h.

Major separation events have an impact on the strength of the electrical parameters of the grid, requiring additional intervention by the market operator to stabilize the grid. Separation events lead to price volatility and additional system cost. For the NEM, the system cost is related to (1) FCAS, (2) Direction compensation, (3) the Reliability and Emergency Reserve Trader (RERT) function, and (4) Variable renewable energy curtailment (AEMO 2020). While system cost is recovered from retailers and generators, generators also receive some of it themselves.

The two separation events in South Australia, together with one separation event in NSW on April 1, 2020, caused by bushfires, contributed AUD229 million to the system cost, or 74% of total system cost for the January–March 2020 quarter (AEMO

2020). Total system cost for the quarter thus amounted to 8% of the energy cost, well above the typical 1–2% range.

3.4 Negative Wholesale Electricity Prices

The renewable transformation has caused many extended periods of negative spot prices and increased uncertainty and variability of electricity prices in South Australia and other NEM regions (Grozev et al. 2022; Havyatt et al. 2022). Negative price frequencies in South Australia and Victoria reached record high values in October 2021, as shown in Fig. 5. While the proportion of negative prices eased in 2022 due to the NEM introducing the aforementioned 5-min settlement period on October 1, 2021, and renewables firms accumulating more bidding experience, such events still occur frequently.

Electricity price volatility assessment and management had been a major challenge for the NEM even before the energy crisis that began in 2022. It relates to the intermittent character of VRE, which makes it harder to balance the variable by nature demand with frequently changing supply. While renewable energy generation has zero-fuel cost and thus helps to reduce electricity spot prices, higher spot price volatility can result in higher wholesale contract prices and therefore higher prices for end consumers, offsetting some or all initial price reductions.

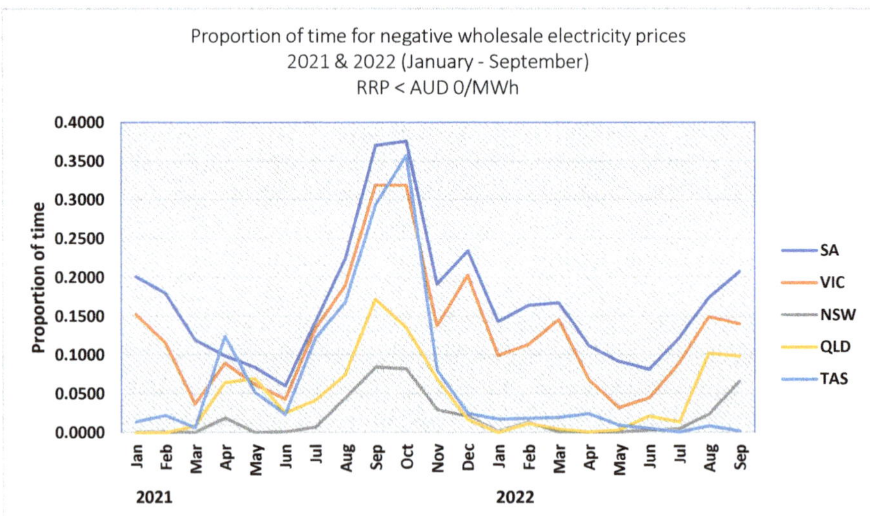

Fig. 5 Proportion of time with negative wholesale electricity prices in the NEM, 2021 and Q1–Q3 2022

3.5 The Role of the Synchronous Condensers in System Cost

Successful testing and commissioning of the four synchronous condensers in SA was completed in November 2021, allowing the grid to operate with fewer synchronous generators (at least two gas generators in SA) and leading to lower system strength curtailment. System strength curtailment in the region fell from 62 MW in Q3 2021 to zero in Q4. System security direction cost in South Australia stood at AUD6 million for Q2 2022, the lowest level since Q2 2019 (AEMO 2022d).

Another important phenomenon associated with renewable generation is restricting renewable power quantities from time to time for grid security and stability. On several occasions, VRE curtailment in SA reached more than 1000 MW, a substantial proportion of renewable generation there. In case of curtailment energy is wasted, however, upgrading the grid to accommodate all possible renewable energy could be very expensive.

4 Costs of Renewable Generation Technologies and Storage in Australia

Renewable energy technologies provide the fastest growing energy sources in Australia and globally alike. They currently represent some of the least expensive abatement opportunities for reducing greenhouse gas emissions, and their role is likely only to increase significantly over the next several decades (Graham et al. 2022).

In Australia, the Commonwealth Scientific and Industrial Research Organisation (CSIRO) and AEMO have established a project to annually estimate and update electricity generation and storage cost data (Graham et al. 2022). Aurecon has supported this work by providing characteristics of current generation technologies and electricity storage (Aurecon 2021). The project applies a scenario modeling approach combined with technology learning rates to estimate future generation technology and storage costs. Learning rates based on historical data aim to determine cost reductions for each doubling of cumulative capacity deployed (Graham et al. 2022). The report was finalized in response to feedback from a wide range of Australian stakeholders and experts.

The future costs of generation technologies and storage are modeled based on the following scenarios:

- Business as Usual (BaU);
- Global Net Zero Emissions by 2050 (Global NZE by 2050); and
- Global Net Zero Emissions post 2050 (Global NZE post-2050).

The BaU scenario is characterized by a slow uptake of renewable energy and having the highest technology cost. The Global NZE by 2050 scenario is consistent with the International Energy Agency's (IEA) report "Net Zero by 2050" (IEA 2021), which defines the most technically feasible and cost-effective roadmap to reach net zero emissions by 2050. The Global NZE post-2050 scenario sits between the other scenarios.

Australia and other countries are currently in an inflationary cycle. The uncertain nature of the inflation cycle, in terms of duration, scale, and coverage, makes it a challenging factor to account for in capital cost modeling of renewable energy generation and storage. The approach that CSIRO's report (Graham et al. 2022) takes is to assume that the real cost of technologies in the first projection year (2022) would be flat, without high inflation, instead of decreasing under normal conditions. The report does not assign a more specific level of change after 2022 due to uncertain future inflationary impact.

Obviously, technology costs depend on local as well as global conditions. Technology cost reductions due to learning by doing could be larger for some regions with greater uptake of given generation technologies. One example is China, where such costs can be substantially lower (Graham et al. 2022). Including local as well as global learning models in CSIRO's approach allows the cost of deployment of new technologies in a given region or country to quickly approach the cost of similar technologies in other regions with larger-scale investment experience. In that sense, the cost projections for Australia are a good starting point for Southeast Asian countries with similar conditions.

Table 6 summarizes current (2021) and projected (2030, 2040 and 2050) capital costs for the following renewable generation technologies per the abovementioned scenarios (Graham et al. 2022):

Table 6 Current and projected renewable generation technology capital costs in 2021–22 AUD/kW

Year	Large-scale solar PV	Rooftop solar PV	Wind	Offshore wind	Scenario
	2021–22 AUD/kW	2021–22 AUD/kW	2021–22 AUD/kW	2021–22 AUD/kW	
2021	1441	1333	1960	4649	Business as usual
2030	1013	949	1897	4545	
2040	733	691	1868	4482	
2050	644	606	1828	4431	
2021	1441	1333	1960	4649	Global NZE by 2050
2030	785	752	1633	2967	
2040	578	557	1553	2653	
2050	521	500	1521	2506	
2021	1441	1333	1960	4649	Global NZE post 2050
2030	1046	977	1778	4437	
2040	689	653	1648	3772	
2050	530	508	1546	3168	

- Large-scale solar PV;
- Rooftop solar PV;
- Onshore wind; and
- Offshore wind.

These estimates are shown graphically in Fig. 6, while Table 7 and Fig. 7 present current and projected total battery costs for 1-h, 2-h, 4-h, and 8-h storage per to the abovementioned scenarios. Total capital cost for batteries includes battery cost and balance of plant cost, i.e., the cost of support components.

Levelized cost of electricity (LCOE) is an important comparison metric when evaluating investment in generation technologies. It is the total cost a generator must recover to meet all its costs, including return on investment (ROI) over its lifetime. It is usually measured in dollars per MWh (AUD/MWh) produced by large generator

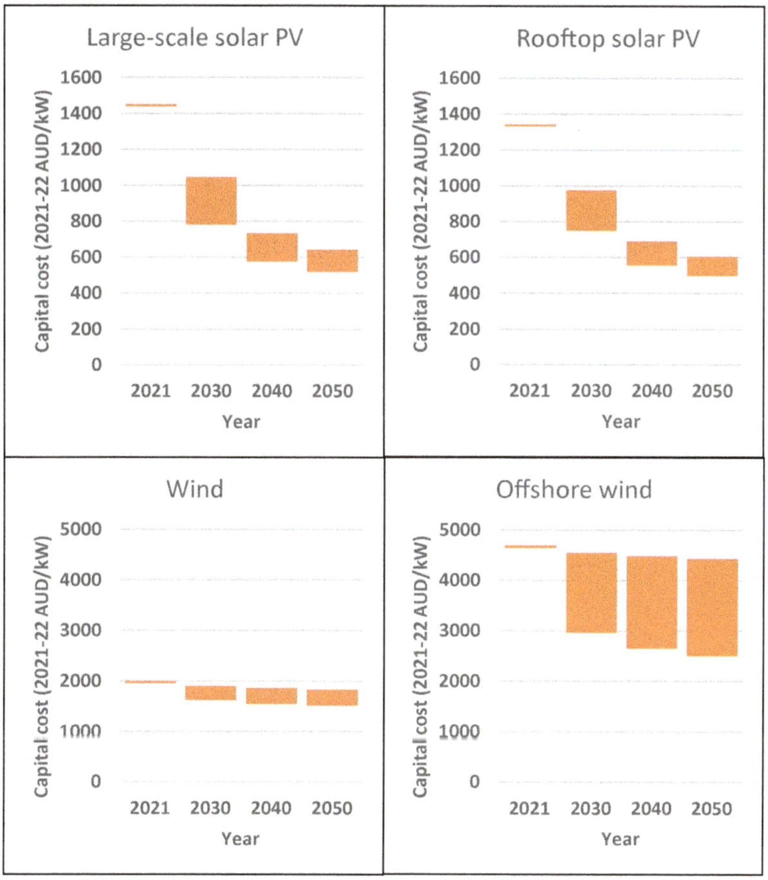

Fig. 6 Current and distribution of projected renewable generation technology capital costs by scenario in 2021–22 AUD/kW

Table 7 Current and projected total capital cost for batteries by scenario in 2021–22 AUD/kWh

Year	Battery storage (1 h)	Battery storage (2 h)	Battery storage (4 h)	Battery storage (8 h)	Scenario
	2021–22 AUD/kWh	2021–22 AUD/kWh	2021–22 AUD/kWh	2021–22 AUD/kWh	
2021	790	527	407	357	Business as usual
2030	687	452	343	298	
2040	565	363	269	230	
2050	485	315	236	203	
2021	790	527	407	357	Global NZE by 2050
2030	553	344	242	200	
2040	436	272	194	161	
2050	337	220	167	144	
2021	790	527	407	357	Global NZE post 2050
2030	608	390	287	244	
2040	483	309	227	193	
2050	385	255	196	172	

Total capital cost = Cost of battery plus Balance of plant cost

units. One contribution of CSIRO's report is estimating the additional integration cost of variable renewables (Graham et al. 2022). The cost to support a combination of solar PV and wind generation in 2030 is estimated at AUD16–28/MWh, depending on the level of renewables. The LCOE for solar PV, wind, and offshore wind, as the report estimates for 2021, 2030, 2040, and 2050, is presented in Table 8, and graphically in Fig. 8. For these technologies, the LCOE of solar PV is lowest, while the LCOE of offshore wind is 2–3 times higher than that for onshore wind. Despite the higher cost of offshore wind, it could play a crucial role for countries with good wind resources, relatively shallow coastal depth, and competition for onshore land use. In August 2022 Australia's federal government selected the first six offshore wind energy zones, with consultation underway for the first wind zone off the Gippsland coast in Victoria.

5 Public Policy

5.1 Renewable Energy Target

The Renewable Energy Target (RET) policy has been the most successful and enduring climate change policy for stimulating renewable technologies uptake in Australia (Nelson et al. 2021; Martin and Rice 2015; Byrnes et al. 2013). The federal

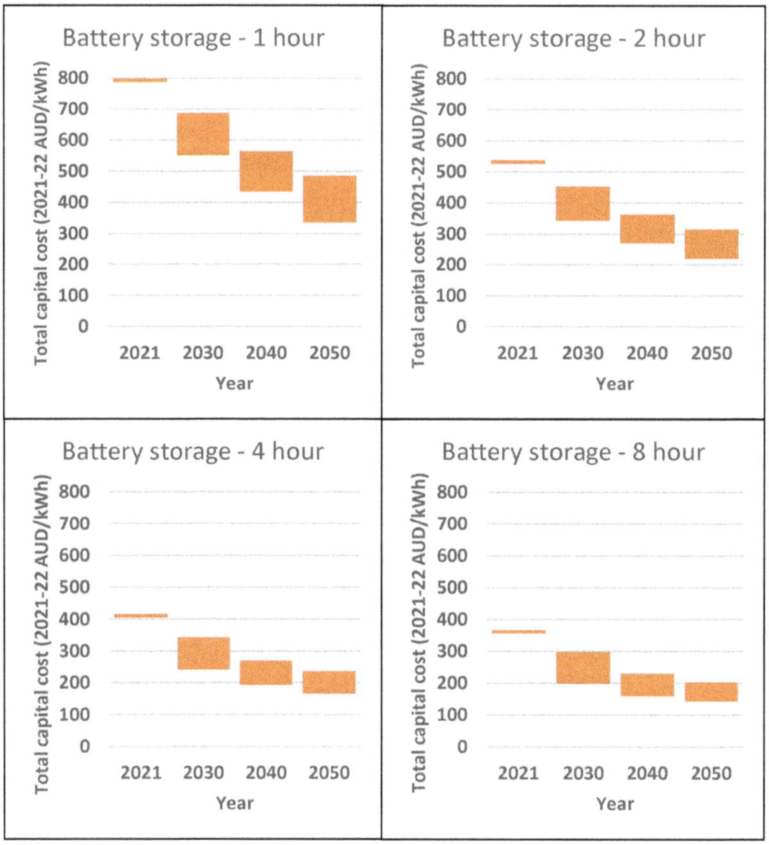

Fig. 7 Current and projected total capital cost for batteries by scenario in 2021–22 AUD/kWh; Total capital cost = Cost of battery plus Balance of plant cost

Table 8 Current and projected renewable generation technology LCOE (2021–22 AUD/MWh)

Year	Solar PV		Wind		Offshore wind	
	2021–22 AUD/MWh		2021–22 AUD/MWh		2021–22 AUD/MWh	
	Low	High	Low	High	Low	High
2021	44	65	49	61	128	166
2030	27	56	40	59	90	163
2040	21	43	37	59	79	162
2050	20	39	34	58	72	160

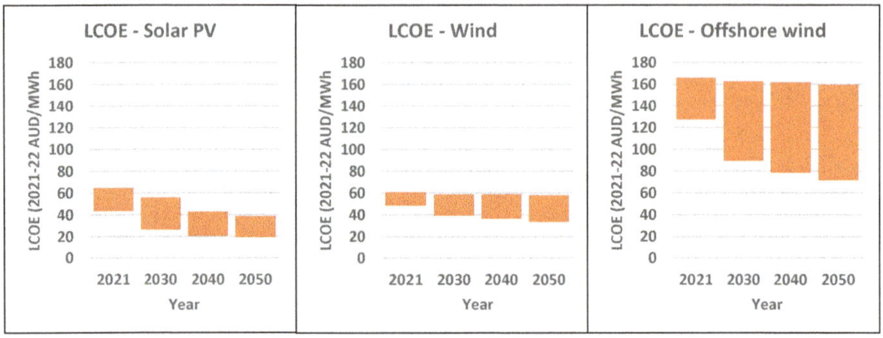

Fig. 8 Current and projected renewable generation technology LCOE (2021–22 AUD/MWh)

government adopted it in 2000 as the Renewable Energy (Electricity) Act 2000, and it applies from January 2001 to December 2030 (Australian Government 2022).

Initially the Act was introduced at the national level as the Mandatory Renewable Energy Target (MRET), aiming to produce an increase of 2% or 9.5 TWh per annum of renewable electricity supply by 2010 from a 1996–97 baseline of 10.5% (Simpson and Clifton 2014). After substantially exceeding the initial target, in 2009 the Australian government expanded it to 20% of Australia's electricity by 2020, or approximately 41 TWh (Clean Energy Regulator 2022a).

In 2011, important modifications in the scheme were implemented, splitting the 10.5% into Large-scale RET (LRET) and Small-scale RE scheme (SRES) (Australian Government 2022). Under the LRET scheme, large-scale generation certificates (LGCs) are created relating to electricity generation by accredited power stations. Renewable energy power station from 19 energy sources can be accredited to create tradable LGCs, one for every 1 MWh generated. These energy sources include hydro, wind, solar, wave, tidal, and others as specified in the Renewable Energy (Electricity) Act 2000 (Australian Government 2022).

Under the SRES scheme, small-scale technology certificates (STCs) are created relating to installation of small generation units, e.g., solar PV, and solar water heaters. Wholesale purchasers of electricity, mainly electricity retail companies and some major electricity users, are required to source a percentage of their electricity from renewable sources annually. "Liable entities" do this by buying LGCs and STCs based on defined percentages by regulator. These companies must surrender these certificates to the Clean Energy Regulator (CER) annually, in quantities based on a percentage of the volume of purchased electricity each year or pay a penalty.

According to the CER, in January 2021 the RET of 33 TWh of additional renewable energy was achieved on a 12-month rolling basis (Clean Energy Regulator 2022a). Achieving this target has not slowed renewable energy investment since 2020. Between January 2016 and July 2022, the CER accredited 15.6 GW renewable capacity, and an additional 5.4 GW capacity was committed.

5.2 Carbon Pricing

In 2011, the federal government introduced the Clean Energy Act 2011, which applied to Australia's bigger emitters of GHG emissions (Clean Energy Regulator 2022b). While designed as an emission trading scheme, for the first several years it introduced a fixed carbon price for large emitters, i.e., liable entities. The Act was only active in FY2012–13 and FY2013–14, as the next government repealed it effective July 1, 2014.

The Act covered approximately 60% of Australia's total GHG emissions and a range of businesses and industrial facilities from several sectors, including electricity generation, stationary energy, wastewater, industrial processes, and fugitive emissions.

For each fiscal year, liable entities had to surrender one carbon unit for every tonne of carbon dioxide equivalent (CO_2-e) emissions they produced. These carbon units could be purchased from the Clean Energy Regulator for a fixed price, which this price was AUD23/unit in FY2012–13 and AUD24.15 in FY2013–14. If a liable entity did not purchase and surrender enough carbon units, it was penalized for 130% of the price of the carbon unit multiplied by the number of units in deficit.

5.3 Feed-In Tariffs, Rebates

Feed-in tariffs (FiTs) is another popular stimulus that governments in many parts of the world use to promote residential solar photovoltaic (PV) system installations, reduce GHG emissions, and improve energy security (Li et al. 2021; Poruschi et al. 2018; Chapman et al. 2016; Martin and Rice 2013). With the highest solar radiation per square meter of any continent, Australia has some of the best solar energy resources in the world (Geoscience Australia 2022). Australia leads the world with total installed solar PV capacity of 1 kW per capita, ahead of the Netherlands and Germany which have less than 800 W per capita (Australian PV Institute 2022). According to the Australian PV Institute (2022), there are over 3.19 million PV installations in Australia with combined capacity of 27.2 GW as of June 2022, including large commercial and utility-scale installations. Small-scale solar led renewable energy growth in 2021, setting a record for new installed capacity for the fifth year in a row with 3.3 GW new capacity (Clean Energy Council 2022). New large-scale solar and wind capacity stood at 3 GW in 2021.

In Australia, state and territory governments implement FiTs, in contrast to the RET policy, which the federal government carries out. The first FiTs, introduced in 2008 varied across states and territories by design and payments. By definition, a FiT is a payment that electricity customers receive from retail companies or governments for the electricity they send into the grid using small-scale generation, comprising solar PV, wind, hydro, biomass, or battery equipment (Essential Services Commission 2022).

There are two main types of FiTs operated in Australia chiefly implementing net and gross FiTs (Chapman et al. 2016). With the latter, all electricity generated in a household is purchased by a retail company at a set tariff, while with the former, only electricity generated in excess of household consumption is purchased. Under the gross FiT scheme consumers pay for all electricity they consume. Net FiTs are prevalent in Australia. Only NT provides gross FiTs currently. NSW and ACT have discontinued them.

Table 9 summarizes the current FiTs in the states and territories of Australia. As mentioned, net FiTs dominate. In contrast with the initial payment rates, which were significantly above retail electricity prices for residential customers, these rates are currently only a fraction of these prices. More information about the history of FiTs in Australia can be found in Australian PV Institute (2021), Clean Energy Council (2018), Li et al. (2021), Poruschi et al. (2018), Chapman et al. (2016), and Martin and Rice (2013).

Advantages

Residential solar PV plays an important role in many countries, generating renewable energy and offsetting fossil fuel-based generation, thereby reducing GHG emissions (Chapman et al. 2016). It is particularly beneficial to Australia, where electricity generation is marked by high levels of GHG emissions. In addition to some of the best solar energy resources, Australia has some of the highest per capita uptake of residential solar PV. Many Australian households have realized significant financial and energy GHG benefits by installing residential solar PV, supported by the federal and state renewables policies. In the NEM, a 1 kW residential solar PV system has average generation potential of 1460 kWh per annum (Chapman et al. 2016). Many of the initial FiT contracts with such high rates as AUD0.60/kWh or AUD0.44/kWh are still effective. The average size of rooftop solar system increased to 8.5 kW in 2021, a more than threefold increase over the previous decade (Clean Energy Council 2022). The installation price of solar PV systems has also decreased significantly over the past two decades. The price for solar systems between 1.5 and 3 kW in 2004 was as high as AUD15/W installed, decreasing to AUD3/W in 2012 (Chapman et al. 2016). In 2021, the cost of a typical 5–10 kW roof-mounted, grid-connected PV system was on the order of AUD1.5/W (Australian PV Institute 2021).

Annual direct full-time employment (FTE) in roof-top solar PV systems in FY2018–19 is estimated at more than 13,000 jobs, including jobs related to hot water systems and small-scale batteries (Australian Bureau of Statistics 2020), an increase of almost 90% versus the same category in FY2009–10. According to Australian PV Institute (2021), there were more than 25,000 FTE positions in the PV industry, with many newly created jobs in installation and maintenance, followed by sales, design, and engineering, and significantly fewer in manufacturing, research, and development (Chapman et al. 2016).

Disadvantages

Sudden changes in renewable energy policy, particularly changes to FiTs in the early stages of their implementation, have not well served the interests of Australia's PV

Table 9 Current state and territory government feed-in tariff schemes in Australia

State	Eligible generation size	Type	Period	Current FiT rate (AUc/kWh)	Comment	Previous schemes	References
ACT					No minimum FiT currently	Gross FiTs for old customers in range 30.16–50.05 AUc/kWh. These small and medium FiTs for system size 10–200 kW closed for new customers in July 2011	ACT Government (2020)
NT	≤ 30 kW	Gross	Since 1/07/2022	9.13, incl. GST		Previous residential and commercial premium FiTs were available. Since 1/7/2022 customers who have been on premium FiTs for four years will be transferred to the standard FiT	Jacana Energy (2022)
NSW		Net	2022–23Y	• 6.2–10.4	• All-day FiT benchmark. Time-dependent FiT benchmark is also specified. Retailers are not required to pay FiT to customers	NSW Solar Bonus Scheme commenced on 1/1/2010 as gross metering FiT with 60 AUc/kWh and later 20 AUc/kWh payments. It closed for new applications in 2011	IPART (2022)
				• 5.6–27.4	• Time-of-the day benchmark FiTs for seven time periods	All payments on the scheme closed on 31 December 2016 (IPART 2021)	
Queensland	≤ 30 kW	Net	2022–23Y	• 9.3—regional Queensland • South-East Queensland—market FiTs		The Solar Bonus Scheme was closed to new customers on 30/06/2014 (Queensland Government 2018). It provides 44 AUc/kWh FiT and will expire on 01/07/2028	Queensland Government (2022)

(continued)

Table 9 (continued)

State	Eligible generation size	Type	Period	Current FiT rate (AUc/kWh)	Comment	Previous schemes	References
SA					SA Government does not set a minimum amount for purposes of the FiTs since December 2016	The Distributor-paid FiT was introduced in 2008 (44 AUc/kWh) The Retailer-paid FiT was introduced on 27/01/2012, for which a minimum FiT was set by ESCSA prior to December 2016	Essential Services Commission of South Australia (2022)
Tasmania	≤ 10 kW for a single-phase system ≤ 100 kW for three-phase system	Net	2022–23 FY	8.883	Minimum rate that retailers must pay to eligible customers	Aurora Energy voluntarily offered the Net Metering Buy Back Scheme before the end of August 2013. The offer was "one-to-one" for residential and small business customers, matching the rate for buying electricity	Office of the Tasmania Economic Regulator (2022)
Victoria	< 100 kW	Net	2022–23 FY	• 5.2	• Minimum flat rate FiT, excl. GST	Premium FiT scheme (60 AUc/kWh) started in 2008, followed by Standard and Transitional FiT schemes	Essential Services Commission (2022)
				• 5.0–7.1	• Minimum time-varying FiT range, excl. GST		

(continued)

Table 9 (continued)

State	Eligible generation size	Type	Period	Current FiT rate (AUc/kWh)	Comment	Previous schemes	References
WA	Max 50 kWh per day	Net	1. From 1/7/2022	• 10.0 • 2.5	Distributed Energy Buyback Scheme (DEBS): 1. For Synergy customers • Peak time 3 pm – 9 pm • Off-peak times	The Renewable Energy Buyback Scheme (REBS) was launched on 1/8/2010 and was replaced by the Distributed Energy Buyback Scheme (DEBS) in September 2020	WA Government (2022)
			2. From 1/7/2021	• 10.0 • 3.0	2. For Horizon Power customers • Peak time 3 pm – 9 pm • All other times		

industry. Lacking significant PV manufacturing, Australian PV-related employment was lower than in Europe or the United States, although Australia does have similar levels of installation and maintenance jobs per MW installed to some European manufacturing nations (Chapman et al. 2016). Total PV-related jobs in Germany were on the order of 20 FTE/MWp installed in 2012, almost twice that of Australia (Chapman et al. 2016).

There are several, sometimes adverse aspects related to the energy-social justice nexus, which pertains to the impact of FiTs and other energy policies on different groups of electricity consumers (Poruschi et al. 2018). While FiTs as subsidies can benefit early adopters, it is essential to consider latecomers as well. An undesirable aspect of FiTs that many authors cite is cross-subsidization from non-solar households to solar households in the form of increased electricity prices and bills for non-participants (Poruschi et al. 2018; Chapman et al. 2016; Nelson et al. 2011). This was particularly significant in the early stages of implementation, when a majority of non-solar PV owning customers supported premium FiTs. It has also been more difficult for customers who rent to install solar PV panels and receive FiT benefits, although some jurisdictions have recently introduced options for them as well (Solar Victoria 2022). Solar PV and battery storage distributed generation options provide opportunities for some customers to disconnect from the grid, potentially leaving grid-dependent customers to pay more for the essential service of delivering electricity (Poruschi et al. 2018).

The diversity of FiTs across Australia, combined with the lack of unified datasets on FiTs, hinders efforts to deriving comprehensive knowledge that could be used to tune the parameters of FiT policy (Poruschi et al. 2018). Martin and Rice (2013) provide a critical analysis of the seven-year Solar Bonus Scheme (SBS) that the NSW government initiated in 2010, with a fixed FiT rate AUD0.60/kWh in gross metering arrangements for systems with 10 kW maximum capacity. In the first 6 months of the SBS, more than 28,500 investors had installed solar PV systems with 53 MW total capacity. Subsequent reviews suggested that the SBS scheme would achieve 1000 MW installed capacity by the end of 2016 at a cost to the government of AUD2.6 billion. In October 2010, the NSW government decided to reduce the FiT rate for new participants to AUD0.20/kWh beginning November 18, 2010. By the new deadline, there were many new investors. Due to continuously surging demand for new PV systems and rising cost to government, in April 2011 the NSW government decided to close the SBS to new participants beginning June 2011. Due to poor initial financial modeling, the NSW government underestimated investor participation by a factor of 2.2. The SBS also lacked such simple operational controls as caps on total capacity and cost.

6 Conclusion and Policy Recommendations

Reaching high levels of renewables in the power system brings many systems integration issues that require a comprehensive policy approach (Browne 2017). The transition from fossil fuel-based power to renewables requires changes to technology, policy, markets, consumer practices and culture, infrastructure, and science knowledge (McGreevy et al. 2021). While such overwhelming transitions happen frequently, in human history, they typically take 50 or more years. Australia, especially South Australia, has demonstrated a highly successful, sustainable transition to a low-GHG emission energy system. This raises the question of what policy lessons the Australian experience over the last two decades might be learned to guide this change into the future and help other countries, such as those in Southeast Asia, aiming to transform their power systems. Here we group some of these recommendations into technical, economic, political, and social implications.

6.1 Technical Implications

It is possible to transform a power system from a traditional centralized, one-directional grid to accommodate intermittent VRE and more distributed energy resources. Australia's NEM has achieved 64% instantaneous renewable generation in 30-min time intervals, while renewable generation in South Australia routinely exceeds operational demand. AEMO is planning to be ready to run the NEM at 100% renewable energy generation by 2025.

While high levels of renewable energy are achievable, doing so requires new system integration approaches. One example is the aforementioned installation of the four synchronous condensers in South Australia in November 2021, which allows the grid to operate with fewer synchronous generators. The NEM has well demonstrated the valuable role of storage systems with the installation of several big batteries. They help integrate renewables, storing excess renewable energy and fulfilling multiple roles in grid stability. Snowy 2.0 is a pump-hydro bulk storage project under construction in Australia that is expected to play a significant supporting role for renewables when it comes online. Many more battery storage projects are also planned for the near future. The Australian Renewable Energy Agency (ARENA) will provide up to AUD100 million competitive funding to new battery energy storage projects for grid support (ARENA 2021).

Solar PV generation, both residential and utility-scale, has seen significant uptake in Australia, and it is accordingly playing a greater role in overall electricity supply. South Australia has powered its grid entirely by solar energy at several times. Solar generation offsets operational demand, particularly on days with high solar irradiation. This impact must be considered in grid operations, in terms of reduced baseload generation and reduced firming generation supporting the grid.

Network stability and security management is more difficult to achieve with intermittent and distributed generation. The grid must be able to supply electricity even when wind and solar conditions are not favorable for renewable energy. Improving diversity of supply, including geographical diversity of renewable generators, may help, together with increased battery storage and demand-side response. Countries like Australia and Southeast Asian states with frequent extreme climatic and weather events must develop more resilient networks.

The growth of distributed energy resources cannot fully replace the energy provided by utility-scale solar and wind. Nor does it diminish the critical role of transmission lines. To properly integrate utility-scale renewable generators, new transmission lines must be built, connecting to zones with high wind and solar resources which are frequently at distance from major population centers and consumers. Appropriate planning and assessment are required for such capital-intensive investment projects to extend transmission grids.

6.2 Economic Implications

Investment in renewable power systems is highly capital-intensive. Successful renewable projects must satisfy many conditions, especially if they are to attract private investors. These include policy stability, long-term revenue certainty, government support, and market transparency. South Australia is a good example of a privatized, market-based system receiving financial support from state government and how public policy may help investors (McGreevy et al. 2021). The SA state government frequently used bulk purchasing agreements for its own energy requirements to underwrite private investment. It was also the first Australian state government to support wind farm development. Renewable technologies were not competitive with fossil fuel-based technologies in the early stages of uptake and required such government support as RETs and FiTs to become attractive to investors and households.

A good example of focused support that the Australian government is providing for renewable energy projects is the creation of ARENA in 2012. Since its establishment, it has supported more than 600 projects with close to AUD2 billion in grant funding and attracting additional AUD7 billion funding (ARENA 2022b). A recent battery project supported by ARENA is the AGL Broken Hill grid-forming battery (50 MW/ 50 MWh) (ARENA 2022a).

Long-term investment requires comprehensive information about current market performance and long-term understanding of electricity demand. The NEM provides a good example of how to implement transparency well, with AEMO regularly publishing extensive market data daily, weekly, monthly, quarterly, and annually. As market operator, AEMO has comprehensive reporting and planning duties and procedures, publishes regular reports on the NEM, including "Electricity Statement of Opportunities", with 5–10 year estimates and forecasts, "Quarterly Energy Dynamics", with recent market dynamics and trends from the previous quarter, and

"Integrated System Plan", with roadmaps through 2030, 2040, and 2050 (AEMO 2022a).

6.3 Political Implications

The discontinuity of climate change policy in Australia has been a major weakness and an obstacle to steady investment in new, low-emission generation technologies (McGreevy et al. 2021; Simshauser and Gilmore 2022). Climate change policy is the responsibility of the Commonwealth (federal) government and the two main parties have been very confrontational on this issue. Several policies have been established only to be repealed or substantially modified within a short period of time. In contrast to climate change policy, energy policy is the main responsibility of the states and examples of poor working relationships and different ideologies between the state and federal governments led to suboptimal results in integration of climate change and energy policies. The renewable energy target policy is one of these examples with too many changes and modifications implemented, specifically at the early stages of its lifetime. Establishing long-term support from the main parties and stakeholders is critical for the success of deep societal changes such as the low-emission, energy transition.

6.4 Societal Implications

Renewable energy transitions require complex engagement with all stakeholders, with civil society perhaps most important (Browne 2017). Current and future users must be educated about and engaged with sustainable energy practices if they are to accept and adopt new technologies, consumption patterns, tariffs, and new ways to buy and sell energy.

Closing coal-fired power stations may create significant regional unemployment and other social dislocations. This is particularly important for regions such as Latrobe valley in Victoria, where two of four coal-fired power stations and one gas-based generator remain operational. Government support for new industries, possible renewable energy projects, and engagement with local populations may mitigate social impact. A complex environmental task following the closure of coal-fired generators is rehabilitating open-pit mines and other areas used by power stations.

As discussed regarding FiTs, some policies may reward early adopters excessively, while penalizing late comers. New policies must consider impact variations on higher and lower socioeconomic groups (Chapman et al. 2016). Transition processes must acknowledge and mitigate fuel poverty and energy injustice (Poruschi et al. 2018). It is crucial not to futher degrade vulnerable groups with new energy policies, as such populations have limited capacity to adapt to climate change.

Acknowledgements The authors thank Assoc. Prof. R. Nepal, Dr. H. Phoumin, and Prof. F. Kimura for their support and guidance. The authors also thank the participants at the first meeting of ERIA's Working Group on "Economic analysis of large-scale deployment of solar PV with battery storage in ASEAN" on June 20, 2022, for sharing their experience pertaining to the subject of this paper.

References

ABC (2021) SA's statewide blackout was five years ago—here's how energy supply has evolved since then. Retrieved 14 Oct 2022, from https://www.abc.net.au/news/2021-09-28/sa-statewide-blackout-anniversary-energy-impacts/100496564

ACT Government (2020) 2019–20 Annual feed-in tariff report. Retrieved 12 Sept 2022 from https://www.environment.act.gov.au/__data/assets/pdf_file/0004/1678810/annual-feed-in-tariff-report-2019-20.pdf

AEMO (2017) Black system South Australia, 28 September 2016. Retrieved 29 Sept 2022 from https://www.aemo.com.au/-/media/Files/Electricity/NEM/Market_Notices_and_Events/Power_System_Incident_Reports/2017/Integrated-Final-Report-SA-Black-System-28-September-2016.pdf

AEMO (2020) Quarterly energy dynamics Q1 2020. Retrieved 30 Sept 2022 from https://aemo.com.au/-/media/files/major-publications/qed/2020/qed-q1-2020.pdf?la=en

AEMO (2022a) 2022 integrated system plan (ISP). Retrieved 3 Nov 2022 from https://aemo.com.au/energy-systems/major-publications/integrated-system-plan-isp/2022-integrated-system-plan-isp

AEMO (2022b) Australian energy market operator. Retrieved 10 Nov 2022 from https://aemo.com.au/

AEMO (2022c) NEM registration and exemption list. Retrieved 6 June 2022 from https://www.aemo.com.au/-/media/Files/Electricity/NEM/Participant_Information/NEM-Registration-and-Exemption-List.xls

AEMO (2022d) Quarterly energy dynamics Q2 2022. Retrieved 3 Nov 2022 from https://aemo.com.au/-/media/files/major-publications/qed/2022/qed-q2-2022.pdf

AEMO (2022e) Quarterly energy dynamics Q3 2022. Retrieved 3 Nov 2022 from https://aemo.com.au/-/media/files/major-publications/qed/2022/qed-q3-2022.pdf?la=en

ARENA (2021) $100 million to support the next generation of grid scale batteries. Retrieved 3 Nov 2022 from https://arena.gov.au/news/100-million-to-support-the-next-generation-of-grid-scale-batteries/

ARENA (2022a) AGL broken hill grid-forming battery. Retrieved 6 July 2022 from https://arena.gov.au/projects/agl-broken-hill-grid-forming-battery/

ARENA (2022b) ARENA—10 years. Retrieved 3 Nov 2022 from https://arena.gov.au/about/

Arraño-Vargas F, Shen Z, Jiang S, Fletcher J, Konstantinou G (2022) Challenges and mitigation measures in power systems with high share of renewables—the Australian experience. Energies 15(2):429. https://doi.org/10.3390/en15020429

Aurecon (2021) 2021 costs and technical parameter review. Retrieved 14 Oct 2022 from https://aemo.com.au/-/media/files/major-publications/isp/2022/iasr/aurecon-2021-cost-and-technical-parameters-review-report.pdf?la=en

Australian Bureau of Statistics (2020) Employment in renewable energy activities, Australia. Retrieved 18 Sept 2022 from https://www.abs.gov.au/statistics/labour/employment-and-unemployment/employment-renewable-energy-activities-australia/latest-release

Australian Bureau of Statistics (2021) Snapshot of South Australia: high level summary data for South Australia in 2021. Retrieved 13 Oct 2022 from https://www.abs.gov.au/articles/snapshot-sa-2021#:~:text=Housing-,Population,Census%20counted%201.1%20million%20people

Australian Energy Regulator (2019) South Australian wind farms in court over compliance issues during 2016 black out. Retrieved 14 Oct 2022 from https://www.aer.gov.au/news-release/south-australian-wind-farms-in-court-over-compliance-issues-during-2016-black-out

Australian Government (2022) Renewable Energy (Electricity) Act 2000. Retrieved 31 Aug 2022 from https://www.legislation.gov.au/Series/C2004A00767

Australian PV Institute (2021) National survey report of PV power installations in Australia. Retrieved 18 Sept 2022, from https://apvi.org.au/wp-content/uploads/2022/08/PViA-Report-2022-AUS_v3a.pdf

Australian PV Institute (2022) Australian PV market since April 2001. Retrieved 16 Sept 2022 from https://pv-map.apvi.org.au/analyses

Batten D, Grozev G (2006) NEMSIM: finding ways to reduce greenhouse gas emissions using multi-agent electricity modelling. In: Perez P, Batten D (eds) Complex science for a complex world: exploring human ecosystems with agents. ANU E Press, pp 227–252

Bergero C, Rich MJ, Saikawa E (2021) All roads lead to Paris: the eight pathways to renewable energy target adoption. Energy Res Soc Sci 80:102215. https://doi.org/10.1016/j.erss.2021.102215

Browne A (2017) Shifting policy focus from renewable energy targets towards grid reform and system integration. Renewable Energy Law and Policy Review 8(1):33–41. Retrieved from https://www.jstor.org/stable/26377519

Byrnes L, Brown C, Foster J, Wagner LD (2013) Australian renewable energy policy: barriers and challenges. Renewable Energy 60:711–721. https://doi.org/10.1016/j.renene.2013.06.024

California ISO (2016) What the duck curve tells us about managing a green grid. Retrieved 3 Jan 2023 from https://www.caiso.com/documents/flexibleresourceshelprenewables_fastfacts.pdf

Chapman AJ, McLellan B, Tezuka T (2016) Residential solar PV policy: an analysis of impacts, successes and failures in the Australian case. Renewable Energy 86:1265–1279. https://doi.org/10.1016/j.renene.2015.09.061

Clean Energy Council (2018) Feed-in tariffs: state by state. Retrieved 7 Sept 2022 from https://www.energycouncil.com.au/media/12974/feed-in-tariffs-state-by-state.pdf

Clean Energy Council (2022) Clean energy Australia report 2022. Retrieved 7 Sept 2022 from https://assets.cleanenergycouncil.org.au/documents/resources/reports/clean-energy-australia/clean-energy-australia-report-2022.pdf

Clean Energy Regulator (2022a) Large-scale renewable energy target. Retrieved 31 Aug 2022 from https://www.cleanenergyregulator.gov.au/RET/About-the-Renewable-Energy-Target/How-the-scheme-works/Large-scale-Renewable-Energy-Target

Clean Energy Regulator (2022b) Carbon pricing mechanism. Retrieved 6 Sept 2022 from https://www.cleanenergyregulator.gov.au/Infohub/CPM/Pages/About-the-mechanism.aspx

Crowley K, Jayawardena O (2017) Energy disadvantage in Australia: policy obstacles and opportunities. Energy Procedia 121:284–291. https://doi.org/10.1016/j.egypro.2017.08.029

DCCEEW (2022) National greenhouse gas inventory quarterly update: December 2021. Retrieved 2 Nov 2022 from https://www.dcceew.gov.au/climate-change/publications/national-greenhouse-gas-inventory-quarterly-update-december-2021#:~:text=Emissions%20for%20the%20year%20to,%25%3B%207.0%20Mt%20CO%E2%82%82%2De

DELWP (2022) Victorian big battery. Retrieved 20 Nov 2022 from https://www.energy.vic.gov.au/renewable-energy/batteries-energy-storage-projects/victorian-big-battery

Essential Services Commission (2022) Minimum electricity feed-in tariff to apply from 1 July 2022: final decision, 24 February. Retrieved 7 Sept 2022 from https://engage.vic.gov.au/minimum-feed-tariff-review-202223

Essential Services Commission of South Australia (2022) Solar feed-in tariff scheme. Retrieved 10 Sept 2022 from https://www.escosa.sa.gov.au/consumers/energy/solar-feed-in-tariff-scheme

Geoscience Australia (2022) Solar energy. Retrieved 16 Sept 2022 from https://www.ga.gov.au/scientific-topics/energy/resources/other-renewable-energy-resources/solar-energy

Government of SA (2017) Our energy plan. Retrieved 17 Oct 2022 from https://apo.org.au/sites/default/files/resource-files/2017-03/apo-nid74435.pdf

Government of SA (2022) Leading the green economy. Retrieved 26 Sept 2022 from https://www.energymining.sa.gov.au/industry/modern-energy/leading-the-green-economy
Graham P, Hayward J, Foster J, Havas L (2022) GenCost 2021–22
Grozev G, Havyatt D, Nepal R, Christopher T, Perez P (2022) Analysis of historical wholesale electricity spot price volatility in South Australia and their projections in 2030 and 2040. Technical report, 44 pp, University of Wollongong. Retrieved 12 Dec 2022 from https://www.sapc.sa.gov.au/inquiries/inquiries/south-australias-renewable-energy-competitiveness/commissioned-research-to-support-the-inquiry/Question-C-Projections-of-spot-price-volatility-UoW.pdf
Havyatt D, Grozev G, Nepal R, Christopher T, Perez P (2022) Future wholesale and retail electricity prices in SA in 2030 and 2040 under select ISP scenarios. Technical report, 38 pp, University of Wollongong. Retrieved 12 Dec 2022 from https://www.sapc.sa.gov.au/inquiries/inquiries/south-australias-renewable-energy-competitiveness/commissioned-research-to-support-the-inquiry/Question-D-Wholesale-and-retail-price-projections-UoW.pdf
Hornsdale Power Reserve (2022) South Australia's big battery. Retrieved 30 Sept 2022 from https://hornsdalepowerreserve.com.au/
Hu X, Grozev G, Batten D (2005) Empirical observations of bidding patterns in Australia's national electricity market. Energy Policy 33(16):2075–2086. https://doi.org/10.1016/j.enpol.2004.04.003
IEA (2021) Net zero by 2050: a roadmap for the global energy sector. Retrieved 30 Sept 2022 from https://iea.blob.core.windows.net/assets/deebef5d-0c34-4539-9d0c-10b13d840027/NetZeroby2050-ARoadmapfortheGlobalEnergySector_CORR.pdf
IPART (2021) Solar feed-in tariff benchmarks: Final report. Retrieved 16 Sept 2022 from https://www.ipart.nsw.gov.au/sites/default/files/cm9_documents/Final-Report-Solar-feed-in-tariffs-benchmarks-2021-22-June-2021.PDF
IPART (2022) Solar energy. Retrieved 12 Sept 2022 from https://www.ipart.nsw.gov.au/Home/Industries/Energy/Retail-prices/Solar-Energy
Jacana Energy (2022) Solar update: changes to the premium solar feed-in tariff. Retrieved 16 Sept 2022 from https://jacanaenergy.com.au/your-home/solar-power/solar-update
Jones S (2010) The national renewable energy target: an example of post-Machiavellian policy-making? The Australian Journal of Public Administration 69(2):165–177. https://doi.org/10.1111/j.1467-8500.2010.00678.x
Lang K, Miller G (2011) Large-scale renewable energy target: a breath of fresh air for windfarms? Australian Environment Review 26(3):77–78
Li XH, Zhang Y, Li Y, Huang J, Costin G, Zhang P (2021) Exploring payback-year based feed-in tariff mechanisms in Australia. Energy Policy 150:112133. https://doi.org/10.1016/j.enpol.2021.112133
Martin N, Rice J (2013) The solar photovoltaic feed-in tariff scheme in New South Wales, Australia. Energy Policy 61:697–706. https://doi.org/10.1016/j.enpol.2013.06.038
Martin N, Rice J (2015) Improving Australia's renewable energy project policy and planning: a multiple stakeholder analysis. Energy Policy 84:128–141. https://doi.org/10.1016/j.enpol.2015.04.034
Martin N, Rice J (2021) Power outages, climate events and renewable energy: reviewing energy storage policy and regulatory options for Australia. Renew Sustain Energy Rev 137:110617. https://doi.org/10.1016/j.rser.2020.110617
McGreevy M, MacDougall C, Fisher M, Henley M, Baum F (2021) Expediting a renewable energy transition in a privatised market via public policy: the case of south Australia 2004–18. Energy Policy 148:111940. https://doi.org/10.1016/j.enpol.2020.111940
Nelson T, Simshauser P, Kelley S (2011) Australian residential solar feed-in tariffs: industry stimulus or regressive form of taxation. Economic Analysis and Policy 41(2):113–129. https://doi.org/10.1016/S0313-5926(11)50015-3

Nelson T, Nolan T, Gilmore J (2021) What's next for the renewable energy target—resolving Australia's integration of energy and climate change policy? Australian Journal of Agricultural and Resource Economics 66:136–163. https://doi.org/10.1111/1467-8489.12457

Nepal R, Foster J (2016) Testing for market integration in the Australian national electricity market. The Energy Journal 37(4):215–237. https://doi.org/10.5547/01956574.37.4.rnep

Office of the Tasmania Economic Regulator (2022) Feed-in tariffs. Retrieved 10 Sep 2022, from https://www.economicregulator.tas.gov.au/electricity/pricing/feed-in-tariffs

Open NEM (2022) Retrieved 14 Nov 2022, from https://opennem.org.au/

Pablo-Romero MP, Sanchez-Braza A, Galyan A (2021) Renewable energy use for electricity generation in transition economies: evolution, targets and promotion policies. Renew Sustain Energy Rev 138:110481. https://doi.org/10.1016/j.rser.2020.110481

Poruschi L, Ambrey CL, Smart JCR (2018) Revisiting feed-in tariffs in Australia: a review. Energy Reviews 82:260–270. https://doi.org/10.1016/j.rser.2017.09.027

Praveen RP, Keloth V, Abo-Khalil AG, Alghamdi AG, Eltamaly AM, Tlili I (2020) An insight to the energy policy of GCC countries to meet renewable energy targets of 2030. Energy Policy 147:111864. https://doi.org/10.1016/j.enpol.2020.111864

Queensland Government (2018) Queensland solar bonus scheme. Retrieved 13 Sept 2022, from https://www.qld.gov.au/__data/assets/pdf_file/0032/168944/solar-bonus-scheme.pdf

Queensland Government (2022) Feed-in tariffs. Retrieved 10 Sept 2022 from https://www.qld.gov.au/housing/buying-owning-home/energy-water-home/solar/feed-in-tariffs

Renew Economy (2020) Tesla big battery in South Australia delivers stunning windfall profits. Retrieved 14 Oct 2022 from https://reneweconomy.com.au/tesla-big-battery-in-south-australia-delivers-stunning-windfall-profits-77644/#:~:text=The%20windfall%20returns%20from%20the,a%20little%20over%20%24A90%20million

Simpson G, Clifton J (2014) Picking winners and policy uncertainty: stakeholder perceptions of Australia's renewable energy target. Renewable Energy 67:128–135. https://doi.org/10.1016/j.renene.2013.11.038

Simshauser P (2022) Australia's national electricity market: an analysis of the reform experience 1998–2021. In: Phoumin H, Nepal R, Kimura F, Uddin GS, Taghizadeh-Hesary F (eds) Revisiting electricity market reforms: lessons for ASEAN and East Asia. Springer, pp 75–115. https://doi.org/10.1007/978-981-19-4266-2_4

Simshauser P, Gilmore J (2022) Climate change policy discontinuity & Australia's 2016–2021 renewable investment supercycle. Energy Policy 160:112648. https://doi.org/10.1016/j.enpol.2021.112648

Shi X, Zhang L, Wang K, Chen W, Phoumin H (2022) Have competitive electricity markets rewarded flexible gas-powered generation? Australia's lessons for ASEAN. In: Phoumin H, Nepal R, Kimura F, Uddin GS, Taghizadeh-Hesary F (eds) Revisiting electricity market reforms: lessons for ASEAN and East Asia. Springer, pp 207–233. https://doi.org/10.1007/978-981-19-4266-2_9

Snowy Hydro (2022) Snowy Hydro 2.0. Retrieved 20 Nov 2022 from https://www.snowyhydro.com.au/snowy-20/

Solar Victoria (2022) Information for renters. Retrieved 19 Nov 2022 from https://www.solar.vic.gov.au/information-renters

WA Government (2022) Energy buyback schemes. Retrieved 12 Sept 2022 from https://www.wa.gov.au/organisation/energy-policy-wa/energy-buyback-schemes

Young S, Bruce A, MacGill I (2019) Potential impacts of residential PV and battery storage on Australia's electricity networks under different tariffs. Energy Policy 128:616–627. https://doi.org/10.1016/j.enpol.2019.01.005

Chapter 8
Effects of Digital Technologies on Renewable Energy Development: Empirical Evidence and Policy Implications from China

Xuemei Zheng, Lu Wang, Rabindra Nepal, and Han Phoumin

Abstract Despite widespread employment of digital technologies in renewable energy generating, transmitting, distribution, storage, and pricing, there is a lack of empirical investigation into the effects of digital technologies on renewable energy development. In this context, this paper estimates the influence of digital technologies on renewable energy market integration in China. This study conducts a series of regressions based on provincial data from 2003 to 2020 and an index of digital technologies measured with the entropy weight method, and finds that digital technologies have significantly bolstered renewable energy development in China. To analyze how to overcome specific barriers to renewable energy expansion, this paper also examines the case study of Qinghai province, which has the potential to power itself with 100% renewable energy. These findings provide valuable policy guidance for ASEAN countries regarding achieving carbon–neutral energy transitions.

Keywords Digital technologies · Renewable energy · China

JEL Classification Q48 · C13 · C54

X. Zheng
Institute of Western China Economic Research, Southwestern University of Finance and Economics, Chengdu, China
e-mail: xm.zheng@hotmail.com

L. Wang
School of Marxism, Chengdu University, Chengdu, China

R. Nepal (✉)
School of Business, Faculty of Business and Law, University of Wollongong, Wollongong, Australia
e-mail: rnepal@uow.edu.au

H. Phoumin
ERIA, Jakarta, Indonesia

H. Phoumin et al. (eds.), *Large-Scale Development of Renewables in the ASEAN*, Economics, Law, and Institutions in Asia Pacific,
https://doi.org/10.1007/978-981-99-8239-4_8

1 Introduction

China aims to hit peak emissions by 2030 and achieve carbon neutrality by 2060. Notwithstanding, given that China is the world's largest emitter of carbon dioxide and that 80% of China's energy comes from fossil fuels, it faces challenges in achieving these goals. From Fig. 1, which shows annual power generation from renewable energy in each province between 2003 and 2020, we see that China has made much progress in using renewable energy over the last two decades. However, to become carbon-neutral, it is crucial that China makes further progress in transitioning to renewable energy, e.g., solar and wind power, and invest in projects that absorb carbon dioxide.

Given that renewable energy, such as solar and wind power, is intermittent, and that the demand side is far removed from suppliers, much of China's renewable energy has gone to waste, particularly solar and wind power in the northwest and hydropower in the southwest. Some 17.1% of total wind generated power was lost in 2017 alone. Although such losses have been reduced since 2019 due to rising energy demand and lower renewable energy prices, much renewable energy is still being wasted at the national level, given its large installed capacity. In the first half of 2021, 12.64 billion KWh wind power and 3.32 billion KWh solar energy was lost.

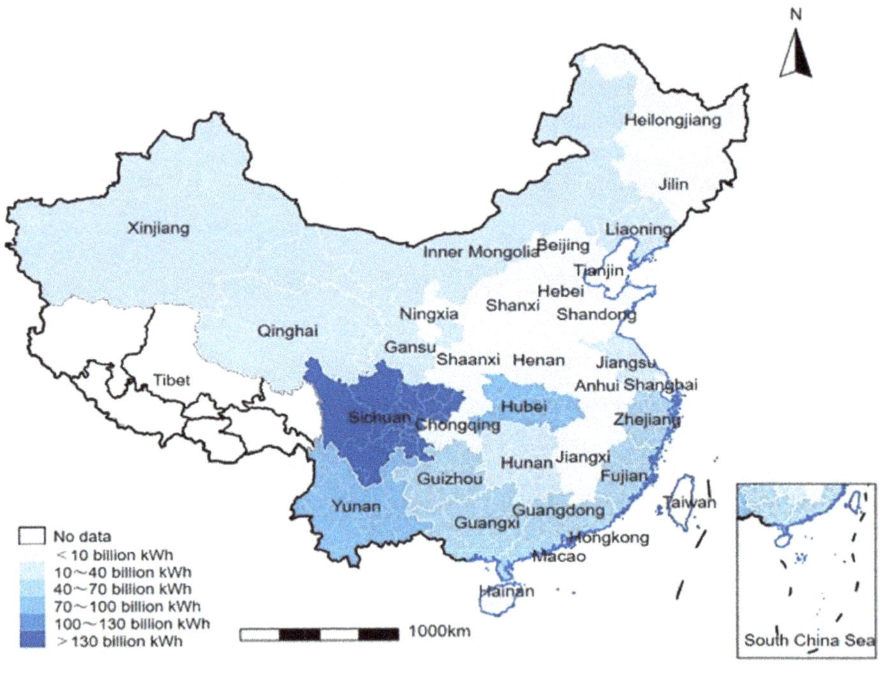

Fig. 1 Renewable power generation in China, 2003–2020

China has taken several measures to address this waste and make renewable power a greater part of the country's energy mix, with digital technology application being crucial. Digital technology has bolstered renewable energy development in many ways, with big data, blockchain, artificial intelligence, fifth-generation (5G) cellular networks, and cloud computing widely used in renewable energy generation, transmission and distribution, storage, and pricing.

The current literature has not paid sufficient attention to the impact of digital technology on renewable energy, however. To date, many studies have looked at the effects of digital technology in terms of social welfare (Shivendu and Zhang 2019), employment (Domini et al. 2021), technological innovation (Feng et al. 2022), factor misallocation (Shen and Zhang 2022), and industrial productivity (Wu and Yu 2022). Notwithstanding, most have analyzed how digital technology transforms the economy overall, ignoring the impact on renewable energy, which plays an important role in energy security, economic growth, and environmental protection (Bhattacharya et al. 2017; Nguyen and Kakinaka 2019). Although literature specializing in renewable energy has explored a number of factors driving renewable energy development, as shown in Sect. 2, literature review, the role of digital technology has not been comprehensively examined. Studies of the relationship between digital technology and energy have mainly investigated the impact of digitization on consumption, with few studies looking at the impact on the supply side of renewable energy.

Accordingly, this study aims to bridge this research gap between renewable energy and digital technology, by empirically estimating how digital technology boosts renewable energy based on evidence from China, and exploring precise mechanisms by which digital technologies facilitate renewable energy. In addition to these estimations and following a heterogeneity analysis of the impact, this paper will take up the case study of Qinghai, a Chinese province that has achieved 100% renewable energy transition for its economy, to further analyze actual steps involved. This paper will then draw on these findings to shed light on how other Chinese provinces and ASEAN member states may find examples of how to achieve their own carbon neutrality goals.

The methodology is as follows. Using Chinese provincial data from 2003 to 2020, we measure China's digital technologies with the entropy weight method and apply the general moment method (GMM) to estimate the impact of said digital technologies on renewable energy development. The results suggest that digital technologies have facilitated renewable energy significantly, through their influence on economic development and industrial structure. The significantly positive relationship between digital technologies and renewable energy development remains robust after a number of robustness checks, including considering spatial spillover of digital technology from neighboring regions and changing the weight of indexes used to calculate the value of digital technologies. The regional heterogeneity analysis reveals that the impact of digital technologies on renewable energy varies across China, with the greatest impact felt in the country's east. This can be explained by such distinctive characteristics as greater digital innovation, more developed market mechanisms, and more efficient administration. These findings, together with Qinghai province's experience of transitioning to 100% renewable energy, provide valuable policy implications for other countries and regions struggling to achieve their own energy transition targets.

The remainder of this paper is organized as follows. Section 2 reviews the literature. Section 3 describes the data used in this study and presents our econometric approaches. Section 4 presents estimation results and tests the mechanisms by which digital technologies affect renewable energy. Section 5 conducts robustness checks and heterogeneity analysis of the primary findings. Section 6 briefly describes experiences in using digital technologies to facilitate renewable energy development in Qinghai province. Section 7 summarizes our main conclusions and provides insights for policy.

2 Literature Review

While many studies examine factors driving renewable energy or the impact of digital technology, there is a lack of investigation into the effects of digital technology on renewable energy development.

2.1 Factors Driving Renewable Energy Development

The literature includes a large number of studies investigating drivers of renewable energy deployment, which find that economic performance and financial development vitally affect renewable energy expansion. Specifically, economic growth rates (Sadorsky 2009a), per capita income (Marques et al. 2010), openness to trade (Omri and Nguyen 2014), FDI inflows (Bhattacharya et al. 2016; Kutan et al. 2018), and economic freedom (Baranes et al. 2017) can positively promote renewable energy demand. Capitalization and growth of stock markets also benefit renewable energy development by financing more clean energy projects and economic activity.

Other factors also affect renewable energy growth, including carbon emissions (Sadorsky 2009b; Marques et al. 2010), oil prices (Sadorsky 2009b; Omri and Nguyen 2014), fossil fuel lobbies, and energy self-sufficiency (Marques et al. 2010). Related polices are fundamental drivers of renewable energy growth, including application of voluntary approaches (Aguirre and Ibikunle 2014). Gozgor et al. (2020) indicate that greater economic globalization promotes renewable energy, while Zheng et al. (2021a, b) find that demand side factors, e.g., consumers' price sensitivity, also closely relates to their support for, and thus overall development of, renewable energy.

Most studies examining renewable energy development determinants are conducted using country-level data, especially from G7 economies (e.g., Sadorsky 2009b), BRICs (e.g., Salim and Rafiq 2012; Kutan et al. 2018), OECD countries (e.g., Gozgor et al. 2020), European countries (e.g., Marques et al. 2010; Baranes et al. 2017), G20 countries (e.g., Bhattacharya et al. 2017), and ASEAN (Association of Southeast Asian Nations) economies (e.g., Nepal and Musibau 2021). Some have examined specific countries, such as China (e.g., Lin et al. 2016; Chen 2018) and Indonesia (e.g., Al-Irsyad et al. 2019).

In sum, while the literature has extensively analyzed factors conducive to renewable energy development from economic, financial, and political perspectives, examination of the role of digital technology is relatively insufficient, despite its extensive employment in generating and using renewable energy. Research on this subject regarding China is particularly limited.

2.2 *Impact of Digital Technologies on Energy*

More and more studies are paying attention digital technology applications to energy. The International Energy Agency (IEA) (2017) points out that digital technologies, such as smart appliances and shared mobility, improve the safety, productivity, efficiency, and sustainability of energy systems. Digitization has the potential to save some 5% of total annual generation costs in electricity in particular, as operation and maintenance costs can be reduced, energy efficiency of generating plants and grids can be improved, and operational lifetimes of assets can be extended. Thanh et al. (2022) empirically analyze the nexus of digitization and energy security in 27 European countries between 2015 and 2019, finding that promoting digitization is beneficial regarding the acceptability and sustainability of energy security, while deleterious on development. Conversely, energy security positively affects digitization, especially in business and the public sector. Baidya et al. (2021) have reviewed the opportunities, challenges, and future directions for energy digitization.

Overall, in existing literature concerning the impact of digital technologies on energy, the role of digitization in energy demand and consumption has attracted the most attention. Bastida et al. (2019) explore the potential of information and communication technology (ICT)-based interventions in households to decrease electricity usage and suggests that such effects on consumer behavior can reduce household final electricity consumption by 0–5%. Lange et al. (2020) estimate the impact of ICT on energy demand across 28 member states of the European Union. They find that overall digitization increases energy consumption, as physical capital and energy complement each other in ICT, which is energy-intensive, and increased energy efficiency thus pays dividends. Ren et al. (2021) examine the situation in China, and find that the relationship between China's internet development and energy consumption is significantly positive, and that internet development promotes energy consumption scaling through economic growth. Husaini and Lean (2022) study the impact of digitization on total and disaggregated energy consumption in five major ASEAN member states, concluding that digitization reduced such consumption by both metrics. Xu et al. (2022) investigate the effects of digitization on energy and related mechanisms from an international perspective, demonstrating that digitization reduces energy consumption, decreases energy intensity, and optimizes energy structure, by promoting technological innovation, accelerating human capital accumulation, and alleviating industrial structure distortions. Digitization also has greater energy savings in low-income and underdeveloped countries.

A number of analytical studies have examined the application and impact of digital technologies to renewable energy. Strielkowski et al. (2021) focus on strategies employing 5G cellular for optimal demand-side response management in future energy systems with large proportions of renewables. They confirm that effective deployment of faster and more reliable cellular networks would allow faster data transfer and processing, including peer-to-peer energy trade markets, Internet of Vehicles markets, and faster smart metering. Hossain et al. (2016) investigate the role of smart grids in renewable energy, concluding that using smart grids may facilitate efficient use of renewables in turn. Ahl et al. (2019) explore potential challenges of blockchain-based peer-to-peer microgrids, and suggest implications thereof for institutional development. Sharifi et al. (2021) analyze the impact of artificial intelligence on energy post-COVID-19 pandemic, and encourage countries whose economies depend on non-renewable energy to develop solar and wind energy, as renewables can reduce the virus's destructive effects and drive economic prosperity.

In summary, in contrast to the increasingly important role of digital technologies in renewable energy development, only a limited number of qualitative studies have been conducted to-date. The current literature has not paid sufficient attention to quantitative analyses based on historical data. Investigations of precise mechanisms by which digital technology plays its role are also few and far between. This paper will accordingly attempt to bridge this gap by conducting empirical analysis to estimate the impact of digital technologies on China's growth in renewables, and examine the case study of Qinghai province to shed light on how to transition to 100% renewable energy.

3 Data and Methodology

3.1 Description

This section presents sources and statistical descriptions of data. The data used to measure the key explanatory variable, i.e., *Digital Technologies*, was extracted from various yearbooks, including *China Statistical Yearbook*, *China Energy Statistical Yearbook*, *China Electricity Statistical Yearbook*, *China Population and Employment Statistical Yearbook*, and *China Technology Statistical Yearbook*. Information about crude oil prices comes from the U.S. Energy Information Administration (EIA), and that about CO_2 emissions from the Carbon Emission Accounts and Datasets (CEADs). Where values are missing, we select data from provincial yearbooks and adjust to match values selected from the abovementioned yearbooks. We supplement data on broadband access ports, which are absent prior to 2006, by backward projecting using the average annual growth rate of this variable.

Table 1 Descriptive statistics

Variables	Obs	Mean	Std.Er	Max	Min
Renewable energy (10^9 kWh)	540	38.9	54.9	365.4	0
Digital techniques	540	0.148	0.122	0.701	0.008
Economic development (10^9 RMB)	540	1215.7	1184.3	7090	36.1
CO_2 emissions (10^6 tons)	540	269	186	950	16
Environmental regulation (10^6 RMB)	540	1879.51	1878.63	14,000	4.76
Government size (10^6 RMB)	540	22.9	10.7	75.8	8.4
Industrial structure (%)	540	0.984	0.321	1.897	0.191
Urbanization rate (%)	540	0.537	0.145	0.938	0.257
Oil price (USD per barrel)	540	6.119	0.330	6.577	5.476
General technology (10^9 RMB)	540	34.751	51.112	309.849	0.121

We measure the dependent variable, renewable energy development, with the difference in electricity generation between the aggregate and that generated from thermal energy. We measure the key explanatory variable, digital technologies, with several approaches, including the entropy weight method in the main analysis and adjusting the weights of related indexes in the robustness check, as shown in more detail in the following section. Economic development is signified by gross domestic product (GDP) per capita by province. We deflate nominal GDP per capita values by the GDP index, using 1998 as the base year. The proxy variable for environmental regulation is costs incurred responding to environmental pollution, while that for government size is the ratio of government spending to GDP. Industrial structure is measured by the ratio of GDP in secondary sectors to that in tertiary sectors in a given province, while urbanization rate is measured by the ratio of urban to total population in a province. General technology is measured with research and development (R&D) investment. In the following regressions, we use the natural logarithms of renewable energy, economic development, CO_2 emissions, crude oil price and general technology. Table 1 gives the descriptive statistics of each variable.

3.2 *Measurement of Digital Technologies*

This paper applies the entropy weight method (EWM), an important information model, to measure the key explanatory variable, that is, digital technologies. It evaluates values by measuring the degree of differentiation in information. The higher the degree of dispersion of the measured value, the higher the degree of differentiation of the index, and the more information that can be derived. Higher weight should be given to the index, and vice versa. Hence, according to the degree of variation of each index, the information entropy tool can be used to calculate the weight of each index and provide comprehensive evaluation of multiple indexes.

The first step in this method is standardizing measured values. Suppose m indexes and n years are set in the evaluation, and x_{ij} denotes the ith sample in year j. Then the standardized value of x_{ij}, which is recorded as X_{ij}, is calculated as follows:

$$X_{ij} = \frac{x_{ij} - min\{x_i\}}{max\{x_i\} - min\{x_i\}}$$

where $min\{x_i\}$ and $max\{x_i\}$ are the minimum and maximum value of the ith sample in all years, respectively. In this study, six indexes are used, and the attributes of all indexes are positive.

The second step is to calculate the weight of index i in year j, and the calculation is given as:

$$w_{ij} = \frac{X_{ij}}{\sum_{j=1}^{n} X_{ij}}$$

Define the entropy value of the ith index, denoted with E_i, as follows:

$$E_i = \frac{\sum_{j=1}^{n} (w_{ij} \times lnw_{ij})}{lnn}$$

Then the range of the entropy value E_i is between zero and one. Given the calculation method of the ith index's weight W_i, which is shown as:

$$W_i = \frac{1 - E_i}{\sum_{i=1}^{m} (1 - E_i)}$$

Then the evaluation score of index i in year j is $S_{ij} = W_i \times X_{ij}$. Financially, the value of digital techniques, denoted as $Digital_j$, is calculated as:

$$Digital_j = \sum_{i=1}^{m} S_{ij}$$

Table 2 gives the indexes used to calculate the values of digital technologies. In total, we have six indexes, which can be grouped into four categories. This means that we select the indexes from four perspectives, including number of employees, outputs, infrastructure, and investment in related fields. The last two columns display the weight and attribute of each index. It can be seen that indexes classified as outputs and infrastructure are assigned higher weights, especially broadband ports and telecommunications business per capita.

Figure 2 shows the trend of China's adoption of digital technologies from 2003 to 2020, as measured by the EWM. It indicates that overall, the level of China's digital technology adoption rises consistently in this period, and even more prominently after 2010. This is consistent with the fact that China acts as one of the world's leading adopters of digital technologies and is shaping the global digital landscape.

Table 2 Indexes used to calculate the value of digital techniques

Classes	Indexes	Weights	Attributes
Number of employees	Ratio of employees in the ICT industry and other information transmission service industry to aggregate employees	0.190	+
Outputs	Telecom business per capita	0.212	+
	Mobile phone switch capacity	0.133	+
Infrastructure	Long-distance optical cable line length	0.074	+
	Broadband access port of internet	0.229	+
Investment	Investment in fixed assets of the whole society in ICT	0.162	+

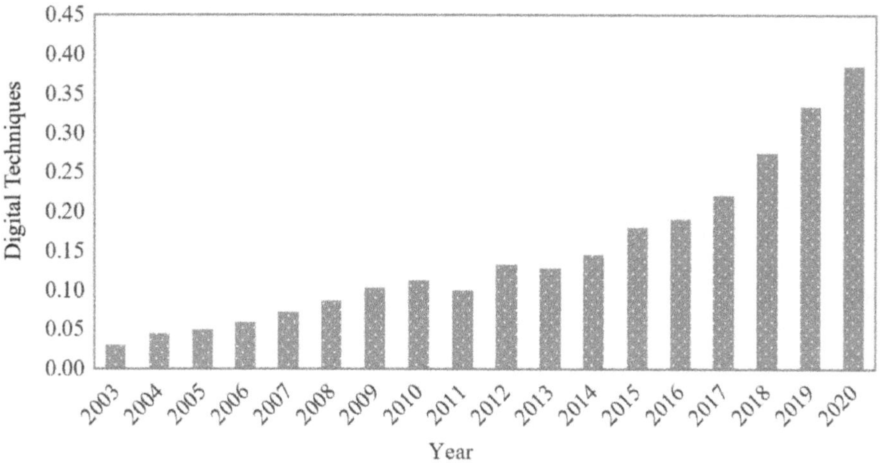

Fig. 2 Trend of digital technology adoption in China, 2003–2020

3.3 The Econometric Model

3.3.1 The Baseline Model

In the baseline model, we apply the general method of moments (GMM) that includes a lagged dependent variable as the instrumental variable to deal with the potential endogeneity problem. We specify the model setting as follows:

$$Renew_{it} = \alpha + \beta Renew_{it-1} + \rho Digital_{it} + \delta X_{it} + V_t + \lambda_i + \varepsilon_{it} \quad (1)$$

where $Renew_{it}$ represents the level of renewable energy development of province i in year t,

$Renew_{it-1}$ is the one-period lagged value of the dependent variable, which is used as the instrumental variable to cope with potential endogeneity. $Digital_{it}$ is the variable that captures digital technologies of province i in time t, and ρ is the associated coefficient. X_{it} indicates a vector of control variables, including economic development, CO_2 emissions, environmental regulation, government size, industrial structure, urbanization rate, crude oil price, and general technology level. β and δ denote the coefficients for the instrumental variable and control variables, respectively. ν_t is the year dummy variable that controls the variables that are constant across provinces but vary over time, i.e., time-fixed effects, λ_i is the dummy variable for provinces that controls the unobserved time-invariant individual effect, i.e., individual fixed effects, and ε_{it} is the error term.

3.3.2 The Spatial Durbin Model in Robustness Check

In one robustness check, we use a spatial econometric model to consider the influence of spatial factors on the development of renewable energy. The main reason to use this method is that neighboring regions, which are based on geographical relationships, share common characteristics in such domains as politics, economics, and culture, implying that there are spatial spillover effects among said neighboring regions. To account for these spatial effects, we apply the spatial Durbin model, which includes a spatially lagged dependent variable and spatially lagged explanatory variables, to estimate the effects of digital technologies on renewable energy. The model is set as follows:

$$Renew = \rho Renew + X\beta + WX\theta + \varepsilon \tag{2}$$

where *Renew* is the renewable energy dependent variable, a $(n \times 1)$ vector, where n is the number of observations included in the model. ρ stands for the effect of renewable energy development of a given region's neighboring regions on the renewable energy development of this specific region. W is a $(n \times n)$ matrix of spatial weighting coefficients. X is a $(n \times k)$ matrix of the independent variables. β is a $(k \times 1)$ vector of parameters associated with explanatory variables. θ is the spatial autoregressive coefficient, which reflects the influence of the spatial factors on the dependent variables. ε is a $(n \times 1)$ vector whose elements follow $\varepsilon \sim (0, \sigma^2 I_n)$.

3.3.3 The Mediation Model in Mechanism Tests

To identify and explain the mechanisms that underpin the relationship between renewable energy development and digital technologies, in Sect. 4.2 we run regressions using the following mediation model:

$$Renew_{it} = \alpha_0 + \alpha_1 Renew_{it-1} + \alpha_2 Digital_{it} + \alpha_3 Z_{it} + V_t + \lambda_i + \tau_{it} \tag{3}$$

$$Mediator_{it} = \beta_0 + \beta_1 RMediator_{it-1} + \beta_2 Digital_{it} + \beta_3 Z_{it} + V_t + \lambda_i + \mu_{it} \quad (4)$$

$$\begin{aligned} Renew_{it} = {} & \gamma_0 + \gamma_1 Renew_{it-1} + \gamma_2 Digital_{it} \\ & + \gamma_3 Mediator_{it} + \gamma_4 Z_{it} + V_t + \lambda_i + \xi_{it} \end{aligned} \quad (5)$$

where Eq. (3) regresses the dependent variable on the independent variable to confirm that the independent variable is a significant predictor of the dependent variable. Equation (4) regresses the mediator on the independent variable to confirm that the independent variable is a significant predictor of the mediator. If the mediator is not associated with the independent variable, it could not possibly mediate anything. Equation (5) regresses the dependent variable on both the mediator and independent variable to confirm that the mediator is a significant predictor of the dependent variable and that the strength of the coefficient of the previously significant independent variable in the first step is now greatly reduced. Equation (3) is similar to Eq. (1); the only difference is the number of control variables included in X_{it} and Y_{it}. In Eq. (3), some variables in X_{it} are excluded, as they are used as mediators in the mediation model, implying that Eq. (5) runs essentially the same regression as Eq. (1). *Mediator*$_{it}$ indicates the possible mediator, and in this study, economic development and industrial structure are tested as mediators. τ_{it}, μ_{it} and ξ_{it} are error terms.

4 Results and Discussion

4.1 Preliminary Results

In the baseline specification, we ran the regression with the full sample using the GMM model, and present the estimation results in column (1) of Table 3. For comparison, we also demonstrate the outcomes estimated with the pooled ordinary least square (POLS) model and random effect (RE) model in columns (2) and (3), respectively. We see that the impact of digital technologies on renewable energy development is significantly positive across all model specifications, at least at the five-percent level.

Renewable energy development in the current period is also positively linked to its level in the last period, suggesting that renewable energy increases are path dependent on resource endowments and infrastructure construction. Economic development level, carbon emissions, environmental regulation, and oil price may all significantly promote renewable energy development in the current period, in line with the literature review in Sect. 2. By contrast, there is a negative relationship between renewable energy development level and government size, industrial structure, and urbanization. Given the measurement of these variables, these findings are economically straightforward.

Table 3 Preliminary estimation results

Variables	GMM	POLS	RE
	(1)	(2)	(3)
Digital technologies	3.831**	4.786***	1.313***
	(1.498)	(0.722)	(0.452)
L. Renewable energy	0.613***		
	(0.110)		
Economic development	0.808**	1.351***	− 0.183
	(0.381)	(0.240)	(0.253)
CO_2	0.330**	0.221*	1.084***
	(0.143)	(0.128)	(0.165)
Environmental regulation	0.080*	0.060	0.063
	(0.042)	(0.075)	(0.039)
Government size	− 4.935**	3.618***	0.151
	(2.097)	(0.580)	(0.675)
Industrial structure	− 0.444***	− 0.844***	− 1.156***
	(0.168)	(0.218)	(0.150)
Urbanization	− 2.989**	− 12.263***	0.072
	(1.394)	(0.815)	(1.038)
Oil price	1.914***	0.139	0.055
	(0.603)	(0.156)	(0.081)
General technology	− 0.716***	0.176**	0.294***
	(0.271)	(0.081)	(0.111)
Constant	0.000	4.843***	− 4.738***
	(0.000)	(1.312)	(1.424)
Time fixed effects	Y	N	N
Individual fixed effects	Y	N	Y
Observations	510	540	540
R^2		0.604	0.750
AR(1)	0.000		
AR(2)	0.425		
Sargan test	0.097		

Robust standard errors are in parentheses. *$p < 0.05$, **$p < 0.05$, ***$p < 0.01$. This applies to all following tables as well

4.2 *Possible Mechanisms*

In this section, we test the potential mechanisms by which digital technologies affect renewable energy development. Considering the established link between GDP and renewable energy consumption (Amri 2017), and the role of industrial policy adjustment in China's energy mix (Liu et al. 2021), we take economic development and industrial structure as mediators in two separate tests, respectively. We test the plausibility of these mechanisms with the mediation model introduced in Sect. 3.3.3, and the associated estimation results are given in Table 4. The third and fourth columns demonstrate the impact mechanism through economic development, and the fifth and sixth columns report the impact channel through industrial structure. Only the estimation results of Eqs. (3) and (4) are presented, as the estimates of Eq. (5) can be found in column (1) of Table 3.

It can be seen that the coefficient of digital technologies in Eq. (5) is smaller than that estimated with Eq. (3), signifying that the presence of the mediator mediates the relationship between digital technologies and renewable energy. The estimation results of Eq. (4) show that the impact of digital technologies on economic development is statistically significant, implying that the changes in digital technologies could predict economic development trends. This is also the case for the impact of digital technologies on industrial structure, as the coefficient of digital technologies in the last column of Table 4 is also statistically significant at the five-percent level.

It is worthwhile to point out that we added the square of digital technologies in the regression of Eq. (4), indicating that the relationship between digital technologies and economic development is non-linear. The estimation results in the third column

Table 4 Possible mechanisms

Variables	Economic development		Industrial structure	
	Equation (3)	Equation (4)	Equation (3)	Equation (4)
Digital technologies	5.539***	− 0.542**	3.999**	− 0.064**
	(1.935)	(0.223)	(1.554)	(0.028)
Square of digital technologies		0.636***		
		(0.244)		
L. Renewable	0.576***		0.676***	
	(0.120)		(0.101)	
L. GDP		0.846***		
		(0.107)		
L. Industrial				1.198***
				(0.067)
Controls	Y	Y	Y	Y
Time fixed effects	Y	Y	Y	Y
Individual fixed effects	Y	Y	Y	Y
Observations	510	510	510	510
AR(1)	0.000	0.033	0.000	0.000
AR(2)	0.392	0.873	0.398	0.054
Sargan test	0.118	0.188	0.113	0.062

of Table 4 show that the sign for digital technologies is negative and that for its square is positive. This implies a U-shaped relationship between digital technologies and economic development, showing that digital technologies exert first a negative, then a positive impact on GDP per capita. The initial negative impact of digital technologies on GDP can be attributed to the phasing-out effect of the investment in digital technologies. To illustrate, at the very beginning, when the investment in digital technology is insufficiently large, the obvious facilitating effects of digital technologies on economic activities cannot be fully unleashed, as economies of scale have yet to be achieved. Instead, as investment in other areas might be affected due to this phasing-out effect, it makes sense that digital technologies might negatively affect GDP at some time in a given place. Increased adoption of digital technologies will however, drive economies of scale sufficient to exceed the phasing-out effect, generating a net positive impact on the economy.

Likewise, the negative coefficient of digital technologies in the last column of Table 4 implies that digital technologies contribute to industrial structure upgrades, as the value of industrial structure in this study is calculated as the ratio of GDP in secondary sectors to that in tertiary sectors, as shown in Sect. 3. It is possible that digital technologies exert these effects through bolstering human capital and technological innovation, which promote overall industrial structure transitions from conventional industry to high-tech.

5 Robustness Checks and Heterogeneity Analysis

In this section, we conduct a serious of robustness checks on the main findings, including applying a different estimation strategy and changing the measurement of the key explanatory variable, i.e., digital technologies. We also investigate the heterogeneity of digital technologies' effects to enrich discussion about these findings.

5.1 Robustness Checks

5.1.1 Spatial Durbin Model

To capture how spatial factors influence the impact of digital technologies on renewable energy development, in this section we run a regression using the spatial Durbin model. To do this, we first construct the spatial weight matrix, W, using the geographic distance spatial matrix. To illustrate, the value of the element w_{ij} in matrix W is assigned with the inverse of the square of the geographical distance between province i and province j. The estimation results of the spatial Durbin model are displayed in column (1) of Table 5.

Table 5 Robustness checks

Variables	Spatial Durbin model	Change measurement of digital techniques
	(1)	(2)
Digital technologies	2.944***	4.236***
	(0.623)	(1.475)
W. renewable	− 0.154**	
	(0.076)	
L. renewable		0.614***
		(0.115)
Control variables	Y	Y
Time fixed effects	Y	Y
Place fixed effects	Y	Y
Observations	540	510
R^2	0.803	
AR(1)		0.000
AR(2)		0.458
Sargan test		0.100

As can be seen, the coefficient of spatially lagged renewable energy, i.e., *W. renewable*, is − 0.154, significantly negative at the five-percent level, indicating that renewable energy development in a given province is likely to be negatively affected by that in neighboring provinces. This can partly be explained by local protectionism pertaining to market segmentation and political contests in the context of political advancement in China (Zheng et al. 2021a, b). Notwithstanding, the coefficient of digital technologies on renewable energy remains robust in terms of both sign and magnitude, showing that advancement in digital technologies effectively facilitates greater renewable energy development, which is consistent with the findings obtained in the baseline models.

5.1.2 Changing Digital Technology Measurements

To further verify the validity of the above findings, we change digital technology measurements and re-run the regressions in the baseline model. More specifically, we standardize each index used to calculate the value of digital technologies, due to their differences in units, and adjust the weight of these indexes, weighting them all equally. The associated estimation results are given in column (2) of Table 5. We can see that the estimated coefficient of digital techniques is significantly positive at the one-percent level, which is consistent with the estimate in the baseline analysis. While the coefficient increases somewhat, it yet remains robust in terms of both sign and magnitude, indicating the reliability of these findings.

Table 6 Heterogeneity analysis

Variables	Eastern regions	The rest regions
	(1)	(2)
Digital technologies	7.118***	4.733***
	(2.430)	(1.581)
L. renewable	0.562***	0.562***
	(0.119)	(0.119)
Control variables	Y	Y
Time fixed effects	Y	Y
Place fixed effects	Y	Y
Observations	510	510
AR(1)	0.000	0.000
AR(2)	0.424	0.424
Sargan test	0.152	0.152

5.2 *Heterogeneity Analysis*

This section analyzes the heterogeneous effects of digital technology on renewable energy in terms of regional differences, taking into account the vast disparity in economic and social development across China. While it is well known that eastern China is much more developed in many aspects, disparities between the inland central and western regions have been considerably reduced due to the efforts of China's prominent place-based policy, that is, the Great Western Development Programme that was instituted in 2000 (Jia et al. 2020). Therefore, in the regional heterogeneity analysis, we divide all samples into two groups, one group in the east regions and the others being the rest of the country. The estimates are displayed in Table 6.[1]

As shown in Table 6, the impact of digital technologies on renewable energy is much greater in the eastern areas than is estimated for the rest of the country. This can be explained by the greater digital innovation, more active market mechanism, and more efficient administrative management in advanced technology delivery in the eastern regions (Jia et al. 2020; Zheng et al. 2022).

[1] As we use the interaction of regional dummy variables and digital technology in this analysis, the number of observations is 510 in both columns of Table 6.

6 Case Study: Qinghai Province

6.1 The Background of Renewable Energy in Qinghai

Qinghai province in Northwest China is renowned for its renewable energy generation. By the end of 2021, its installed power generation capacity reached 41.14 million kilowatts (KW), of which 25.28 million KW were renewable energy and 37.21 million KW clean energy, accounting for 61.5 and 90.45% of its power generation capacity, respectively. Qinghai thus has the highest proportion of renewable clean energy in its energy supply in China. In the first half of 2022, Qinghai's clean energy power generation reached 42.67 billion kilowatt-hours (kWh), accounting for 84.8% of the province's total power generation, with renewable energy power generation of 21.26 billion kWh, accounting for 42.3%.[2]

Figure 3 shows changes in renewable energy in Qinghai versus the Chinese average between 2003 and 2020. A comparison of histograms shows that power generation from renewable energy in Qinghai remains higher than the national average over most of the past two decades. Only in the period from 2014 to 2017 did Qinghai's renewable energy generation fall relatively below the national average, as renewable energy curtailment during that time was too great. Since then, however, power generation from renewable energy in Qinghai has once more surpassed the national average.

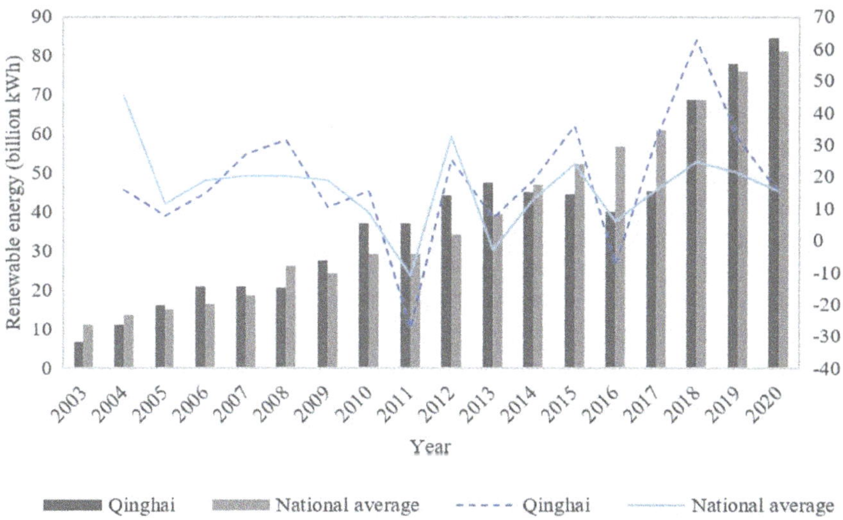

Fig. 3 Renewable energy and digital technology changes between 2003 and 2020 in Qinghai versus the Chinese national average

[2] See http://qh.news.cn/2022-08/13/c_1128912266.htm.

Based on its outstanding energy infrastructure, Qinghai province has carried out a "Green Power" campaign annually since 2017. In 2017, it successfully ran on 100% renewable energy for seven continuous days, as part of a trial conducted by the State Grid Corporation of China. From June 25 to July 29, 2022, Qinghai conducted a 35-day "Green Power 5 Weeks" campaign, using wind, solar, hydro, and other renewable energy sources, to achieve "zero carbon emissions" in both industrial and residential electricity supply. In the past five years, its cumulative clean power supply was 25.156 billion kWh, reducing coal consumption by 11.43 million tons and CO_2 emissions by 20.58 million tons.

6.2 *Applying Digital Technology to Promote Renewable Energy*

Since 2017, Qinghai province has carried out an overall assessment of demand potential and grid flexibility for large-scale renewable energy connections, and comprehensively studied optimal power mixes of various energy sources, e.g., wind, solar, hydro, and thermal energy. They have also analyzed optimal dispatching of multi-energy units and maximum power generation of new energy stations. In the process, a number of programs applying digital technology have been launched, including "Software Demonstration for Optimal Annual/Monthly Electricity Generation Scheduling", "Random Optimal Renewable Energy Generation Scheduling System", and "Complementary and Coordinating Dispatching System for Energy Generation from Multiple Power Sources". Qinghai has leveraged digital technology to maximize renewable energy use while effectively reducing dispatching risk and improving its electricity supply system's safety and reliability. Many entities promote these programs, including top universities, companies in energy or related infrastructure construction, and professional associations. Many projects applying cutting-edge digital technology have been put at the top of their research lists, including interaction between providers and end-users, load modes and assessment systems with renewable energy as the main generation source, and the demonstrated interaction between generation and load. In practical work, the following tasks have also become key work, including network-based energy storage, direct current (DC) collection and networking, electricity-carbon collaborative management, and comprehensive energy planning and operation based on industrial zones.

Figure 3, which shows the trend of digital technology growth rates, also shows via the line graphs that digital technology adoption in Qinghai accelerated in 2017 and have significantly exceeded the national level since, although the gap between Qinghai and the national level is unstable in the previous years. Overall, the change of digital technology in Qinghai goes in line with its outstanding performance in renewable energy. To some extent, this reveals that digital technology in Qinghai has considerably facilitated renewable energy development there.

6.3 Barriers and Policy Responses

In China, almost all provinces are trying to apply digital technologies to facilitate expansion renewable energy generation, and Guizhou and Sichuan, two western provinces with extensive renewables, are rapidly building more computing power to drive digital economies and provide digital services to eastern China. In this context, Qinghai faces tremendous challenges from peer provinces. Qinghai has been making many efforts in many ways to deeply integrate digital technology with renewable energy, particularly in improving policy structure and mechanism design, launching key projects, attracting high-level human capital in related fields, and facilitating social capital participation in digital technology.

Regarding improving policy structure and mechanism design, key provincial leaders direct dedicated personnel in developing Qinghai's digital economy. The Digital Economy Development Bureau, a provincial government agency, manages such relevant affairs as confirmation, openness, circulation, transactions, and security data resources. The bureau has established targeted standards and systematic regulations for said digital economic development, and is exploring better coordination across departments and hierarchies. It has also simplified processes for applying for data use among governments and concerned companies.

Qinghai has constructed several data centers, computing infrastructure and national computing hubs in the course of launching key projects, including the National Qinghai-Tibet Plateau Scientific Data Center—Qinghai Branch, Qinghai-Tibet Plateau Ecological Big Data Center, Huawei Big Data Center, and Big Data Centers of the Three Major Telecom Operators. The bureau has taken a number of measures, to stimulate these data centers to release more potential and thereby promote Qinghai's digital economic development, including sharing project resources, taking advantage of counterparts' assistance to reduce poverty, and striving to become a showcase for investment. In particular, two 10-million KW renewable energy bases built by Qinghai are continuously allocating relevant energy to companies inside and outside the province, to promote the reciprocal model of carrying out crucial projects by sharing resources.

Considering that human capital is one of the core elements for digital technology development, Qinghai tries to attract and employ representative talent in computer science, software engineering, artificial intelligence, data science, and electronic engineering. The province has also adopted innovative policies featuring telework and teleconferencing which allow workers to work remotely for Qinghai regardless of where they actually reside.

Qinghai also grasps the emerging characteristics of digital economy and provides preferential policies for social capital participation. The province compensates for its economic development shortcomings by considering its local economic conditions and relevant social capital investment demands.

7 Conclusion

In this paper, we evaluate the role of digital technologies in renewable energy development, based on China's provincial data from 2003 to 2020, by applying the entropy weight method to measure China's digital technology level and employing the GMM estimation approach in the baseline analysis. We also test potential channels through which digital technologies bolster renewable energy growth, in terms of economic development and industrial structure adjustment. We also conduct a series of robustness checks to ensure the validity of our primary findings, using the spatial econometric method to take possible geographical influences into account and handling the influence of the weighting of the indexes used to measure digital technologies. We also conduct a heterogeneity analysis before depicting Qinghai province's particular measures to how achieve 100% renewable energy. The aforementioned primary findings suggest that the application of digital technologies has significantly facilitated renewable energy development in China, an outcome that is robust across a number of model specifications and assessment methods. Digital technologies exert these effects through affecting economic development and adjusting industrial structure. This impact is particularly prominent in the more developed eastern regions where conditions are more conducive to advancement in digital technologies. These conclusions shows that it is plausible to expand renewable energy application by enhancing digital technology. Moreover, to make digital technologies work better in improving renewable energy, it is necessary to ensure conditions suitable for digital innovation, in both market mechanisms and government efficiency.

The Qinghai province case study shows that a place with extensive renewable resources that nonetheless lags significantly in economic development is able to promote renewable energy development with the help of digital technologies, given reasonable policy and mechanism design. Other important factors to consider include emerging characteristics of digital technologies, representative human capital, and social capital investment.

Acknowledgements Xuemei Zheng acknowledges financial support from ERIA (Grant No. RIA-RD/ RA-1-1-2212/06/FY22) and the National Natural Science Foundation of China (Grant No. 72003156).

References

Aguirre M, Ibikunle G (2014) Determinants of renewable energy growth: a global sample analysis. Energy Policy 69:374–384

Ahl A, Yarime M, Tanaka K, Sagawa D (2019) Review of blockchain-based distributed energy: implications for institutional development. Renew Sustain Energy Rev 107:200–211

Al-Irsyad MI, Halog A, Nepal R (2019) Estimating the impacts of financing support policies towards photovoltaic market in Indonesia: a social-energy-economy-environment model simulation. J Environ Manage 230:464–473

Amri F (2017) The relationship amongst energy consumption (renewable and non-renewable), and GDP in Algeria. Renew Sustain Energy Rev 76:62–71

Baidya S, Potdar V, Ray PP, Nandi C (2021) Reviewing the opportunities, challenges, and future directions for the digitalization of energy. Energy Res Soc Sci 81:102243

Baranes E, Jacqmin J, Poudou J-C (2017) Non-renewable and intermittent renewable energy sources: friends and foes? Energy Policy 111:58–67

Bastida L, Cohen JJ, Kollmann A, Moya A, Reichl J (2019) Exploring the role of ICT on household behavioural energy efficiency to mitigate global warming. Renew Sustain Energy Rev 103:455–462

Bhattacharya M, Churchill SA, Paramati SR (2017) The dynamic impact of renewable energy and institutions on economic output and CO_2 emissions across regions. Renewable Energy 111:157–167

Bhattacharya M, Paramati SR, Ozturk I, Bhattacharya S (2016) The effect of renewable energy consumption on economic growth: evidence from top 38 countries. Appl Energy 162:733–741

Chen Y (2018) Factors influencing renewable energy consumption in China: an empirical analysis based on provincial panel data. J Clean Prod 174:605–615

Domini G, Grazzi M, Moschella D, Treibich T (2021) Threats and opportunities in the digital era: automation spikes and employment dynamics. Res Policy 50(7):104137

Feng S, Chong Y, Li G, Zhang S (2022) Digital finance and innovation inequality: evidence from green technological innovation in China. Environ Sci Pollut Res 29(58):87884–87900

Gozgor G, Mahalik MK, Demir E, Padhan H (2020) The impact of economic globalization on renewable energy in the OECD countries. Energy Policy 139:111365

Hossain M, Madlool N, Rahim N, Selvaraj J, Pandey A, Khan AF (2016) Role of smart grid in renewable energy: an overview. Renew Sustain Energy Rev 60:1168–1184

Husaini DH, Lean HH (2022) Digitalization and energy sustainability in ASEAN. Resour Conserv Recycl 184:106377

Jia J, Ma G, Qin C, Wang L (2020) Place-based policies, state-led industrialisation, and regional development: evidence from China's great western development programme. Eur Econ Rev 123:103398

Kutan AM, Paramati SR, Ummalla M, Zakari A (2018) Financing renewable energy projects in major emerging market economies: evidence in the perspective of sustainable economic development. Emerg Mark Financ Trade 54(8):1761–1777

Lange S, Pohl J, Santarius T (2020) Digitalization and energy consumption does ICT reduce energy demand? Ecol Econ 176:106760

Lin B, Omoju OE, Okonkwo JU (2016) Factors influencing renewable electricity consumption in China. Renew Sustain Energy Rev 55:687–696

Liu X, Zhao T, Chang C-T, Fu CJ (2021) China's renewable energy strategy and industrial adjustment policy. Renewable Energy 170:1382–1395

Marques AC, Fuinhas JA, Manso JP (2010) Motivations driving renewable energy in European countries: a panel data approach. Energy Policy 38(11):6877–6885

Nepal R, Musibau HO (2021) Energy security, renewable, non-renewable energy and economic growth in ASEAN economies: new insights. The Singapore Economic Review 66(02):457–488

Nguyen KH, Kakinaka M (2019) Renewable energy consumption, carbon emissions, and development stages: some evidence from panel cointegration analysis. Renewable Energy 132:1049–1057

Omri A, Nguyen DK (2014) On the determinants of renewable energy consumption: international evidence. Energy 72:554–560

Ren S, Hao Y, Xu L, Wu H, Ba N (2021) Digitalization and energy: how does internet development affect China's energy consumption? Energy Economics 98:105220

Sadorsky P (2009a) Renewable energy consumption and income in emerging economies. Energy Policy 37(10):4021–4028

Sadorsky P (2009b) Renewable energy consumption, CO_2 emissions and oil prices in the G7 countries. Energy Economics 31(3):456–462

Salim RA, Rafiq S (2012) Why do some emerging economies proactively accelerate the adoption of renewable energy? Energy Economics 34(4):1051–1057

Sharifi A, Ahmadi M, Ala A (2021) The impact of artificial intelligence and digital style on industry and energy post-COVID-19 pandemic. Environ Sci Pollut Res 28(34):46964–46984

Shen Y, Zhang X (2022) Digital economy, intelligent manufacturing, and labor mismatch. Journal of Advanced Computational Intelligence and Intelligent Informatics 26(4):655–664

Shivendu S, Zhang RA (2019) The impact of digitization on content markets: prices, profit, and social welfare. Management Information Systems Quarterly

Strielkowski W, Dvoˇrák M, Rovny‘ P, Tarkhanova E, Baburina N (2021) 5G wireless networks in the future renewable energy systems. Frontiers in Energy Research 654

Thanh TT, Ha LT, Dung HP, Huong TTL (2022) Impacts of digitalization on energy security: evidence from European countries. Environment, Development and Sustainability 1–46

The International Energy Agency (2017) Digitalization and energy

Wu HX, Yu C (2022) The impact of the digital economy on China's economic growth and productivity performance. China Economic Journal 15(2):153–170

Xu Q, Zhong M, Li X (2022) How does digitalization affect energy? International Evidence. Energy Economics 107:105879

Zheng X, Li C, Fang X, Zhang N (2021a) Price sensitivity and consumers' support for renewable energy in China. Energy 222:119862

Zheng X, Wu C, He S (2021b) Impacts of market segmentation on the over-capacity of the thermal electricity generation industry in China. J Environ Manage 279:111761

Zheng X, Menezes F, Zheng X, Wu C (2022) An empirical assessment of the impact of subsidies on EV adoption in China: a difference-in-differences approach. Transportation Research Part a: Policy and Practice 162:121–136

Chapter 9
Potential Solar, Wind, and Battery Storage Deployment for Decarbonization in ASEAN

Han Phoumin and Rabindra Nepal

Abstract Achieving carbon neutrality will require multiple approaches to decarbonizing greenhouse gas emissions across all sectors. Accordingly, this study investigates the maximum contributions of solar and wind deployments together with energy storage potentials with the objective of changing such deployments from intermittent supply to more stable load by employing energy storage systems. To this end, we use data generated by a linear programming model to minimize total system under such constraints as CO_2 emissions and supply–demand balance, in order to assess the aforementioned maximum contributions in ASEAN's decarbonization scenario. Our findings provide policymakers a second opinion on how to scale up solar and wind with battery storage to contribute to future significant ASEAN decarbonization.

1 Introduction

At the 26th Conference of Parties (COP26), a roadmap was established for achieving climate goals including phasing out coal, ending fossil fuel subsidies, putting a price on carbon, protecting vulnerable communities, and delivering a USD100 billion climate finance commitment, despite some debate over the timeline for net-zero emissions (NZE) and other climate policies (UN Climate Change Conference, UK 2021). If translated into real policy action, this roadmap will have an enormous impact on investment in clean technologies, renewables, and clean fuels. COP26 has thus influenced national policy the world over toward transitioning to low-carbon societies in order to limit global warming to at least below 2 °C and preferably 1.5 °C compared with pre-industrial levels (Kimura and Han 2023).

H. Phoumin (✉)
Economic Research Institute for ASEAN and East Asia (ERIA), Jakarta, Indonesia
e-mail: han.phoumin@eria.org

R. Nepal
School of Business, University of Wollongong, Wollongong, Australia

H. Phoumin et al. (eds.), *Large-Scale Development of Renewables in the ASEAN*, Economics, Law, and Institutions in Asia Pacific,
https://doi.org/10.1007/978-981-99-8239-4_9

The goal of achieving energy transitions toward carbon neutrality by 2050 or beyond poses a challenge for ASEAN to replace the fossil fuels that it is currently dependent on with clean energy systems. The Economic Research Institute for ASEAN and East Asia (ERIA) predicts that ASEAN as a group will experience continuous rising energy demand through 2050, and that clean energy, especially renewables, will play a critical role in gradually replacing fossil fuels in a stepwise manner. The reason is that ASEAN will still need fossil fuels in 2050 in a Business as Usual (BAU) scenario (Kimura and Han 2020), in which coal, oil, and natural gas will still account for 80% of ASEAN's primary energy supply by 2050. Any clean energy policy, therefore, must redesign systems to accommodate renewable energy (RE) sources, i.e., wind, solar, hydro, geothermal, and biomass. Increased renewables generation must be combined with deployment of clean technology that requires solutions to the problem of its persistent high cost. Greenhouse gas (GHG) emission rates in ASEAN and India also continue to rise in the BAU scenario, contravening the Paris Agreement's NZE target (Han 2022). ASEAN Member States (AMS) and India must thus act immediately to reduce emission rates to achieve the aforementioned goals. It is important to look at the renewable resources that ASEAN must secure to make the transition affordable and ensure seamless use of locally available resources.

Solar has strong potential for inclusion in ASEAN's energy mix, while wind can only work in some countries, including Vietnam, the Philippines, and Indonesia. Notwithstanding, some crucial obstacles remain to deploying wind and solar at scale, the largest ones being that they are intermittent and require battery support and backup from other renewable sources to prevent blackouts, and that their deployment costs are much higher than other resources such as thermal and coal. Faster integration of ASEAN's electricity grids, however, would allow just such use of resources from other regions to bridge the aforementioned solar and wind gaps. For example, countries in the Greater Mekong Subregion (GMS) endowed with hydropower energy resources could provide backup for AMSs utilizing solar and wind. This combination would allow greater solar and wind integration into ASEAN's connectivity infrastructure depending on the status of said infrastructure and policy progress.

The ongoing Russia-Ukraine conflict is a wake-up call for the European community, as well as other countries around the world, that they can no longer depend on fossil fuels for a secure energy source as doing so has created a great dependency on fuel imports. Thus, this war that has raised energy costs might also provide impetus for countries to shift away from fossil fuels altogether in the long-term. The International Energy Agency (IEA) has issued 10 recommendations for measures to reduce the European Union's reliance on Russian natural gas import (IEA 2022a), including jump-starting renewables by accelerating the deployment of new wind and solar projects, and maximizing output from such existing dispatchable low-emissions sources as bioenergy and nuclear. The war in Ukraine may thus drive RE to economies of scale capable of reducing costs in the not-too-distant future. Cohen (2022), author at State of the Planet, wrote, "The only true way to secure real energy independence is to break our dependence on fossil fuels. Renewable energy is the ultimate form of energy independence since no sovereign state owns the sun." While true, fears for the aforementioned supply disruptions and energy security might also keep some

Asian countries dependent on coal beyond their COP26 commitments, owing to coal being reliable and politically stable within ASEAN and East Asia.

Thus, there is cause for concern that these countries may make a priority of extending contracts to secure fossil fuel supplies. Doing so would have an impact on the timeframe for phasing out fossil fuels to meet the Paris Agreement as well as to limit the global warming to the aforementioned below 2 °C, and preferably to 1.5 °C, compared to pre-industrial levels. Oil market concerns may be prolonged if the Russian-Ukraine war continues and no immediate alternative sources of supplies of oil or natural gas come into play. Any decision on investment for new energy projects may take at least a year in the case of solar or wind, and longer still for bioenergy or nuclear. Although we see no signs of reneging on climate change commitments, we should be cautious about policy changes that may distract from investment in renewables. Countries should redesign energy policy to shift away from fossil fuel dependency for the long-term, the sooner the better, utility scale investment, into solar and wind and other clean energy sources.

Combining such policy changes with technological innovations, however, may enable ASEAN to incorporate a greater share of renewables in its energy mix, including solar, wind, biomass, geothermal, and hydropower, as part of achieving NZE by 2050. And as solar is abundant in all AMSs, it is incumbent upon ASEAN to deploy large-scale solar photovoltaic (PV) with battery storage, which this study accordingly thoroughly analyzes, as previously mentioned.

2 Analytical Framework

This study benefits from data collection and analysis of ERIA's study project titled "Decarbonization of Energy System: Optimum Technology Selection Model Analysis up to 2060." We used the Optimum Technology Selection Model developed by the Institute of Energy Economics, Japan (IEEJ), adapted from Otsuki et al. (2019). The study spans the energy systems of all 10 AMSs, taking 2017 as the baseline year and 2030, 2040, 2050, and 2060 as the years for analysis, covering the energy conversion and end-use sectors, including industry, transport, residential, and commercial, incorporating more than 350 technologies (Kimura et al., 2022). It is formulated as a linear programming model, taking the cost and performance of each energy technology to be adopted as input values and outputting a single combination of the scales and operational patterns of such technologies. Doing so minimizes the total cost of the energy systems when accounting for such constraints as CO_2 emissions and power supply–demand balance. The model evaluates combinations of the technologies by applying factors such as capital costs and fuel costs to each technology, in addition to the aforementioned CO_2 emissions.

ASEAN will deploy large Solar PV systems with battery storage, among other clean technologies, to become carbon–neutral. Figure 1 shows ASEAN's solar and wind potential (Global Solar Atlas 2022; Global Wind Atlas 2022), on which we base the model's assumptions.

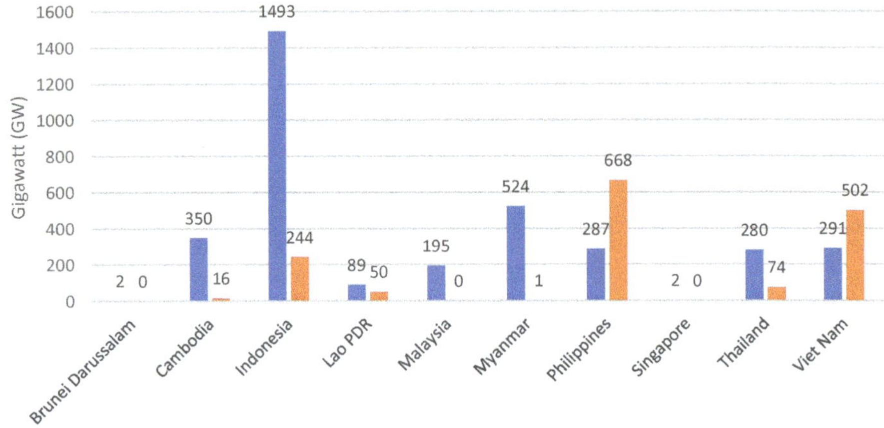

Fig. 1 Solar PV and wind potential in AMSs. *Source* Global Solar Atlas (2022) and Global Wind Atlas (2022)

Other power generation assumptions in terms of Levelized Cost of Energy (LCoE) are also made across different technologies, per Fig. 2.

The model also includes other low-carbon technologies such as hydrogen (H_2), ammonia (NH_3, and negative-emission technologies including direct air capture with carbon storage (DACCS) and bioenergy with CCS (BECCS).

The ERIA study considered the following ASEAN carbon neutrality scenarios: (1) BAU, with no CO_2 emissions targets; (2) Carbon Neutrality Scenario 2050/

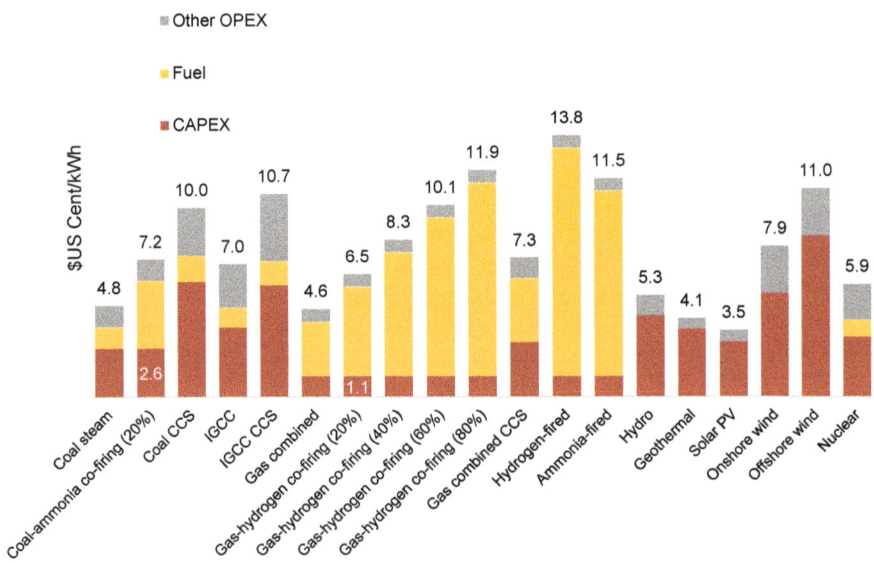

Fig. 2 LCOE assumptions by technology. *Source* Authors' calculation

60 (CN2050/2060), reflecting nationally declared carbon–neutral target years and considering carbon sinks; (3) CN2050/2060_Innovation cases, describing the impact of various technological innovations; (4) CN2050/2060_Stringent2030, tightening CN2050/2060's emission constraints in 2030 to the same level as the IEA Sustainable Development Scenario; and (5) CN2050/2060 without carbon sink (CN2050/2060 w/o Carbon Sink), which assumes that CO_2 emissions become net zero by 2060 and does not consider carbon sinks.

For simplicity's sake, the results of potential solar and wind penetration into the energy mix are reported only for BAU and CN2050/2060.

3 Literature Review

In 2021, the IEA released a report titled "Net Zero by 2050: A Road Map for the Global Energy Sector," which mentioned the need to transform energy systems toward renewables and clean fuels. Such RE technologies as solar and wind are key to reducing CO_2 emissions in the electricity sector, currently the single largest source of such emissions. According to this plan, almost 90% of global electricity generation in 2050 would come from such renewable sources, with solar PV and wind together accounting for nearly 70% (IEA 2021).

The Energy Outlook and Energy Saving Potential in East Asia Region 2022 (Kimura and Han 2022) highlighted the East Asia Summit (EAS) region in various decarbonization pathway scenarios. It evaluated various socioeconomic and political circumstances that may aid countries' efforts to reach carbon neutrality, with the "Low Carbon Energy Transition (LCET)" scenario analyzing the impact of NZE technologies capable of helping countries become carbon–neutral by 2050. The report indicated the need for multiple pathways, including transition technologies, and more renewables and clean fuels in stepwise phases, to minimize cost and ensure energy affordability and security in all countries.

Storage plays an important role in enabling power grids to function with more flexibility and resilience. In the US, electric power markets are undergoing significant structural change to allow more solar and wind integration which will result in the installation between 2021 and 2023 of 10 times the capacity of large-scale battery storage in 2019, capable of contributing 10,000 MW to the grid, (IEA 2021). Grid-scale storage plays a similarly important role in the NZE by 2050 Scenario, providing crucial system services from short-term balancing and operating reserves, ancillary services for grid stability, and deferment of investment in new transmission and distribution lines, to long-term storage and restoring operations following blackouts. However, projected grid-scale storage capacity growth requires greater efforts if the Net Zero Scenario is to be fulfilled (IEA 2022b).

Academic research tests renewables' significant contribution to reducing GHG emissions globally and in Asia. Much of this research finds that such emissions are largely linked to the energy systems of countries that rely heavily on fossil fuels. Abbass et al. (2022) found that while non-renewable energy consumption and economic growth increase long-term CO_2 emissions in South Asia, RE use and trade reduce same in the short term.

Regional institutional reform and energy cooperation will help accelerate clean energy and renewables adoption. Lee (2013) analyzes the dual challenge of depletion of fossil fuels and climate change in Northeast Asia, i.e., China, Japan, and Korea, and their introduction of green energy initiatives in recent years. This paper suggested that green energy cooperation is not free from neo-mercantilist competition, as the current initiatives entail strong industrial policy elements. Such efforts should instead be based on continued domestic momentum as well as building of sub-regional institutions. Regional multilateral institutions, such as Asia–Pacific Economic Cooperation (APEC), ASEAN + 3, and the ASEAN Regional Forum, as well as diverse international organizations, can provide useful venues for the Northeast Asian countries to share information and adopt a common position on the aforementioned green energy cooperation.

Sahoo et al. (2022) use data from 14 developing countries in Asia between 1990 and 2018 to examine the potential impact of environmental innovation on GHG emissions by controlling globalization, urbanization, and economic growth, finding that RE and globalization significantly reduce such emissions, while innovations in environmental technology play a minor role, and even then only when economic growth supports investments in same. They also found that urbanization, oil consumption, and economic growth are detrimental to the environment. They suggested countries should invest accordingly.

Irfan et al. (2021), examine the impact of energy efficiency (EE) and RE on GHG emissions, using panel data from South Asian countries between 1990 and 2014. Their results suggest that a cointegrating link is evident between GHG emissions, EE, and RE in South Asia, especially after controlling for economic growth and trade openness. Their homogenous coefficient estimates reinforce the findings that in the long run, GHG emissions decline as RE increases.

Usman and Hammar (2021), analyze the dynamic relationship between technological innovation, financial development, RE, and ecological footprint in APEC by utilizing balanced longitudinal data between 1990 and 2017. The study found that financial development and RE utilization accelerate environmental quality by 0.0927% and 0.4274% respectively, while greater technological innovation, economic growth, and population size are detrimental to environmental quality in the long run by 0.099%, 0.517%, and 0.458% respectively.

4 Carbon Neutrality Scenario Results

4.1 Primary ASEAN Energy Supply in BAU Versus Carbon Neutrality Scenarios

ASEAN's energy supply was 616 million tonnes of oil equivalent (Mtoe) in 2017, and it is expected to grow to 2006 Mtoe by 2060 in the BAU or Baseline scenario, per Fig. 3 and Table 1. Coal, oil, and natural gas accounted for approximately 80.06% in 2017, and are forecast to reach 85.09% in 2060 in the BAU scenario.

In the CN2050/60 scenario, clean fuels and renewables are forecast to increase significantly from the baseline by 2050 and 2060. Nuclear could be an option in this scenario, with a forecast of providing some 62–63 Mtoe of the primary energy mix by 2050 and 2060 respectively. Per Fig. 4, solar PV and wind capacity are forecast to increase aggressively in this scenario. Other renewables, such as hydropower, geothermal, and biomass, are also forecast to contribute. Clean fuels, such as hydrogen and ammonia, are also forecast to be introduced into the supply mix in this scenario beginning in 2040.

Of all fossil fuels, coal is expected to decline most drastically from peak demand in the CN2050/60 scenario as per Table 1. By contrast, oil and natural gas are predicted to increase slowly in this scenario, chiefly due to their being used to bridge the energy transition. Natural gas is anticipated to be a clean energy source if combined with carbon capture, utilization, and storage (CCUS), while oil will remain important for transportation, especially heavy vehicles such as buses and trucks.

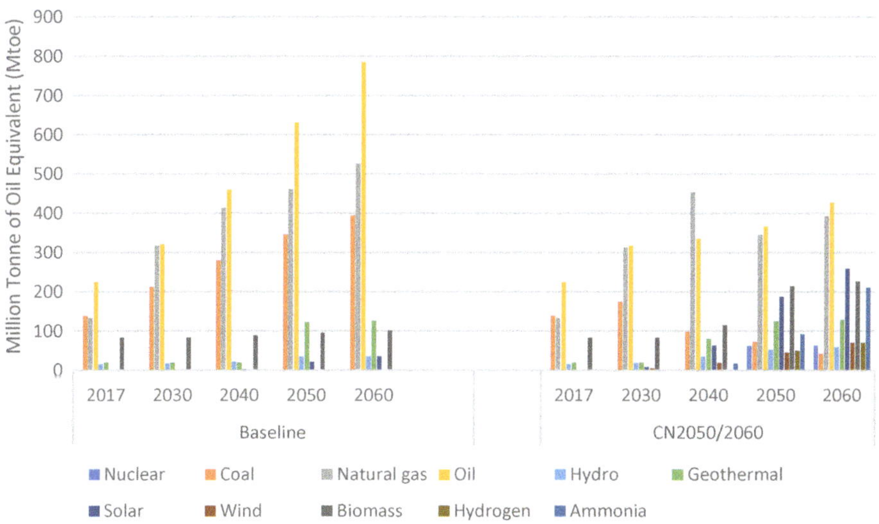

Fig. 3 Total primary energy supply (TPES), BAU versus CN2050/60. *Source* Authors' calculations

Table 1 Primary energy supply by source, BAU versus CN2050/60 scenarios (Mtoe)

Source	BAU					CN2050/2060				
	2017	2030	2040	2050	2060	2017	2030	2040	2050	2060
Nuclear	0.0	0.0	0.0	0.0	0.0	0.0	0.0	0.0	62.4	62.7
Coal	138.8	212.5	279.8	346.8	394.9	138.8	174.3	99.2	72.8	42.8
Natural gas	132.4	317.1	413.2	460.9	526.2	132.4	312.2	453.1	345.2	393.1
Oil	225.7	322.2	460.7	631.9	785.8	225.3	318.9	335.8	368.0	429.3
Hydro	16.2	17.7	22.5	35.6	35.6	16.2	19.6	36.1	53.0	59.4
Geothermal	19.9	19.9	19.9	122.9	126.2	19.9	19.9	79.8	125.4	128.9
Solar	0.5	1.0	2.7	21.6	35.4	0.5	8.6	63.6	187.3	258.7
Wind	0.2	0.2	0.2	0.3	1.2	0.2	5.8	19.2	45.6	70.3
Biomass	82.8	82.9	88.8	94.7	100.2	82.8	82.6	114.9	214.1	226.2
Hydrogen	0.0	0.0	0.0	0.0	0.0	0.0	0.0	0.2	50.6	70.9
Ammonia	0.0	0.0	1.2	0.8	0.4	0.0	0.0	17.2	91.6	210.9
Share of imported H_2/NH_3	0%	0%	0%	0%	0%	0%	0%	1%	9%	14%
Total	616	974	1289	1716	2006	616	942	1219	1616	1953

Source Authors' calculations

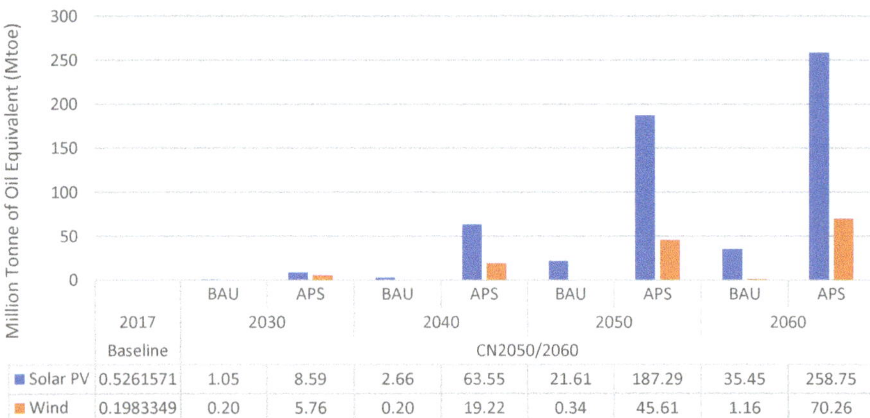

Fig. 4 Potential ASEAN solar PV and wind deployment in CN2050/60 scenario. *Source* Authors' calculations

It is observed that the total energy supply is forecast to fall from 2006 Mtoe in BAU to 1953 Mtoe in CN2050/60, due to renewables and clean fuels comprising the major share of the energy mix in the latter scenario.

4.2 ASEAN Power Generation Mix, BAU Versus CN2050/60

ASEAN's power generation mix is forecast to double by 2060 in CN2050/60 from BAU. Per Fig. 5 and Table 2, by 2060 electricity demand is predicted to reach 6720 TWh in CN2050/60, compared with 3657 TWh in BAU, owing to large-scale introduction of intermittent renewables, i.e., solar PV and wind. Battery storage is critical in this instance for such purposes as back-ups to avoid interruptions or to smooth RE load curves. Surplus wind and solar PV electricity could also be used to produce hydrogen for use in fuel cell vehicles and for power generation, i.e., co-combustion with natural gas (U.S . Energy Information Administration [EIA], 2021).

While a proper real time power mix simulation would be necessary to understand how much battery storage will be needed, for present purposes we apply a rule of thumb ranging from 20 to 25% of solar PV and wind installed capacity, per Figs. 6 and 7. In CN2050/60, it is forecast that there will be some 1627 GW installed wind and solar PV capacity by 2060, with Solar PV having the larger share at 1385 GW, and onshore and offshore wind combined on the order of 331 GW.

Figure 7 shows that while battery storage can back up intermittent wind and solar PV, how much storage is needed depends on the energy mix and load curve, as mentioned above. In this study, it is estimated that in CN2050/60 ASEAN would require battery storage on the order of 1365 GWh.

Scheduling for battery storage deployment is consistent with the increased installed solar PV and wind capacity beginning in 2040. While ASEAN could see further large-scale deployment for solar PV and wind due to the aforementioned regional potential, system integration and cost may pose obstacles, resulting in greater lead times for adoption at scale. Although coal remains in use in CN2050/60, ASEAN moves toward such cleaner power initiatives as CCUS and co-combustion of coal with ammonia.

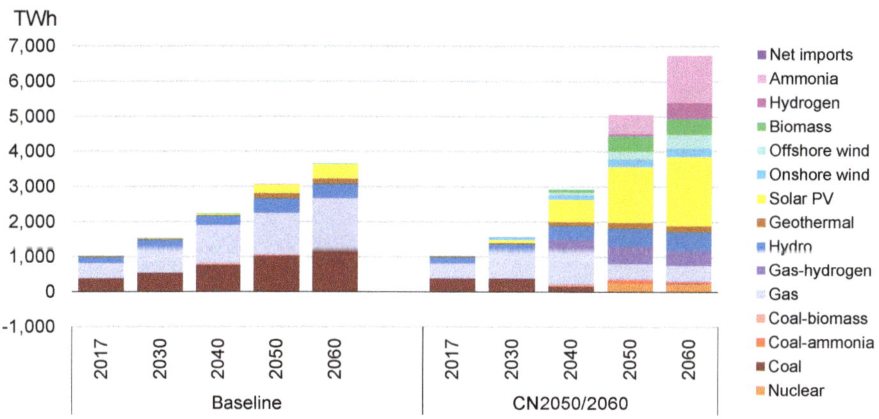

Fig. 5 ASEAN power generation mix, BAU versus CN2050/60. *Source* Authors' calculations

Table 2 Power generation mix by source, BAU versus CN2050/60 (TWh)

Source	BAU					CN2050/2060				
	2017	2030	2040	2050	2060	2017	2030	2040	2050	2060
Nuclear	0.0	0.0	0.0	0.0	0.0	0.0	0.0	0.0	241.9	243.2
Coal	383.7	544.0	757.2	1033.6	1226.9	383.7	383.7	155.4	0.0	5.7
Coal-ammonia	0.0	0.0	24.1	15.9	8.0	0.0	0.0	24.1	100.1	44.3
Coal-biomass	0.0	0.0	48.2	31.9	15.9	0.0	0.0	48.2	31.9	15.9
Gas	437.7	741.1	1079.7	1173.3	1421.0	437.7	767.1	998.0	438.6	459.1
Gas-hydrogen	0.0	0.0	0.0	0.0	0.0	0.0	0.0	272.5	495.7	423.7
Hydro	188.0	205.9	262.2	414.5	414.5	188.0	227.4	404.1	531.0	543.4
Geothermal	23.1	23.1	23.1	142.9	146.8	23.1	23.1	92.8	145.9	149.9
Solar PV	5.9	12.2	31.0	251.4	412.3	5.9	99.9	648.9	1591.2	1985.1
Onshore wind	1.9	1.9	1.9	3.5	13.1	1.9	66.5	123.1	211.1	226.6
Offshore wind	0.4	0.4	0.4	0.4	0.4	0.4	0.4	81.0	230.5	401.0
Biomass	0.0	1.4	1.3	1.5	1.2	0.0	0.4	75.0	432.6	454.6
Hydrogen	0.0	0.0	0.0	0.0	0.0	0.0	0.0	0.0	54.5	446.7
Ammonia	0.0	0.0	0.0	0.0	0.0	0.0	0.0	0.0	554.4	1335.8
Net imports	−2.0	−2.1	− 1.6	− 3.5	− 3.1	− 2.0	− 3.1	− 3.2	− 5.5	− 5.7
Total	1039	1528	2227	3065	3657	1039	1566	2920	5054	6720

Source Author's calculations

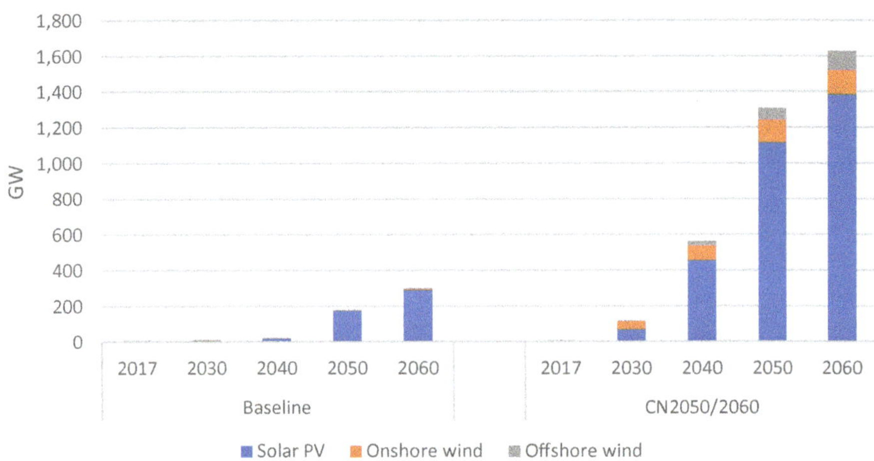

Fig. 6 Total installed solar PV and wind capacity, BAU versus CN2050/60 (GW). *Source* Authors' calculations

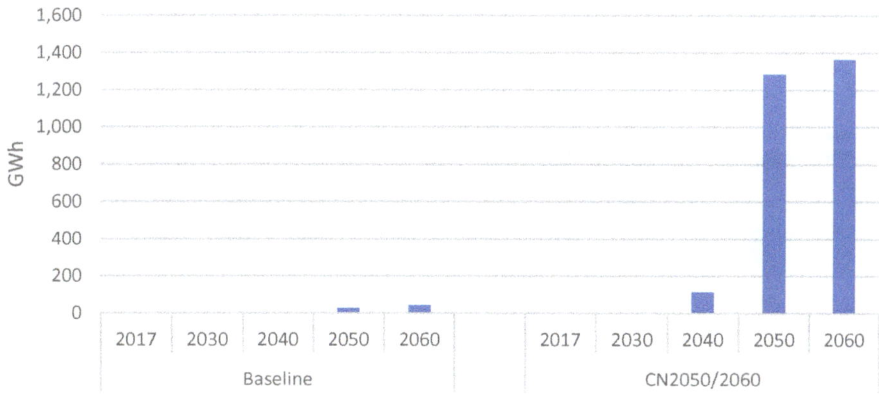

Fig. 7 Total Installed Battery Storage, BAU versus CN2050/60 (GWh). *Source* Authors' calculations

4.3 Final ASEAN Energy Consumption, BAU Versus CN2050/60

Turning again to fossil fuels, coal is forecast to decline significantly by 2060, from 108 Mtoe in BAU to 4 Mtoe in in CN2050/60. Oil is also expected to decline by half by 2060, from 694 Mtoe in BAU to 348 Mtoe in in CN2050/60. On the other hand, natural gas use does not decline significantly by 2060, as it serves as an energy transition bridge, as mentioned above, as well as potentially being viable beyond that stage if combined with such negative emissions technologies as Direct Air Capture (DAC). Figure 8 and Table 3 additionally show that electricity is the most used power source in CN2050/60.

The shift to large-scale electricity use is due to the introduction of electric vehicles (EV) and increasing residential and commercial adoption. As mentioned above, some buses and heavy trucks continue to use oil and hydrogen, while biomass and natural gas are expected to serve some sectors for heating purposes, such as industry. End-use decarbonization must focus on more electricity use in CN2050/60, which could be further accelerated, and decarbonization of power sources will require appropriate technologies and such aforementioned clean energy sources as biomass, hydropower, geothermal, wind, solar PV, and such clean fuels as hydrogen and ammonia.

4.4 ASEAN Emissions by Sector, BAU Versus CN2050/60

Decarbonization by sector will be necessary if ASEAN is to fulfill CN2050/60. In BAU, ASEAN's total CO_2 emissions are expected to rise to nearly 5000 Mt-CO_2 by 2060, with all sectors contributing, including transport, industry, and other, i.e., commercial and residential, as mentioned above. In CN2050/60, however, electricity

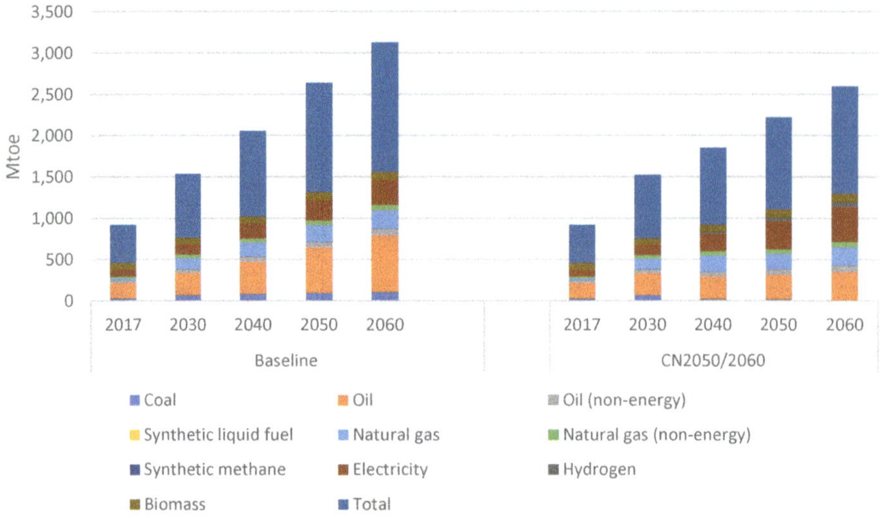

Fig. 8 Total final energy consumption (TFEC), BAU versus CN2050/60 (Mtoe). *Source* Authors' calculations

Table 3 Final energy consumption by source, BAU versus CN2050/60 (Mtoe)

Source	BAU					CN2050/2060				
	2017	2030	2040	2050	2060	2017	2030	2040	2050	2060
Coal	31.9	70.4	84.5	97.4	108.1	31.9	68.9	23.9	20.1	4.1
Oil	187.8	268.7	394.3	554.1	695.4	187.8	266.0	274.7	297.7	348.5
Oil (non-energy)	30.2	45.7	54.7	64.2	74.7	30.2	45.7	54.7	64.2	74.7
Synthetic liquid fuel	0.0	0.0	0.0	0.0	0.0	0.0	0.0	0.0	0.0	0.0
Natural gas	26.6	139.7	181.8	209.3	229.3	26.6	134.7	201.6	190.6	226.1
Natural gas (non-energy)	16.7	37.4	45.2	52.8	59.9	16.7	37.4	45.2	52.8	59.9
Synthetic methane	–	0.0	0.0	0.0	0.0	–	0.0	0.0	0.0	0.0
Electricity	84.8	124.8	182.0	250.3	298.7	84.8	127.8	216.9	344.0	430.6
Hydrogen	0.0	0.0	0.0	0.0	0.0	0.0	0.0	22.9	47.1	55.6
Biomass	82.2	82.5	86.0	92.7	99.1	82.2	82.5	86.0	93.4	100.3
Total	460.3	769.1	1028.6	1320.8	1565.0	460.3	762.9	925.9	1110.0	1299.8

Source Authors' calculations

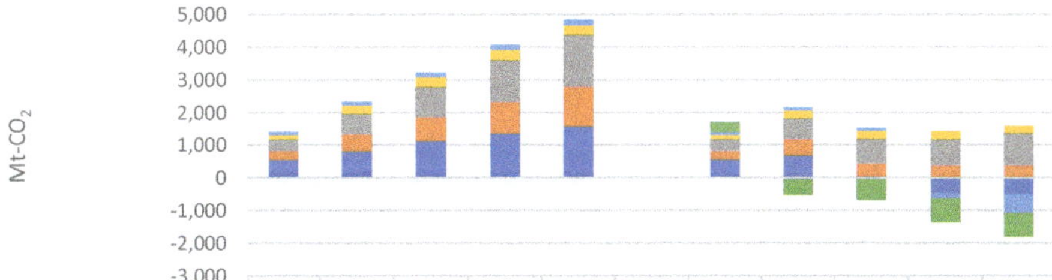

	Baseline 2017	Baseline 2030	Baseline 2040	Baseline 2050	Baseline 2060	CN2050/2060 2017	CN2050/2060 2030	CN2050/2060 2040	CN2050/2060 2050	CN2050/2060 2060
LULUCF						314.1	-519.6	-670.5	-732.8	-732.8
Other including DACCS	98.3	104.1	129.5	149.0	169.4	97.0	91.0	84.5	-148.4	-564.5
Other end-use	138.1	258.8	298.9	312.1	288.0	138.1	243.0	263.8	255.5	236.8
Transport	349.0	640.0	912.4	1266.2	1579.9	349.0	640.0	725.8	790.7	987.0
Industry	266.1	527.2	755.4	996.1	1232.0	266.1	515.1	411.7	386.8	385.7
Electricity	558.4	812.6	1119.6	1350.5	1575.9	558.4	677.7	43.1	-469.2	-502.6

Fig. 9 ASEAN emissions by sector (Mt-CO_2). *Source* Authors' calculations

consumption, chiefly the aforementioned solar PV, wind, biomass, hydropower, and geothermal, and Direct Air Capture with Carbon Storage (DACCS), are forecast to be introduced into ASEAN's energy system to decarbonize emissions. Natural carbon offsets through Land Use, Land Use Change, and Forestry (LULUCF) may also play a part.

As mentioned above, carbon capture, including DACCS, plays a vital role in helping ASEAN cut emissions by some 565 Mt-CO_2 by 2060, with renewables helping decarbonize an additional 503 Mt-CO_2. Solar PV will again be the most significant renewable in ASEAN'S system mix and its role, thus, remains critical to ASEAN fulfilling CN2050/60. While emissions persist in transport and industry, emissions are expected to decline more significantly in these sectors in CN2050/60 than BAU (Fig. 9).

5 Conclusions and Policy Implications

This chapter presents perspectives on greening ASEAN by potential solar PV and wind deployment coupled with battery storage to provide a stable and resilient energy system according to CN2050/60. Key findings are that in this scenario, clean fuels and renewables are forecast to increase significantly in ASEAN'S primary energy supply mix from BAU by 2050 and 2060. Solar PV capacity is forecast to increase aggressively as follows: 8.6 Mtoe by 2030, 63.6 Mtoe by 2040, 187.3 Mtoe by 2050, and 258.7 Mtoe by 2060. Wind is also forecast to increase as follows: 5.8 Mtoe by 2030, 19.2 Mtoe by 2040, 45.6 Mtoe by 2050, and 70.3 Mtoe by 2060. Nuclear may be an option in CN2050/60, potentially providing some 62–63 Mtoe of the primary energy mix by 2050 and 2060 respectively. Other clean energy, such as biomass, geothermal, and hydropower, also contribute significantly, depending on availability in AMSs, helping decarbonize GHG emissions by 2050 and 2060.

ASEAN'S power generation mix in CN2050/60 is forecast to have large-scale intermittent renewables online, i.e., the aforementioned solar PV and wind. Forecasts call for some 1627 GW total installed wind and solar PV capacity by 2060, with solar PV the larger at 1385 GW installed capacity and onshore and offshore wind combined supplying the remainder. Battery storage is critical for such purposes as back-ups or smoothing the renewables' load curves. Surplus renewable electricity is forecast to be used to produce hydrogen for fuel cell vehicles and co-combustion with natural gas.

Electricity, chiefly generated by solar PV, wind, biomass, hydropower, and geothermal will be used more than any other energy source by end users in CN2050/60, due to increased EV adoption and other use in residential and commercial sectors. Said use is forecast to grow and, together with DACCS, drive decarbonization in these sectors on the order of 503 Mt-CO_2 and 565 Mt-CO_2 respectively by 2060 as follows: 127 Mtoe by 2030, 2017 Mtoe by 2040, 344 Mtoe by 2050, and 431 Mtoe by 2060. Natural carbon offsets through LULUCF will also contribute.

Following are key implications for institutional and regulatory issues to be addressed to achieve carbon neutrality with solar PV and wind combined with battery storage and other clean technologies in ASEAN.

Establishing a framework for a regional regulatory and power trading body to create an integrated regional power system: ASEAN should consider forming a regional electricity regulatory and trading institution which is exclusively responsible for the promotion, harmonization, implementation, and enforcement of policies and regulations for an integrated power market and RE technologies within that market. Given the geographical spread of ASEAN, this recommendation is more immediately pertinent for AMSs. These should assess whether the current ASEAN Power Grid (APG) initiative and other ASEAN platforms such as the ASEAN Energy Regulators Network (AERN), which have the commitment of ASEAN heads of state, may provide a foundation for a functioning market with principal regulatory and trading functions including the aforementioned administering and enforcing power market regulation, providing accurate high-quality information, ensuring equal access for all market participants, and guaranteeing all trades and their delivery.

Such an integrated power market that brings RE and clean energy into the regional electricity mix requires cross-border power connectivity infrastructure. A starting point would be the APG initiatives on interconnection with Northeast India and Southwest China. Branching out in this way from a future ASEAN integrated power market to a cross-border pan-regional trading market with EAS will provide the region with resilient and energy security for power supply, in addition to accelerating RE utilization overall.

It is highly recommended that ASEAN's regional power market develop at its own gradual pace, initially forming sub-markets, e.g., country-to-country market coupling, followed by sub-regional market coupling, toward wider regional integration. It is crucial that concerned ASEAN governments and industry coordinate on investment and removing barriers.

Institutional capacity building and financing mechanism development are key to reducing lead time for appropriate deployment, especially of clean technologies and renewables. Policy priorities among ASEAN and advanced countries are financial assistance and capacity development in support of developing AMSs embarking on electricity trading in addition to the aforementioned clean technologies and RE development. The reason is that these face both regulatory and technical risks, making it crucial that governments assess each of these carefully and provide appropriate support.

All renewables and clean technologies may be classified into and pilot and demonstration, early-stage deployment, and mature technologies. This last includes the aforementioned wind, solar PV, geothermal, hydropower, and biomass. Notwithstanding such maturity, they still face market barriers to their large-scale deployment, necessitating policy reform, especially electricity market reform. Pilot and demonstration technologies and early-stage deployment alike confront many risks for any resource mobilization to fund them, again necessitating government spearheading to subsidize such technologies early-stage technologies as CCUS, hydrogen, and battery storage, augmented by some possible private sector sharing of the aforementioned risks.

References

Abbass K, Qasim MZ, Song H, Murshed M, Mahmood H, Younis I (2022) A review of the global climate change impacts, adaptation, and sustainable mitigation measures. Environ Sci Pollut Res 1–21

Cohen S (2022) State of the planet. https://news.climate.columbia.edu/author/steve-cohen/. Retrieved 22 March 2022

Global Solar Atlas (2022) https://globalsolaratlas.info/map

Global Wind Atlas (2022) https://globalwindatlas.info/en

Han P (2022) Decarbonization of ASEAN energy system: the optimum model analysis up to 2060. The presentation at 1st Asia Green Growth Partnership Ministerial. 25th April 2022. https://www.aggpm2022.org/2022ppf/english/. Retrieved 1 March 2023

IEA (2021) Net zero by 2050: a road map for the global energy sector. https://www.iea.org/reports/net-zero-by-2050. Retrieved 10 March 2023

International Energy Agency [IEA] (2022a) A 10 point-plan to reduce the European Union's reliance on Russian natural gas. https://www.iea.org/reports/a-10-point-plan-to-reduce-the-european-unions-reliance-on-russian-natural-gas. Retrieved 10 March 2022

IEA (2022b) Grid-scale storage, IEA, Paris. https://www.iea.org/reports/grid-scale-storage, License: CC BY 4.0

Irfan M, Mahapatra B, Ojha RK (2021) Examining the effectiveness of low-carbon strategies in South Asian countries: the case of energy efficiency and renewable energy. Environ Dev Sustain 23:11936–11952. https://doi.org/10.1007/s10668-020-01150-w

Kimura S, Han P (2020) Energy outlook and energy saving potential 2020. https://www.eria.org/publications/energy-outlook-and-energy-saving-potential-in-east-asia-2020/. Retrieved 30 Nov 2022

Kimura S, Han P (2022) Energy outlook and energy saving potential in East Asia Region 2022. Scheduled for publication in June 2023

Kimura S, Han P (2023) Energy outlook and energy saving potential 2023. Forthcoming from ERIA

Kimura S et al (2022) Decarbonisation of ASEAN energy systems: optimum technology selection model analysis up to 2060. https://www.eria.org/publications/decarbonisation-of-asean-energy-systems-optimum-technology-selection-model-analysis-up-to-2060/. Retrieved 22 Nov 2022

Lee JS (2013) Towards green energy cooperation in Northeast Asia: implications from European experiences. Asia Eur J 11:231–245. https://doi.org/10.1007/s10308-013-0359-1

Otsuki T, Komiyama R, Fujii Y (2019) Techno-economic assessment of hydrogen energy in the electricity and transport sectors using a spatially-disaggregated global energy system model. Journal of the Japan Institute of Energy 98(4):62–72

Sahoo B, Behera DK, Rahut D (2022) Decarbonization: examining the role of environmental innovation versus renewable energy use. Environ Sci Pollut Res 29:48704–48719. https://doi.org/10.1007/s11356-022-18686-1

UN Climate Change Conference, UK (2021) COP 26 Glasgow climate pack. https://ukcop26.org/wp-content/uploads/2021/11/COP26-Presidency-Outcomes-The-Climate-Pact.pdf. Retrieved 12 Nov 2022

U.S. Energy Information Administration [EIA] (2021) Battery storage in the United States: an update on market trends. https://www.eia.gov/analysis/studies/electricity/batterystorage/pdf/battery_storage_2021.pdf. Retrieved 20 March 2023

Usman M, Hammar N (2021) Dynamic relationship between technological innovations, financial development, renewable energy, and ecological footprint: fresh insights based on the STIRPAT model for Asia Pacific economic cooperation countries. Environ Sci Pollut Res 28:15519–15536. https://doi.org/10.1007/s11356-020-11640-z

Chapter 10
India's Cross Border Electricity Trade with BIMSTEC Countries

Sangeeta V. Sharma, Han Phoumin, Vinod K. Sharma, and Rabindra Nepal

Abstract This chapter assesses the present status of India's Cross Border Electricity Trade (CBET) with partners Nepal, Bhutan, Bangladesh, and Myanmar, and its effects on energy security. A mathematical model, consisting of source, trade, and result functions, was developed based on simple energy balance with and without CBET. In the first scenario, India and its trading partners are independently simulated for energy balance to individually assess energy deficit and storage dynamics. Drought season could cause serious deficits for countries dependent on single power sources, undermining energy security in ways that adding renewables to the mix might address. In the second scenario, interconnected grids could reduce storage and generation capacity and curtailment period for renewables. Trading partners dependent on single power sources might avoid deficits as energy could be imported, showing how CBET measurably affects energy security, helping achieve clean energy targets, minimize curtailment periods, and promote renewable energy penetration.

1 Introduction

Environmental pollution from fossil-fuel based energy, particularly thermal electricity generation, is a serious concern across the world. Cross Border Electricity Trade (CBET) among neighboring countries may help achieve sustainable energy targets and overall greater social welfare. One such CBET group is the Bay of Bengal

S. V. Sharma (✉)
NEEF, Mumbai 400053, India
e-mail: sangey2501@gmail.com

H. Phoumin
ERIA, Jakarta Pusat 10270, Indonesia

V. K. Sharma
Indira Gandhi Institute of Development Research, Mumbai 400065, India

R. Nepal
University of Wollongong, Wollongong, Australia

H. Phoumin et al. (eds.), *Large-Scale Development of Renewables in the ASEAN*, Economics, Law, and Institutions in Asia Pacific,
https://doi.org/10.1007/978-981-99-8239-4_10

Initiative for Multi-Sectoral Technical and Economic Cooperation (BIMSTEC), including Bangladesh, Bhutan, India, Nepal, Sri Lanka, Thailand, and Myanmar. India benefits from sharing borders with South Asian and BIMSTEC countries in terms of effectively utilizing diversified energy resources in the region, thus facilitating CBET.

Renewable energy (RE) plays an important role in transitioning to a low-carbon economy by reducing CO_2 emissions in the power sector. However, some RE technologies, particularly those based on intermittent wind and solar energy, can vary significantly over short periods, introducing instability into electricity systems, which can be overcome by using lithium battery-based storage systems or CBET. The former is not very cost-effective, as storage systems can significantly increase the levelized cost of electricity. CBET, conversely, is a promising solution with economic advantages that can include electricity produced in diverse forms across a broad geographical area.

As per BIMSTEC Energy Outlook-2030, the total primary energy supply in the BIMSTEC region may increase from 772 million tonnes of oil equivalent (Mtoe) in 2008 to 1758 Mtoe by 2030. Similarly, the total primary energy consumption is estimated to increase from 539 Mtoe in 2008 to 1210 Mtoe by 2030. Despite being endowed with abundant natural resources, most BIMSTEC countries depend upon imports to meet their primary energy needs. Increased CBET could thus be an important component of an integrated strategy for ensuring regional energy security (ORF 2020).

Bangladesh Power Management Institute (BPMI 2022) reports that CBET can assist in reducing energy prices, mitigating power shortages, facilitating decarbonization, and providing opportunities for market integration. Bangladesh CBET statistics indicate that energy transfers may increase from 1160 to 4500 MW by 2030 and 9000 MW by 2041, mostly between India and Bangladesh. Some of the challenges in implementing an active and robust network for CBET and restricting uncontrolled use of energy resources include lack of political will, insufficient grid interconnection, lack of legal framework, and regulations varying by country.

2 Context

Nepal is the first country to reap the benefit of buying day-ahead power from the Indian Energy Exchange (IEX) since IEX commenced CBET in April 2021. An analysis based on the economic factors affecting CBET between Nepal and India indicates that maximizing the value of any new transmission intended to enhance CBET value may depend on bilateral tariff reform and greater operational coordination between the two countries. India aims to have a regional power grid among Nepal, Bhutan, Bangladesh, Myanmar, and Sri Lanka, and has accordingly notified the concerned parties of planned changes to cross-border trading regulations.

Currently, India is connected with Nepal, Bhutan, Bangladesh, and Myanmar, and CBET between India and these other countries stands at approximately 18 billion units (BU), conducted through medium to long-term bilateral contracts. As per the Central Electricity Authority (CEA) and the Central Electricity Regulatory Commission (CERC), India imported 8.7 BU of electricity from Bhutan and exported 2.37 BU to Nepal and 7 BU to Bangladesh in 2021. Power trade with these countries is expected to increase to some 70 BU by 2027 (MoP 2022; Mint 2022).

The BIMSTEC countries inked a Memorandum of Understanding in 2018 on establishing a regional power grid, and are working toward establishing a power grid spanning approximately 3000 km from Myanmar-Thailand to India for more efficient capacity utilization to meet member states' demand and supply. India has bilateral connections with Nepal, Bhutan, Bangladesh, and Myanmar via high-voltage high-capacity transmission projects including 400 kV and 765 kV AC, as well as HVDC systems (MoP 2022). Presently, India imports 1.5 GW of hydropower from Bhutan and exports approximately 500 MW to Pakistan, 120–150 MW to Nepal, and some 500 MW to Bangladesh. Options for power import/export to and from Myanmar, Bangladesh, Pakistan, and Sri Lanka are also being explored. Greater CBET cooperation is thus seen as a long-term solution to energy deficits. Infrastructure is required for the region's countries to expand their electricity trade, as are regulations defining terms and conditions of electricity exchanges. Enhanced CBET among BIMSTEC countries will ensure uninterrupted power supply in the region, as well as reduce electricity prices. It will also help achieve India's Intended Nationally Determined Contribution (INDC) targets, as well as the SDG-7 (affordable and clean energy) and SDG-13 (climate action) sustainable energy targets, showing that it fulfills social needs of energy security, economic needs of increased GDP, environmental needs of reduced pollution, and policy needs of geopolitical stability.

3 Literature Review

An Asian Development Bank (ADB) study (ADB 2015) of South Asian countries, specifically Afghanistan, Bangladesh, Bhutan, India, Nepal, Pakistan, and Sri Lanka, showed that Nepal and Bhutan have large potential to develop energy transmission infrastructure for transferring their hydroelectric surpluses to India, resulting in significant improvements in fossil fuel use, power shortages, and carbon dioxide (CO_2) emissions in the region. The study concluded that India could be a central hub driving power trading in the region, owing to its geographical location, the size of its power system, and the amount of power it consumes. An evolving electricity market in India combined with a shift toward RE generation could increase CBET opportunities in the region. RE growth in India and Sri Lanka also offers grid interconnection benefits between these countries (NREL 2022).

Hurlbut (2019) examined the economic factors affecting CBET between Nepal and India, finding that the future of CBET between them will depend on the cost of alternate power sources in each country. India and Nepal both use energy banking as an interim CBET measure, wherein excess hydropower in Nepal during the monsoon season is exported to India, with the volume credited against future flows from India to Nepal during the dry season when hydropower production in Nepal falls. In July 2019, Nepal provided as much as 200 MW to India in accordance with this agreement (NREL 2019).

Nepal and Musibau (2021) examined long-term impact among energy security, renewable energy, and non-renewable energy on ASEAN member states' economic growth during 1980–2018. Their results confirm a feedback relationship between renewable energy and economic growth in ASEAN, suggesting that ASEAN governments must prioritize renewable energy funding and investment.

Pu et al. (2021) show that while the scale of the electricity trade network is expanding, many economies still have yet to participate. CBET may help reduce CO_2 emissions, achieve renewable energy transformation, and reduce power supply and demand mismatch, by improving coordination among Asian countries. Issues to be addressed regarding increased CBET within South Asian economies are large gaps in GDP, electricity prices, industrial structure, geographical distance, and institutional support.

Do and Burke (2022) review progress toward establishing an ASEAN Power Grid and the key barriers to multilateral cross-border electricity trade in ASEAN across political, technical, institutional, economic, environmental, social, and time dimensions. Using a policy sequencing framework, they conclude that it is premature for ASEAN to pursue a strong form of power sector market integration, due to the sizeable barriers that currently prevail, especially economic and institutional. They suggest that focusing on bilateral power purchase agreements and large-scale investments in solar and wind power between 2022 and 2030 would foster stronger foundations for ASEAN to move toward greater integration thereafter, while also being consistent with renewables adoption goals.

Timilsina (2018) quantified, under various scenarios, the importance of improving cross-border transmission interconnections and regional electricity trade to promote South Asian hydropower, and the potential for hydropower development and trade if such improvements were made. The study found that if the region could facilitate an unrestricted flow of electricity across South Asian borders, hydropower development in the region would increase 2.7 times over the next two decades.

Singh et al. (2018) stressed increased CBET by capitalizing on complementarities in electricity demand patterns, diversity in resource endowments for power generation, and market access gains. Despite increased bilateral cooperation in the region's electricity sector, broader regional cooperation and trade initiatives have lagged in the face of regional barriers and domestic inefficiencies. Their key findings are that, while greater electricity market reforms are not required for continued cross-border electricity trade, slow progress in removing domestic and regional barriers will limit the scope of the market and the benefits it can provide.

Agostini et al. (2019) proposed a framework of basic conditions for import and export of energy from available surpluses. Based on simulations, empirical analysis of regulatory proposals shows that energy transmission from Chile with its neighbouring countries is feasible clearly and transparently, lowering marginal energy costs and total cost of operation while keeping average generation cost relatively constant.

Sharma et al. (2020) studied the nexus amongst energy, economy, trade, and the environment by estimating the determinants of CO_2 emissions through empirical analysis (ARDL model) and found a causal relationship amongst energy, economy, trade, and the environment—both in the long-run and the short-run. The study suggests increasing energy efficiency by closing the gap between the end-user tariffs and the cost of supply and reduction of wasteful consumption of energy due to losses from pilferage, non-billing, and non-payment; and transmission and distribution.

4 Objective and Research Questions

The objective of this study is to assess the impact of India's CBET with South Asian and BIMSTEC countries. The research will make the following inquiries, with an emphasis on the efforts of the Government of India (GoI) in promoting CBET:

- The present status and prospects of CBET between India and other South Asian and BIMSTEC countries;
- The effect of CBET on energy security and social welfare in India and its trading partners;
- How CBET is facilitating region RE development;
- Challenges associated with CBET and possible solutions;
- Role of government in policies and market reforms promoting CBET; and
- Policy recommendations for promoting CBET.

5 Data and Methodology

A mathematical model (MATLAB) was developed consisting of simple energy balance for two scenarios, with and without CBET. Secondary storage was integrated into the model for a realistic simulation. Following Agostini (2019), we have designed a similar but much simpler approach to study the effects of CBET on energy security. The effects of RE penetration such as wind, solar, and combinational have been analyzed in the context of trading partners. The major limitation of this research is the availability of trusted data sources.

We created solar and wind distribution data to simulate any rated power plant capacity to study RE integration into the trading partners' grids. Distribution file and capacity factor were created by software simulations using System Advisor Model (SAM) for solar and Windographer for wind. Data for the simulation was obtained from an inbuilt data finder, the NREL database for solar and Windographer database for wind.

Distribution data refers to a nation's electricity consumption pattern and its electricity generation pattern for such various power sources as hydro, wind, or solar. For accurate analysis and to determine the influence of CBET in satisfying peak load power, hourly or t-15 data is required. We have instead used monthly data, however,

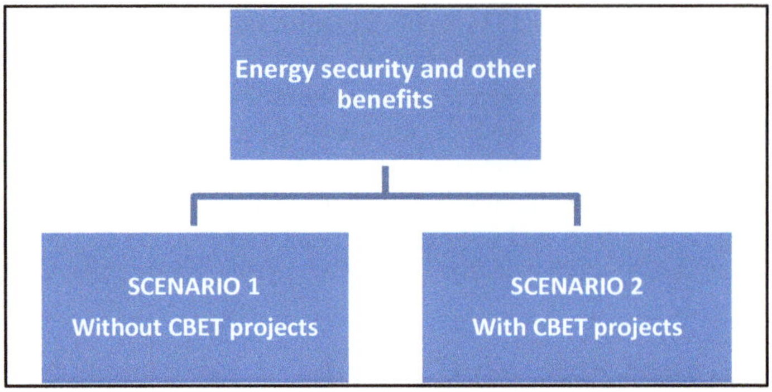

Fig. 1 Energy security scenarios

owing to a lack of t-15 data for some power sources. We developed distribution data on the assumption that consumption is directly proportional to the ambient temperature as C α T, where C is the electricity consumption of the trading nation and T is the ambient temperature.

As mentioned above, we generated two scenarios for assessing the effects of CBET between India and her BIMSTEC trading partners. Scenario 1 assessed energy security and other socio-economic benefits without CBET, while scenario 2 assessed these characteristics with CBET, as per Fig. 1.

In the first scenario, India and its trading partners are independently simulated for energy balance to assess individual country's energy deficit and storage dynamics. To balance the load, it is assumed that each country needs to expand its own generation and storage. If renewables are added to the mix, curtailment of these power sources to balance the grid could be unavoidable. For countries dependent on a single power source, e.g., Bhutan on hydropower, drought seasons could cause serious power deficits, which, in turn, undermines energy security that could be addressed by adding renewables to the mix.

The second scenario, with CBET, is anticipated to solve the problems of the first scenario. As the grids are interconnected, storage and generation capacity and curtailment periods for renewables could be reduced. Trading partners dependent on single sources of energy supply may not face serious deficit as they can import more energy. The results of this simulation could show how CBET affects energy security in a measurable way, helps all trading partners achieve clean energy targets, minimizes curtailment periods, and promotes RE penetration. The model is as follows:

1. Total annual energy generation of any power source is defined by Eq. (1):

$$E = c.f. * P * 8760 \tag{1}$$

where, E is annual energy generation, c.f. is the capacity factor of the power generation source, P is the rated power of the power generation source, and 8760 is the interval for hourly data—365 days × 24 h.

2. Energy surplus or deficit is defined by Eq. (2)

$$\Delta E = G - C \tag{2}$$

where ΔE is the energy difference, G is total power generation, and C is total consumption.

3. Storage dynamics could be simplified as in Eq. (3). The energy surplus or deficit is either stored or obtained from storage:

$$\Delta S = \int \Delta E \tag{3}$$

where, ΔS is the energy storage dynamics and ΔE is the energy difference.

CBET occurs when the following conditions are satisfied:

(i) Nation A has an energy surplus during any time interval, while nation B has an energy deficit; and

$$\Delta EA > \Delta EB$$

$$\Delta EA > 0$$

$$\Delta EB < 0$$

(ii) Nation B's storage system is unable to meet the deficit. In such a scenario trade occurs from country A to country B, with surplus energy in storage the rest of the time.

The model simulates two conditions. The first is where storage is not used and CBET takes place, likely result in reduced storage capacity. The second is when solar and wind are introduced and affect CBET. Storage capacity is assumed to be half-full at the start of the year. The simulation interval is between 01-01-2021 and 31-01-2022.

6 Results and Discussion

6.1 Cross Border Electricity Trade Without Storage

The simulation was carried out for different countries as detailed below.

India–Nepal

Figure 2 represents India's energy surplus or deficit curve, while Fig. 3 represents the same for Nepal. This energy excess or deficit is balanced to reduce stress which may cause grid failure, usually by such direct strategies as storage and such indirect strategies as demand side management, requiring strategic and costly investment and policy adjustments. Unexpected increases in consumption or generation remain a problem, however. CBET is a promising solution to such issues, in that both parties involved benefit in terms of energy security, reduced storage cost, and promoting renewables penetration in the grid by reducing storage requirements.

These figures show summer season is when India experiences the majority of its surplus, which is when Nepal experiences the majority its deficit, a condition presumably related to average monthly precipitation for Nepal, which relies heavily on hydropower. In such a scenario, if India and Nepal did not trade, a pumped-storage hydroelectricity (PSHE) system would be necessary, driving increased investment.

In a scenario in which the two nations conduct mutual energy trade, Nepal has the option to sell electricity to India in periods of surplus and import energy from India during periods of deficit, per Fig. 4. In this scenario, India has ported 872.70 GWh of energy to Nepal during the summer months.

This has reduced stress on both grids, increasing the energy security of both nations. After energy trading began, India utilized energy from Nepal more than from storage, and Nepal has commensurately reduced its energy surplus significantly.

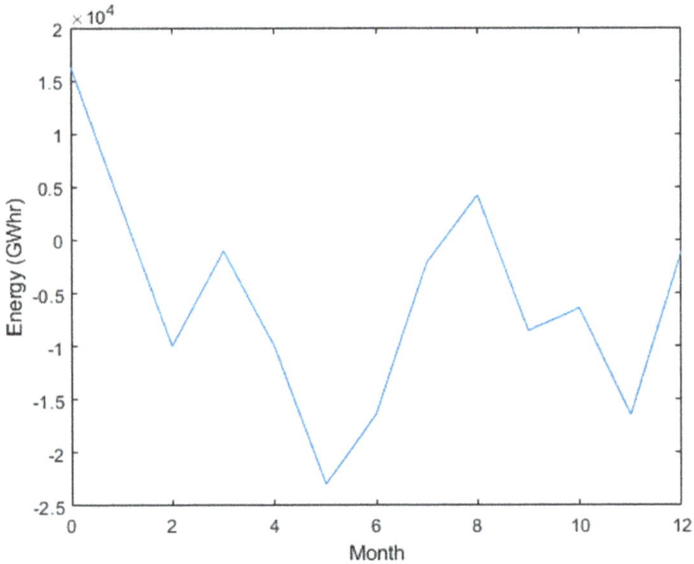

Fig. 2 Energy surplus/deficit for India

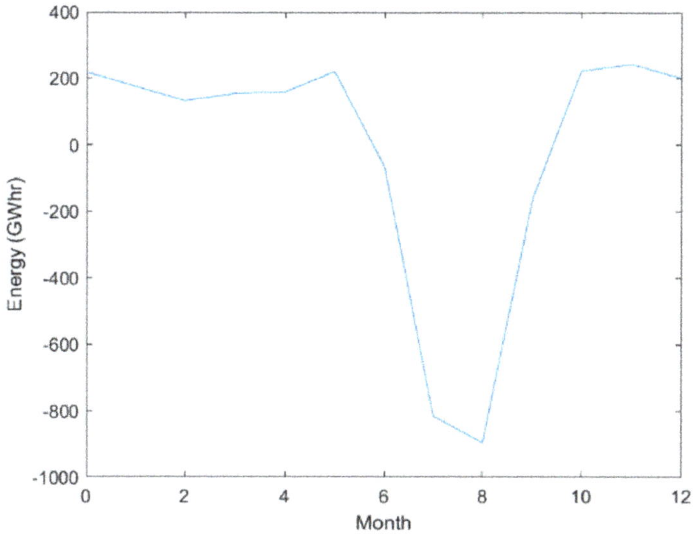

Fig. 3 Energy surplus/deficit for Nepal

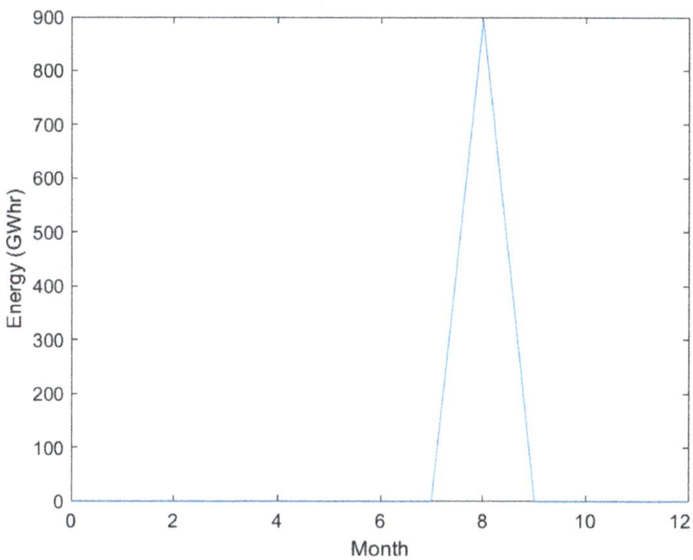

Fig. 4 India's energy exports to Nepal

India–Bhutan

Figure 5 represents Bhutan's grid balance, generation versus consumption. We see that Bhutan has predominantly excessive generation, albeit during periods of heavy rainfall. Bhutan could export 467 GWh of energy to India monthly during this period, thereby reducing its storage requirements as seen in Fig. 7 while Fig. 6 shows that India has excessive generation at the start of the year, during Butan's deficit period, and was able to export 211 GWh, again reducing grid load for both countries.

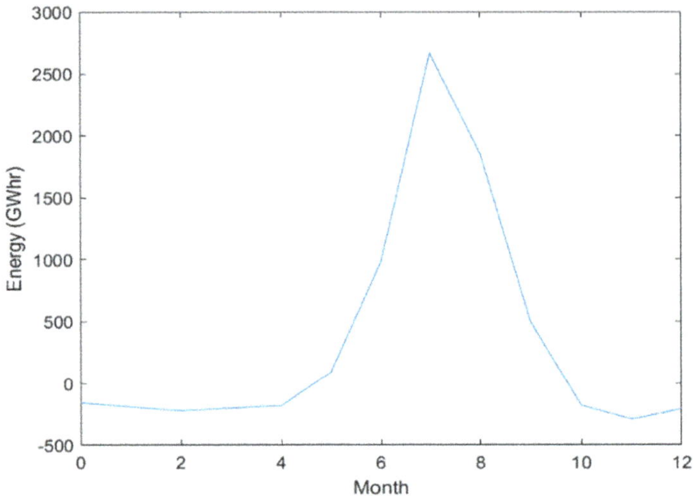

Fig. 5 Bhutan's energy surplus/deficit

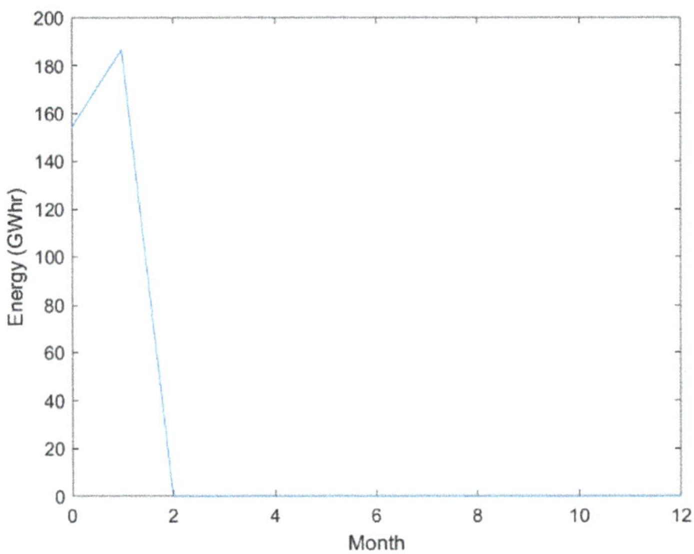

Fig. 6 Energy exported by India

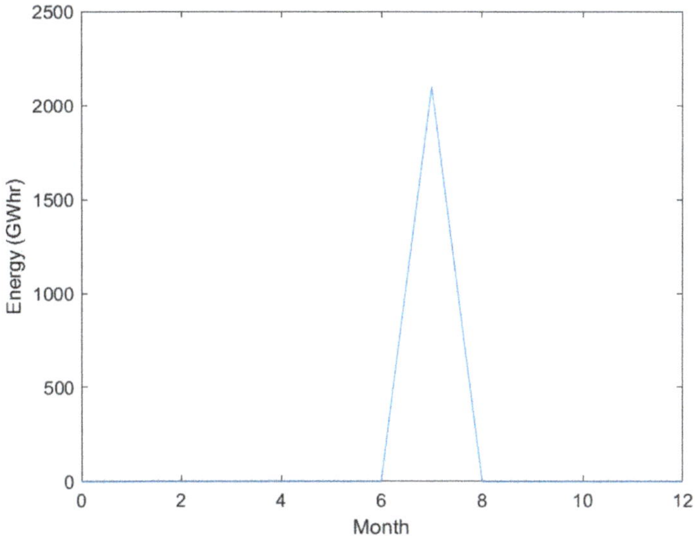

Fig. 7 Energy imported by India from Bhutan

India–Bangladesh

Figure 8 shows that Bangladesh has a grid balance similar to India, with surplus energy at the start of the year followed by months of deficit. After trade began, not much difference was observed except in August, when India was able to transfer surplus energy to Bangladesh.

6.2 RE Adoption Effect

India–Nepal

Wind: Fig. 9 shows a simulated grid with 100 MW of wind turbine power in Nepal, leading to energy exports to India rising from 871.1 to 842 GWh.

Solar: Fig. 10 shows that 100 MW solar penetration causes energy exports from India to rise from 871.7GWh to 861GWh.

Wind + Solar: Fig. 11 shows that 100 MW each of wind and solar reduces export requirements from 871.7 to 831 GWh throughout the year.

India–Bhutan

Wind: Fig. 12 shows that wind penetration in Bhutan of 100 MW increases its export capacity to 1.39 TWh with no import requirements.

Solar: Figs. 13 and 14 show that Solar penetration of 100 MW increases India's energy exports from 419 to 467 GWh and reduces Bhutan's energy exports from 211

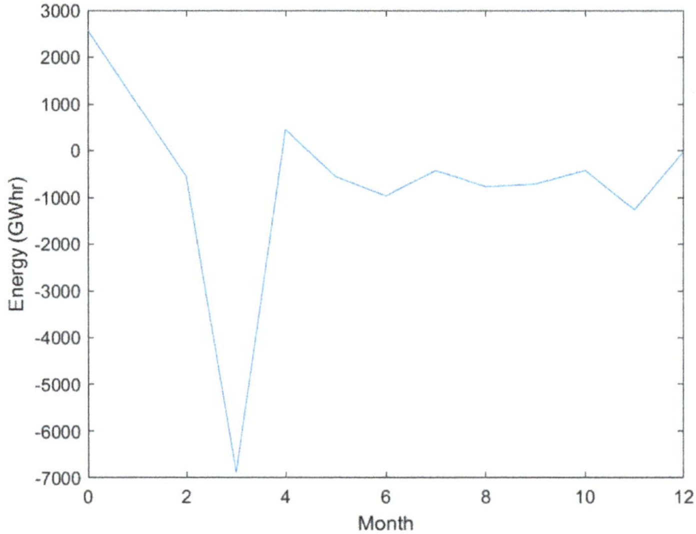

Fig. 8 Bangladesh grid balance

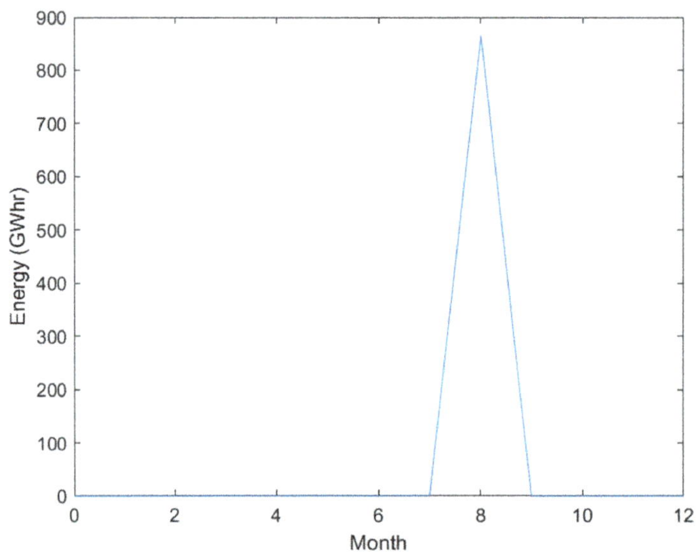

Fig. 9 Energy exported to India

to 198.2 GWh. The significant difference is in surplus energy available for export, which is economically feasible as it avoids costly storage which, as mentioned above, is needed in the absence of CBET, in turn, facilitate solar penetration in Bhutan,

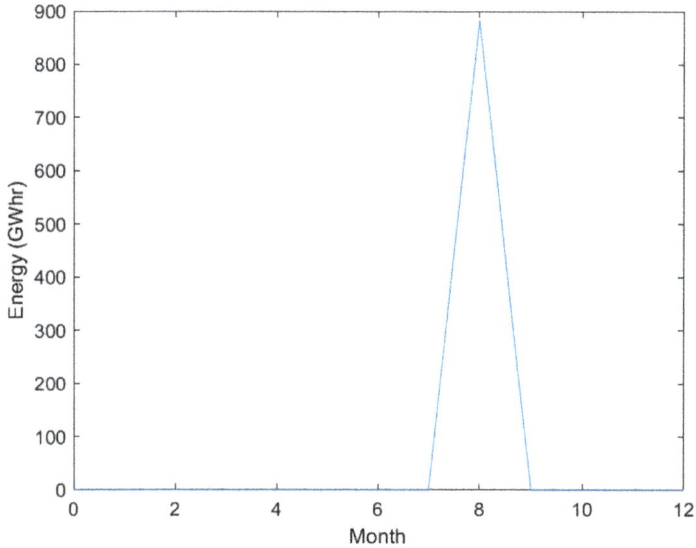

Fig. 10 Energy exported from India

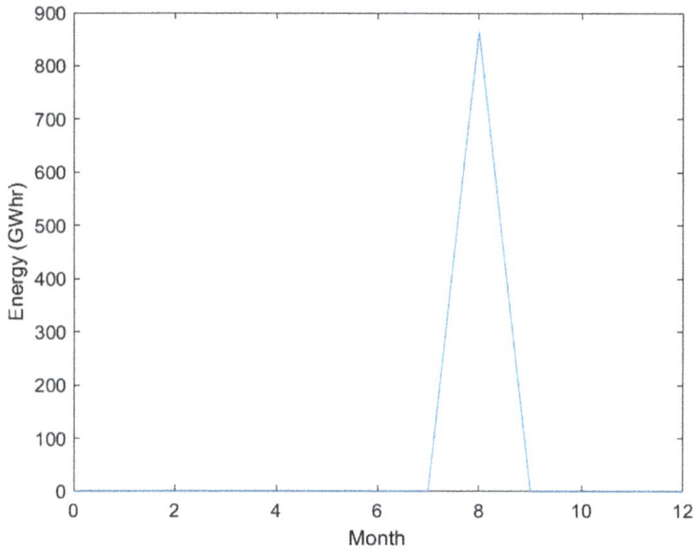

Fig. 11 Nepal's Import requirements

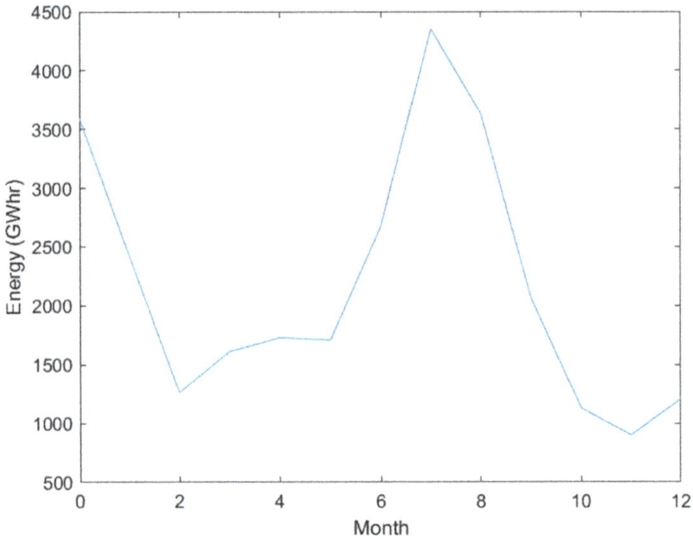

Fig. 12 Energy imported from India

especially with CBET. The sharp fall to zero indicates energy traded to India, which can only be stored in June and August satisfying demand during deficit periods.

Wind + Solar: The combination of wind and solar increase exports to India to 1.228 TWh and negate any need for imports, particularly benefiting India. Bhutan

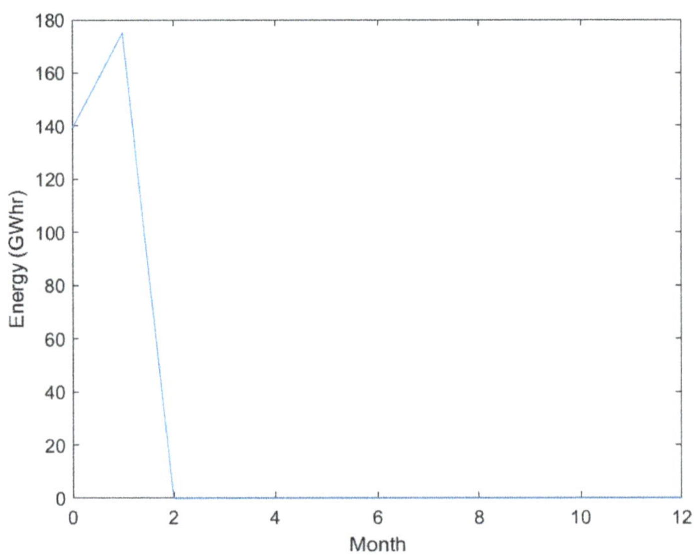

Fig. 13 Energy exported by India

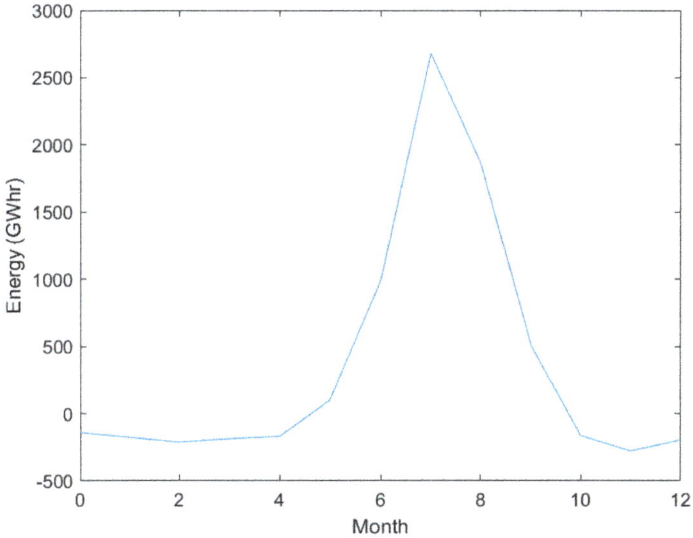

Fig. 14 Energy imported to India

has some periods of deficit, however, coinciding with Indian surpluses. This could be explained by similar climatic conditions. CBET could thus have greater benefits for trade among nations with complementary climatic conditions.

India–Bangladesh

Wind: Fig. 15 shows that the addition of 100 MW wind system decreased energy imports from India from 604 to 561.7 GWh.

Solar: Penetration of solar has decreased import requirements to 558.8 GWh. Figure 16 shows Bangladesh's post-CBET grid balance. The difference in energy trade could not be shown as relative trade is too low to be noticeable.

Wind + Solar: Wind and solar introduction of 100 MW each into the Bangladesh grid decreases import requirements to 561.7 GWh.

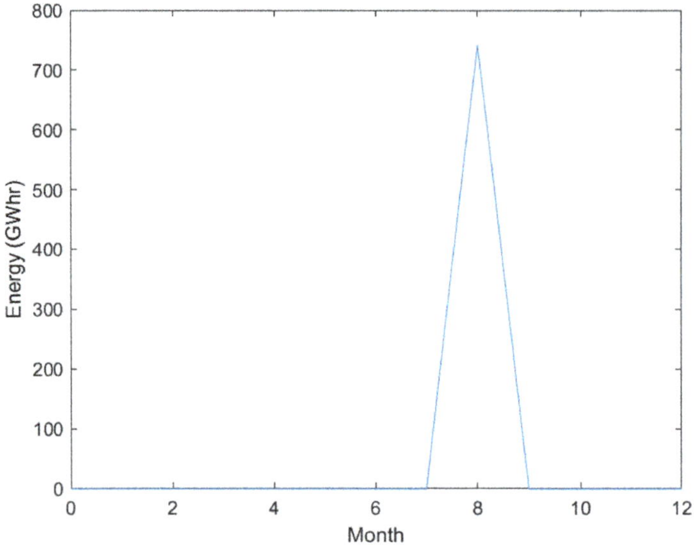

Fig. 15 Energy exported from India

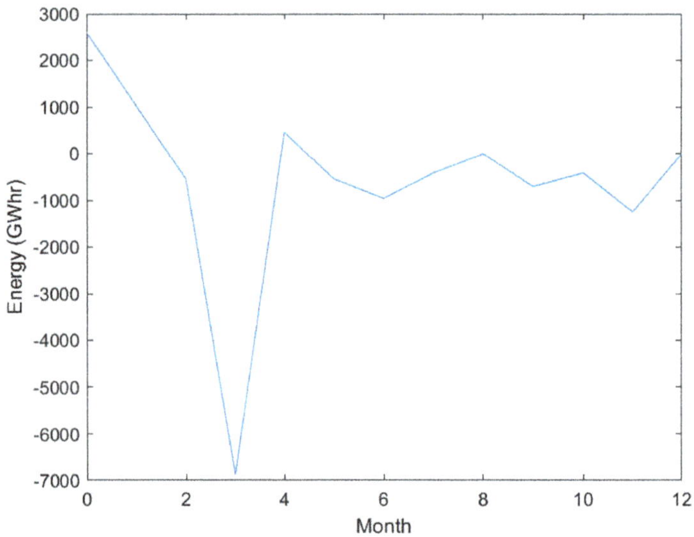

Fig. 16 Bangladesh grid balance post-CBET

7 Conclusion

In this study, we have simulated CBET in South Asia, especially India, Bhutan, Bangladesh and Nepal. Initially, the simulation was done without any storage in the energy network to determine the effects of CBET, with especially positive results in terms of grid balance. Further analysis is done with solar, wind, and combined solar and wind penetration in the grids of India's trading partners, which drive decreased imports and increased exports of energy. We observed the following outcomes:

(i) When trading nations' generation patterns do not coincide with consumption patterns and deficit periods, heavy grid stress results; and
(ii) When trading nations' generation patterns do not coincide with consumption patterns but do coincide with deficit, opportunities arise for trading non-renewable and renewable energy.

These results show that CBET has significant impact in grid balancing, thereby decreasing large-scale storage requirements. While energy security increases significantly due to renewables adoption, which is evident from consequent reductions in import requirements and increases in energy exports, if, as mentioned above, trading nation's generation patterns do not coincide with consumption patterns or deficit patterns, excessive load stresses the grid, leading to periods of curtailment or increased storage requirements.

Lastly, as also mentioned above, CBET is observed to be especially beneficial among nations with complementary climatic conditions.

8 Policy Implications

The study suggests the following policy recommendations:

- To boosting multilateral CBET, South Asian and BIMSTEC governments may need to reform their policies on energy generation, transmission, and utilization;
- A mutually agreed legal framework, which could be established in consultation with all energy trade partners and may follow other successful frameworks, such as in Europe or Africa, is necessary; and
- Participation and cooperation of countries in multilateral CBET may have a positive impact on economic development, social welfare, and environmental outcomes for all trading partners.

References

ADB (2015) Cross-border power trading in South Asia: a techno economic rationale. In: ADB South Asia working paper series, No. 38. Available via DIALOG. https://www.adb.org/sites/default/files/publication/173198/south-asia-wp-038.pdf. Retrieved 6 May 2022

Agostini CA, Guzmán AM, Shahriyar N, Silva C (2019) A surplus based framework for cross-border electricity trade in South America. Energy Policy 128:673–684. ISSN 0301-215, https://doi.org/10.1016/j.enpol.2019.01.053

BMPI (2022) Challenges of cross border power trade with neighboring countries and interest of Bangladesh. In: Bangladesh Power Management Institute (BPMI) Report, Available via DIALOG. https://bpmi.portal.gov.bd/sites/default/files/files/bpmi.portal.gov.bd/page/bc0f0c3c_c487_463b_8873_6effb42cc584/2020-09-21-23-45-0efdcfa44e9dbb06e97f0a5fd45b9b7c.pdf. Retrieved 10 May 2022

Do TN, Burke PJ (2022) Is ASEAN ready to move to multilateral cross-border electricity trade? J Asia Pacific Viewpoint. https://doi.org/10.1111/apv.12343

Hurlbut DJ (2019) Cross-border energy trade between, Nepal and India: trends in supply and demand. In: National renewable energy laboratory, technical report, NREL/TP-6A20-72345, April 2019. Available via DIALOG. https://www.nrel.gov/docs/fy19osti/72345.pdf. Retrieved 12 April 2022

Mint (2022) Cross border electricity trade through power exchange begins. In: The Mint epaper, 19 April 2022. Available via DIALOG. https://www.livemint.com/industry/energy/cross-border-electricity-trade-through-power-exchange-begins-11618827281432.html. Retrieved 1 May 2022

MoP (2022) Interconnection with neighbouring countries. In: Ministry of power, India website. Available via DIALOG. https://powermin.gov.in/en/content/interconnection-neighbouring-countries. Retrieved 1 May 2022

Nepal R, Musibau H (2021) Energy security, renewable, non-renewable energy and economic growth in Asean Economies. J New Insights, The Singapore Econ Rev 66(02):457–488. https://doi.org/10.1142/S0217590819430045

NREL (2019) Cross-border power trade to support resilience. In: National renewable energy laboratory report, US Department of Energy. Available via DIALOG. https://www.nrel.gov/docs/fy20osti/75153.pdf. Retrieved 12 May 2022

NREL (2022) South Asia cross-border electricity trade and cooperation study. In: National renewable energy laboratory, US Department of Energy, Available via DIALOG. https://www.nrel.gov/international/south-asia-cross-border-electricity-trade-and-cooperation-study.html. Retrieved 6 May 2022

ORF (2020) Manifestation of SDG goals and scope for cross-border energy cooperation in BIMSTEC. Available via DIALOG. https://www.orfonline.org/expert-speak/manifestation-sdg-goals-scope-cross-border-energy-cooperation-bimstec/. Retrieved 6 May 2022

Pu Y, Li Y, Wang Y (2021) Structure characteristics and influencing factors of cross-border electricity trade: a complex network perspective. J Sustainability 13(11):5797. https://doi.org/10.3390/su13115797

Sharma SV, Le TH, Phoumin H, Sharma VK (2020) Energy, income, trade and environment nexus: dynamic data analysis for India. In: Phoumin H, Farhad T-H, Kimura F (eds) Energy sustainability and development in ASEAN and East Asia. Routledge, Abingdon

Singh A, Jamasb T, Nepal R, Toman M (2018) Electricity cooperation in South Asia: barriers to cross-border trade. Energy Policy 120:741–748. ISSN 0301-4215, https://doi.org/10.1016/j.enpol.2017.12.048

Timilsina GR (2018) How would cross-border electricity trade stimulate hydropower development in South Asia? In: Policy research working paper no. 8513. World Bank, Washington, DC. Available via DIALOG. http://hdl.handle.net/10986/29986

Chapter 11
Toward a Coherent Policy Approach to Solar Uptake in Southeast Asia: Insight from Indonesia and Vietnam

Muyi Yang, Achmed Shahram Edianto, Thi Anh Phuong Nguyen, Rabindra Nepal, and Han Phoumin

Abstract This chapter examines Indonesia and Vietnam's experiences with adopting utility-scale solar power, finding that despite landscape pressures, such as the need to address energy security concerns and policy commitments to emissions reduction, challenges in local contexts and incumbent electricity regime often hindering translating these pressures into action. It also highlights the need for a coherent policy framework that can address both the emergence and wider adoption of niche electricity technologies and reconfigure the incumbent regime. Developing such a framework requires careful planning and consideration of cross-cutting issues, however, which can take time. A key strategy to reconcile the need for rapid transitioning to address the climate crisis with the usually prolonged transition process is to focus initial efforts on promoting clean technologies that already play a significant role in the energy mix, which could reduce immediate demand for major regime change and ensure a quick start to the transition, while buying time to plan to reconfigure the incumbent regime for a clean electricity future.

M. Yang (✉) · A. S. Edianto
International Society for Energy Transition Studies (ISETS), Sydney, Australia
e-mail: muyi.yang@uts.edu.au

T. A. P. Nguyen
Tona Syntegra Solar, Ho Chi Minh, Vietnam

R. Nepal
Wollongong University, Wollongong, Australia

H. Phoumin
Economic Research Institute for ASEAN and East Asia (ERIA), Jakarta, Indonesia

H. Phoumin et al. (eds.), *Large-Scale Development of Renewables in the ASEAN*, Economics, Law, and Institutions in Asia Pacific,
https://doi.org/10.1007/978-981-99-8239-4_11

1 Introduction

Southeast Asia is one of the fastest growing and economically dynamic regions of the world. From 2010 to 2020, before the COVID-19 pandemic, it sustained strong annual per capita GDP growth of approximately 3.7%, outperforming many other parts of the world (ADB 2022). In tandem, its demand for electricity, much of which comes from fossil fuels, especially coal, surged by an annual average of almost 6% (Foong 2022), resulting in a substantial rise in carbon emissions from the electricity sector. According to ACE (2022), these emissions reached 1815 Mt $CO_{2\text{-eq}}$ in 2020, up from 1039 Mt $CO_{2\text{-eq}}$ in 2005. Southeast Asia has thus become the world's fourth largest emitter, ranking behind only China, at 10,707 Mt, the United States, at 4817 Mt) and India, at 2456 Mt (World Bank 2022a).

With electricity demand recovering as economic recovery takes hold, Southeast Asia faces a challenge in balancing its need to secure electricity supply to support economic expansion with the imperative to carry out decarbonization. At the core of this challenge is the uptake of solar power, particularly utility-scale solar power, which is considered a crucial element of a clean electricity supply, per Fig. 1. Despite its importance, solar power uptake has been negligible in most Southeast Asian countries, accounting for only some 2% of the region's electricity in 2020 (Handayani et al. 2022). Progress has recently stalled even in Vietnam, which is often cited as a solar uptake success story in the broader Southeast Asian context (Asia News Network 2022).

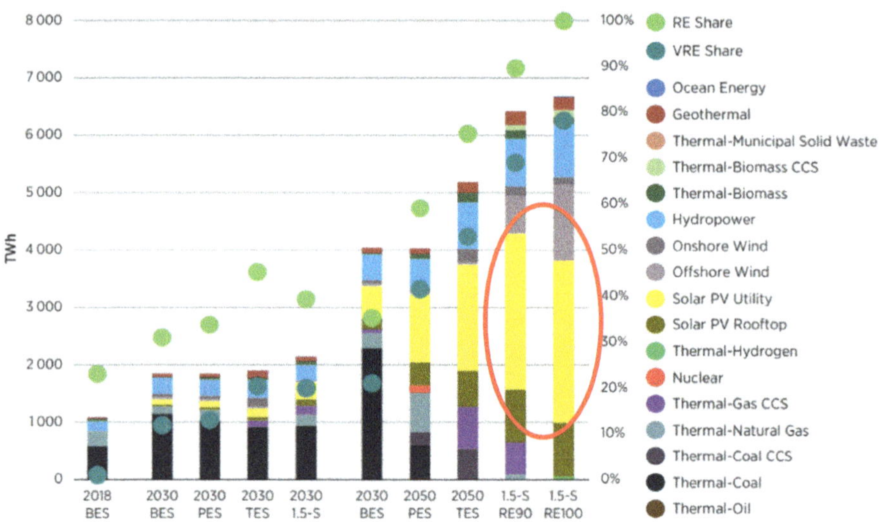

Fig. 1 Electricity generation and renewables share, 2018–2050. *BES* Baseline Energy Scenario; *PES* Planned Energy Scenario; *TES* Transforming Energy Scenario; *1.5-S* 1.5°C-aligned energy pathways for Southeast Asia; *RE90* 90% renewable generation; *RE100* 100% renewable generation. *Source* IRENA (2022a, b)

There is significant commentary on possible reasons for the slow progress of solar uptake in Southeast Asia, ranging from insufficient network infrastructure (Do et al. 2021) to complex administrative procedures (Do et al. 2020a, b), high upfront costs (Jayaraman et al. 2017; Setyawati 2020), uncertain financial support (Barroco and Herrera 2019; Koerner et al. 2022; Rababah et al. 2021), and incumbent resistance (Fathoni et al. 2021). By implication, this suggests that solar power uptake is a complex phenomenon requiring multiple inputs, such as the aforementioned network infrastructure, regulatory reform, financial support, and socio-political acceptance. A coherent policy framework, able to drive these factors in a concerted manner, is therefore essential for solar uptake progress. Despite some credible initiatives and programs introduced in the past few years, however, Southeast Asian countries have largely failed to develop such a framework, which also helps explain the region's slow solar uptake progress.

In this context, the first essential step to rectify the slow progress of utility-scale solar power uptake is gaining insight into how coherent and effective policy frameworks can be developed to drive necessary changes to support utility-scale solar power expansion. The need for such insight is heightened by the region's growing net-zero energy commitments and the role of utility-scale solar power in achieving these commitments. This chapter attempts to fulfil this need by analyzing Indonesia and Vietnam's experiences with solar uptake. Both countries have extensive solar resources and have introduced various policy measures in recent years to support the solar uptake, with varying degrees of success. An analysis of the dynamics and outcomes of solar uptake in these countries would indeed provide valuable insight for other Southeast Asian countries seeking to improve their solar uptake promotion policies. This analysis focuses on utility-scale solar power, in contrast to many other studies analyzing rooftop solar PV in the region (Fathoni et al. 2021; Jayaraman et al. 2017; Potisat et al. 2017; Rababah et al. 2021; Setyawati 2020; Tongsopit et al. 2016).

This chapter is organized as follows: Sect. 2 provides an overview of Indonesia and Vietnam's electricity sectors for context, emphasizing key initiatives and programs that have been implemented in these countries to support the renewable energy expansion, including utility-scale solar power. Section 3 outlines the interview-based approach adopted in this chapter to identify key issues affecting utility-scale solar power uptake in Indonesia and Vietnam. Section 4 presents empirical results of these interviews, which are discussed in Sect. 5 to draw some general insight into factors affecting utility-scale solar power deployment. Section 6 draws key conclusions, including messages for Southeast Asian policymakers as they endeavor to attain net-zero by mid-century.

2 Context

This section provides a brief introduction to the electricity sectors of Indonesia and Vietnam, as well as their recent initiatives and programs to support renewable generation, including utility-scale solar power. The information presented here aims to enhance the reader's understanding of the nuances of the arguments presented in this chapter.

2.1 Indonesia

Indonesia nationalized all electricity assets in the 1950s into Perusahaan Listrik Negara (PLN) (McCawley, 1971). It initiated market reform of its electricity sector in the early 1990s, emphasizing a greater role for the private sector in the generation business in the form of Independent Power Producers (IPP). Notwithstanding, PLN continues to dominate the sector, which acts as a single buyer, purchasing electricity from IPPs under long-term contracts. It also owns and operates approximately 70% of the country's generating capacity and maintains an effective monopoly over network businesses (PLN 2022).

Coal-fired power has been the mainstay of Indonesia's energy mix. As shown in Fig. 2, its share has increased in recent years from some 45% in 2011 to more than 60% in 2021. The renewables share has also been rising since the mid-2010s, especially after the release of the National Energy Policy that stipulated a structural shift in primary energy mix to at least 23% of renewable energy by 2025 (IEA 2015). To implement this policy, the draft National Electricity Plan (RUKN) 2015–2034 included a target of 25% renewable energy by 2025, necessitating a more than fivefold increase in renewable capacity to 45 GW from 8.7 GW in 2015 (IRENA 2017).

To support the expansion of renewable capacity, the Ministry of Energy and Mineral Resources (MEMR), Indonesia's main energy sector governing body, introduced several regulatory changes between 2014 and 2016, to promote uptake of small-scale renewable projects by providing technology-specific feed-in-tariff (FiT) schemes for project developers. Large-scale solar projects were, however, not included (Kennedy 2018). The Indonesian government also launched other initiatives during this period to complement the FiT schemes, including establishing a task force for renewable energy development, creating the Centre of Excellence on Clean Energy, and initiating the Bright Indonesia program (Maulidia et al. 2019).

Following changes in MEMR leadership in 2016, renewable energy policy support shifted toward reducing renewable project costs. In 2017, MEMR Decree No. 12 established a new FiT scheme for all renewable projects. This scheme, later revised by MEMR Decree No. 50, included a cap on tariffs for renewable projects of 85% of the average cost of generation for the local grid, or 100% if said average cost was lower than the national average (IRENA 2017). This new scheme was considered

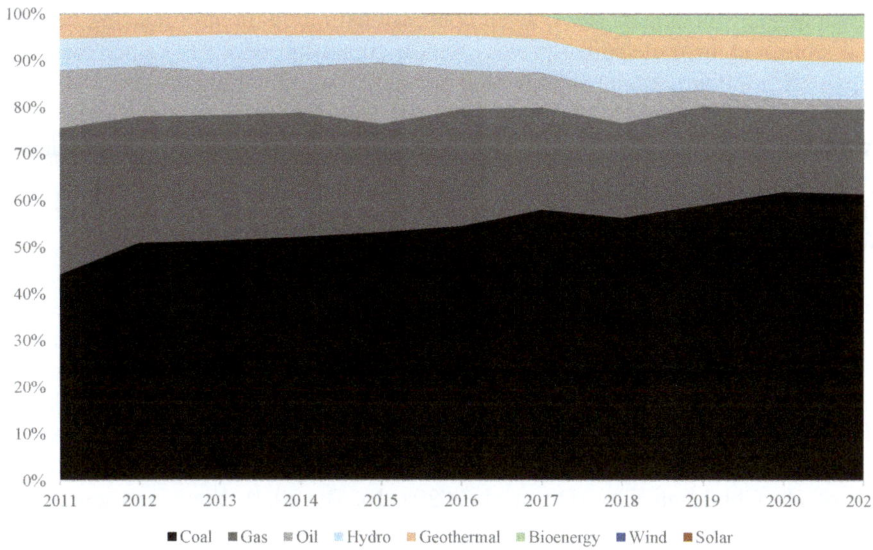

Fig. 2 Energy mix in Indonesia, 2011–2021. *Source* MEMR (2022)

unattractive to private investors, particularly for large-scale solar projects, which are more expensive than other renewable projects in Indonesia. As a result, compared to other renewable energy sources, solar power has seen the smallest expansion between 2015 and 2021, with only 132 MW added. By contrast, wind power saw an increases of 153 MW, bioenergy 178 MW, geothermal 839 MW, and hydro 1280 MW (IRENA 2022a).

Renewable energy expansion has gained momentum in recent years. In September 2022, Presidential Regulation No. 112 (PR 112/2022) was enacted to promote renewable investment and to expedite phasing out coal-fired power plants. Later that year, the government launched the Just Energy Transition Partnership (JETP), with Indonesia committing to peak carbon emissions from electricity generation by 2030. The agreement emphasizes generating at least 34% of electricity from renewable energy sources by 2030 (The White House 2022). A new energy and renewable energy bill (*Rancangan Undang Undang tentang Energi Baru dan Terbarukan*) is currently under discussion in Indonesia to support the renewable energy expansion. How to translate this rising momentum into concrete action and progress is therefore a priority.

2.2 Vietnam

In 1986, Vietnam initiated the Đổi Mới reform, with the objective of transitioning the country's economy from centrally planned to market-oriented. The transition gained further impetus in 1993, when concessional international financing became accessible, and the trade embargo was lifted. Vietnam subsequently gained membership in several international and regional organizations, including the Association of Southeast Asian Nations (ASEAN) in 1995, the Asia–Pacific Economic Cooperation (APEC) in 1998, and the World Trade Organization (WTO) in 2007. These developments have significantly affected the country's economic growth, lifting it from one of the world's poorest nations to a middle-income economy in one generation (World Bank 2022b).

Vietnam's impressive economic growth has led to a significant surge in electricity demand, making supply capacity expansion a priority. Between 2010 and 2020, electricity demand grew at an average annual rate of 15%, primarily driven by an industrial boom (IEA 2022). Supply capacity did not keep up with demand, however, causing widespread security concerns. It was reported that the Ministry of Industry and Trade (MOIT), the governmental body responsible for managing Vietnam's energy sector, expected power shortages to occur as early as 2020, especially in the manufacturing hub of Ho Chi Minh City (Do et al. 2020a, b). By 2030, Vietnam's generation capacity shortfall is projected to exceed 10GW (International Trade Administration 2022), roughly equivalent to 13% of the country's total installed capacity in 2021.

To address these concerns, Vietnam embarked on an ambitious plan to expand its coal-fired power capacity in the mid-2010s, as outlined in the National Power Development Plan (PDP) 7 (Gallagher et al. 2021), making coal the country's primary source of electricity in 2016, surpassing hydropower. Vietnam ratified the Paris Agreement the same year, signaling the Vietnamese government's intention to prioritize addressing climate change. Since then, policy machinery has been attuned to promoting renewable energy as the means of satisfying the country's rising appetite for electricity, with the government adopting several initiatives and programs to support renewable energy uptake, including utility-scale solar power. In 2017, the Vietnamese government introduced highly favorable FiTs, where utility-scale solar plants commissioned before June 30, 2019, would be eligible for a 20-year preferential FiT, selling electricity to the grid at US$93.5/MWh. In April 2020, said FiTs were reduced to between US$70.9 and $83.8 per MWh (Do et al. 2020a, b). Notwithstanding, investors still had ample profit potential, especially considering that the levelized cost of energy (LCOE) for solar PV in Vietnam was in the range US$66 to US$76 per MWh between 2019 and 2020 and is expected to fall further as technology advances (Do et al. 2021). The government has also offered a range of incentives to solar project developers, including tax breaks and equipment import tariff exemptions.

Such policy support has driven a solar boom. Between 2017 and 2021, Vietnam's solar generation rose from practically nothing to nearly 26 TWh, accounting for approximately 11% of total electricity, per Fig. 3, and making Vietnam the world's tenth-largest solar power producer. Its expansion of renewable generation is likely to accelerate in coming years, as it endeavors to further wean itself off fossil fuel-based electricity. As part of the JETP, it has committed to peak carbon electricity emission by 2030, as well as peak coal of 30.2 GW, down from 37 GW contemplated in the current (PDP), and at least 47% electricity from renewable sources by 2030.

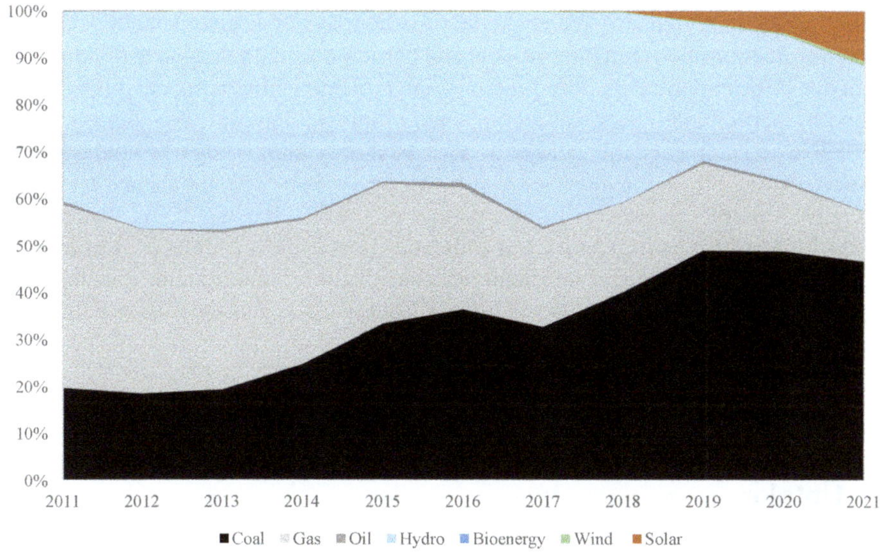

Fig. 3 Energy mix in Vietnam, 2011–2021. *Source* IEA (2022)

The role of utility-scale solar power in Vietnam's future expansion of renewable generation will be negotiated in the context of growing grid constraints to integrated variable solar generation, however. More than 16 GW solar capacity came online between 2018 and 2020 alone, overwhelming grids and forcing EVN, the national utility, to curtail solar generation in order to maintain system reliability and stability. In 2021, plans called for approximately 500 GWh of solar power to be curtailed in Vietnam (Sang 2021), with the curtailment rate at the country's largest solar power plant in Thuan Nam, with a capacity of 450 MW, in the range of 40% (Vu 2022).

3 Interview-Based Methodology

To gain a more nuanced understanding of solar uptake and factors influencing same in Indonesia and Vietnam, we conducted semi-structured interviews with relevant experts between September and December 2022. Sixteen experts, selected for their deep involvement in solar power development in Indonesia and Vietnam, participated online in sessions lasting between 30 min and an hour. They included specialists in Indonesia and Vietnam's electricity sectors, domain experts from leading local and regional think tanks, senior managers from local solar companies and industry associations, and former officials with first-hand knowledge of their country's power sectors.

The interviews were exploratory, with participants asked to share their opinions and viewpoints on issues influencing utility-scale solar power development in Indonesia and Vietnam. We chose this methodology because it facilitated flexible conversations between interviewers and participants, allowing us to obtain more in-depth information. This posed the challenge of managing subjective bias introduced by participants' personal feelings and opinions, however. We addressed this challenge and enhanced the validity of our analysis by comparing and linking information obtained from the interviews with secondary data collected from a review of publicly available documents from a variety of sources (Huang 2021; Wang et al. 2021). To maintain data accuracy and authenticity, we gave higher priority to sources from national governments and regional organizations, multilateral development agencies including the Asian Development Bank (ADB) and the World Bank, and peer-reviewed journals.

4 Empirical Results: Issues Affecting Utility-Scale Solar Uptake

4.1 Indonesia

Most participants agreed that lack of competitiveness is the main factor impeding solar uptake in Indonesia. One interviewee from the power sector noted that "in 2013, the Ministry of Energy and Mineral Resources (MEMR) conducted a bidding process that procured several solar projects in Kupang, Gorontalo, and Sumba. The contracted prices for these projects were very high. Later, PLN directly procured two solar PV projects in Lombok and Likupang. These two projects have lower costs than the earlier ones but are still more expensive than coal power."

As the former MEMR Minister mentioned in 2016, "the government supports (the change in) energy fuel mix in a bid to address climate change issues. However, the price must be affordable" (Kennedy 2018). As previously mentioned, MEMR Decree No. 12 was issued in 2017, introducing a new FiT program for all renewable projects, later revised by MEMR Decree No. 50. A key aspect is that it caps prices paid to renewable generators based on PLN's average costs of electricity provision (i.e., Biaya Pokok Penyediaan, BPP) rather than on generators' cost of production (IRENA 2017). As one interviewee noted, this tariff restriction is a major obstacle to renewable energy development, including solar, because "it makes renewable energy unattractive in regions that heavily reply on cheap coal for power generation. A clear example of this is that only one solar project has passed PLN's bidding process from 2017 until more recent times." It is, however, anticipated that the Presidential Regulation (PR 112/2022) enacted in 2022 will create more space for renewable energy uptake by, for example, replacing the BPP with a ceiling price-based scheme, as well as streamlining the renewable project procurement process.

We asked participants why solar power is as expensive as it is in Indonesia. Several interviewees noted that a significant contributing factor is the local content requirement, which mandates that project developers must procure certain amounts of materials and services used in the project from local sources, increasing renewable project costs. This is particularly challenging in the context of solar PV, where Indonesia's small domestic manufacturing base means that solar panels produced locally are often of lower quality and significantly more expensive than those available in international markets (IESR 2023). One interviewee noted that "the local content requirement was introduced by the Ministry of Industry with good intention of reducing import dependency. But the Ministry of Industry did not fully understand the solar development issues; therefore, this regulation also affects the progress of solar deployment in the country."

Two interviewees from private utility companies cited land procurement as another factor contributing to the high cost of solar power in Indonesia. As one explained, "Large-scale solar projects require significant amounts of land. At least one hectare of land is needed for every one MW solar capacity. There is currently no incentive to support land acquisition." The other argued that "if incentives to address these issues are not available, the acceleration of solar uptake in Indonesia will be challenging."

Interviewees also mentioned the procurement process as another key factor, with one noting that "in some cases, addressing social and environmental considerations causes significant delays in the permitting process and makes solar investment less attractive." Another issue associated with the procurement process, as some interviewees mentioned, is that "only companies included in the so-called selected supplier list can join the auction process led by PLN and this potentially reduces the scope for private participation." One interviewee, however, pointed out that "this is because PLN learned from its experience that there were some companies that act as intermediaries and obtained PLN's quote for a solar project and sold it to another company, which sometimes did not have sufficient capacity for project development."

According to one interviewee from the power sector, another important obstacle to solar uptake in Indonesia is unfavorable power purchase agreements (PPA) with investors forced to take excessive risk. This interviewee explained that "the PPAs sometime allow PLN to not take the electricity from solar project for any days in a month without the need to provide a reason. PLN can default for maximum of two days a month." Another issue with the PPAs, according to this interviewee, is that "only PLN can claim carbon credit from the solar projects. This affected a recent solar bidding. About 120 companies showed their interest in joining the bidding, but only four joined in the end. In another bidding to acquire solar projects to replace diesel power plants, over 100 companies showed interest, but only three companies in Java and one company in Kalimantan finally joined the bidding process."

Some interviewees cited excessive supply capacity as another obstacle to solar uptake in Indonesia. This is understandable if one notes that in 2015, the Jokowi administration introduced the 35 GW program to increase supply capacity, offering private investors long-term PPAs with take-or-pay and guaranteed rate-of-return clauses (Hamdi 2021). The program and less-than-expected demand growth resulted, however, in PLN, having excess supply capacity supported by expensive take-or-pay PPAs, imposing payment obligations PLN whether it needs electricity or not. This also explains, as one interviewee noted, why PLN has opposed a new regulation introduced in 2021 to give incentives to investment in commercial and industrial rooftop solar PV. According to this interviewee, this regulation enables solar owners to sell all excessive electricity generated to the grid, which could cause PLN significant financial losses given the existing market glut. The amount of surplus electricity that could be sold was later reduced to a mere 15%. Another interviewee added that this is because "PLN needs to gain a profit as a state-owned company, as required by the Ministry of State-Owned Enterprises."

4.2 Vietnam

As several participants indicated, the abovementioned concerns about electricity supply security were the main driver behind the Vietnamese government's support for solar power. One explained that "electricity is widely considered an important ingredient for economic growth, better living standards, and industrialization in Vietnam, and supply shortfall is therefore often perceived as a threat to the country's socio-economic progress." According to some interviewees, solar power was an attractive option for addressing Vietnam's looming power shortage, mainly due to it being inexpensive. One interviewee explicitly mentioned that "cost is not a problem...-solar power has already proven that it is cost comparable to coal and gas." Another interviewee added that "the recent surge in gas prices further enhance the cost competitiveness of solar power." The International Renewable Energy Agency (IRENA) conducted a study that supports this viewpoint, finding that solar PV projects in Vietnam have among the lowest average investment costs in the Southeast Asian region, at approximately US$690/kW, versus as much as US$2000/kW elsewhere in the area (IRENA 2022a). Another study found that the levelized costs for solar PV are also low in Vietnam, at approximately US$64/MWh, compared to US$80/MWh in Thailand and over US$200/MWh in Indonesia (Lee et al. 2020).

It came out during the interviews that short construction times are another contributing factor to the attractiveness of solar power as a quick fix to perceived supply shortfalls. One interviewee noted that "it takes roughly nine months or even less to complete a solar project in Vietnam...this is quick when compared to coal and hydro power plants." External influences are another important factor in solar's attractiveness in Vietnam. One interviewee noted, "International lenders have started to cease coal financing...this makes coal power less attractive." Another interviewee indicated that "large foreign companies have demanded to use more clean energy in their manufacturing factories in Vietnam."

One of the biggest challenges that several participants noted to solar uptake in Vietnam is complex administrative procedures that lack transparency. One interviewee, from a private solar company, explained that "investors need to clear several administrative steps, including environmental impact assessment, construction license, grid connection approval and so on...there are different governmental entities involved in the approval process...the procedures for obtaining these approvals are not clear and often lack details...the investors often doesn't know what procedures to follow."

This poses significant risks to solar investors. It also somewhat explains the provision of generous FiTs for solar power. To further reduce risks, one interviewee mentioned that "some investors sought to work with well-connected local partners and used low-quality materials." MOIT is currently investigating the policies that led to the 2019 Vietnamese solar power boom, which could lend credence to this claim. Another strategy for risk mitigation employed by some investors, as solar industry interviewees mentioned, was to build small-scale, less efficient solar projects that were easier to implement. As Do et al. (2020a, b) noted, only 12 out of 87 approved solar projects had capacities greater than 50 MW.

Another major challenge that some interviewees cited is limited grid capacity to handle increasing solar power, resulting in solar curtailments and significant delays in grid connections. One interviewee pointed out that "high FiT attract developers and make a lot of projects, but there are some problems on the transmission lines." Another suggested that "transmission system is not strong enough to deal with intermittency…investing in transmission or using battery storage could help."

According to one interviewee, planning inertia was the main cause of such grid constraints on solar uptake in Vietnam, stating, "Improper planning process is one of the main barriers for solar utility-scale uptake in Vietnam. The recent case of limited grid capacity is one of the examples, where in 2022 completed solar projects need to wait to come online until 2030 when the grid expansion is finished." This is partly due to "the lack of experience with managing new technologies, like solar power," as another interviewee noted, adding that "government needs to adapt…they need to learn how to deal with variable renewable energy, because previously the power sector is dominated by baseload coal and hydro power." This perspective is supported by the fact that Vietnam had a total solar capacity of 16.5 GW by the end of 2020, up from almost nothing in 2017 (IRENA 2021), exceeding its 2030 target for solar uptake a decade ahead of schedule. However, this surge in solar power had not been incorporated into the grid capacity expansion plan in time.

Several participants from the power sector highlighted land as an important factor contributing to Vietnamese solar grid constraints. One pointed out that "solar irradiation is high in central Vietnam, but the demand is in south and north…land is expensive, and developer sometimes face lengthy negotiations with local communities." Another added that "land clearing is an issue in grid expansion. Firstly, state project may not be able to pay above market prices. Secondly, most of the projects are in remote areas, and sometimes in forest areas." Addressing these issues will take significant time and effort, as one interviewee noted, which is why the Vietnamese government is now prioritizing offshore wind and rooftop solar PV. This interviewee explained that "the potential (for offshore wind) is more equally distributed in the north, central and south…on-site solar power, like rooftop solar, does not need much effort on grid augmentation."

5 Discussion

Section 2 discussed the electricity landscapes in Indonesia and Vietnam, and Sect. 4 gave an overview of key issues affecting utility-scale solar uptake in these countries. In this section, we extend the discussion, complemented by the transition literature, to find some general insight into factors affecting utility-scale solar deployment.

5.1 Landscape Pressures

In both Indonesia and Vietnam, landscape pressures, particularly the need to address perceived power shortages and growing public demand for decarbonization, created windows of opportunity for renewable energy the of, as discussed in Sect. 4. This aligns with the transition literature, which views the transition as a co-evolutionary process shaped by a myriad of context-specific interactions between technology niches, incumbent regimes, and changing landscapes (Geels 2002, 2005, 2018). In the context of electricity transition, these comprise niche electricity technologies, e.g., solar power, emerging in protected spaces, incumbent electricity regimes consisting of engineering practices, market rules, regulations, and planning processes that impose selection pressures on new technologies and other innovations, and landscape pressures involving sets of deep structural factors that create imputes for change (Geels 2002; Rip and Kemp 1998; Smith et al. 2010; Yang et al. 2022). Transition scholars find that transitions start when landscape pressures, e.g., public concern about climate change, create the aforementioned windows of opportunity for niche electricity technologies to thrive, facilitated by various policy measures Raimed at protecting these technologies from the selection pressures in the incumbent electricity regime, including &D support, FiTs, and tax benefits (Bergek et al. 2008).

5.2 Local Contexts

Despite facing similar landscape pressures, Indonesia and Vietnam have taken different approaches to exploiting renewable energy development opportunities, as discussed in Sect. 2. Indonesia devoted much attention to conventional renewables, such as hydro and geothermal, which already occupy an important place in the country's electricity technology mix. Between 2012 and 2021, total renewable capacity in Indonesia increased 30%, from 7489 MW in 2012 to 11,157 MW in 2021. More than 90% came from hydropower, at 67%, and geothermal, at 26%, with solar accounting for only approximately 5% (IRENA 2022b). By contrast, Vietnam took a different approach, emphasisizing utility-scale solar, leading to exceptional growth in this segment as part of an overall solar boom that saw it surpass Thailand in 2019 to achieve the largest solar capacity in Southeast Asia.

Local contextual factors could help explain these countries' divergent response to landscape pressures. One such factor is energy endowment. Indonesia has abundant low-cost coal reserves, especially in Kalimantan and Sumatra. Such availability, combined with coal price subsidies, e.g., a coal price cap for domestic users, has adversely affected competitiveness of solar there (Bridle et al. 2019). By contrast, rising demand and rapidly depleting indigenous resources in Vietnam have led to a widely held belief that the country is likely to become dependent on imports to satisfy its energy needs (Minh Do and Sharma 2011). Vietnam's imports of coal, the country's main source of electricity, have increased considerably since the mid-2010s, from 72 GJ in 2014 to 1211 GJ in 2020 (IEA 2022). The country's central leadership has acknowledged its growing dependence on coal imports as a strategic concern (Dorband et al. 2020), which helps explain its prioritizing solar power, as it provides Vietnam an opportunity to reduce said dependence on coal imports by taking advantage of its plentiful solar potential. It also suggests that local contextual factors could moderate landscape pressures on choices of generation technologies to clean up electricity sectors.

5.3 Regime Inertia

Early transition studies were often challenged for viewing electricity transitions as an outcome of top-down landscape pressures and bottom-up development of niche electricity technologies, such as solar PV, while largely ignoring incumbent regimes (Turnheim and Sovacool 2020). In this view, as these niche technologies mature and become ready for wider adoption, they will start to challenge the dominant fossil fuel-based electricity regime, which will naturally lead to a gradual emergence of a new regime with clean technologies as its backbone that replaces the old one (Köhler et al. 2019). In recent years, some transition scholars have called for more attention to be paid to the incumbent regime, particularly how to facilitate regime change, the so-called 'flip side' of the transition (Steen and Weaver 2017; Turnheim and Geels 2013, 2012). In response, a growing body of studies has been undertaken that highlights the need to destabilize the regime by addressing lock-in factors (Smith and Raven 2012) and resistance from incumbent actors (Geels 2014; Roberts et al. 2018; Ting and Byrne 2020). Other studies also found that incumbent actors do not always resist change but may also pursue different strategies, leading to regime fragmentation that accelerates regime destabilization and decline (Steen and Weaver 2017; Turnheim and Geels 2013).

Indonesia's experience with solar power, as discussed in Sect. 4.1, highlights the importance of regime factors in shaping utility-scale solar uptake. BPP pricing that ties renewable prices to the average cost of electricity provision, often determined by subsidized coal prices, made large solar projects unattractive. Stringent local content requirements also affected the attractiveness of solar projects. PLN, the incumbent national utility, signed long-term supply contracts with excessive risk foisted off on solar project developers and opposed regulatory changes that would allow rooftop solar owners to sell all surplus electricity to the grid. In Vietnam, the absence of a strong incumbent utility with interests tied to the status quo has created room for rapid of solar deployment. When major regime changes are needed to further its progress, however, inertia becomes a major concern. An example is difficulties involved in acquiring enough land for grid expansion to accommodate greater solar penetration, as discussed in Sect. 4.2.

6 Conclusions and Policy Implications

In analyzing Indonesia and Vietnam's solar uptake experiences, this chapter showed that landscape pressures, such as the need to address energy security concerns and ambitious policy commitments to emissions reduction, could not always be translated into concrete action due to challenges posed by local contexts and incumbent electricity regimes. This highlights the need to create a coherent and effective policy framework capable of driving transitions toward a clean and more sustainable electricity future, in addition to raising the ambitions for such transitions. This is not to say that making more ambitious transition commitments is unimportant. Rather, these commitments, once made, are important steps in promoting electricity transitions. Their achievement, however, depends on whether effective policy frameworks can be developed to drive the transitions. The need for such frameworks is heightened if one notes that electricity transitions are gaining momentum across Southeast Asia, with most countries committing to becoming carbon–neutral between 2050 and 2065.

Such frameworks must address two dimensions of these transitions: (1) the emergence and wider adoption of niche electricity technologies and such supplementary innovations as battery storage; and (2) reconfiguring incumbent electricity regimes to be more accommodating to these technologies. As this chapter suggests, these dimensions are closely connected, particularly when the aforementioned niche technologies, such as utility-scale solar PV, are mature and ready for wider adoption. This presents additional challenges to policymaking, as policy support is needed to address techno-economic issues affecting the uptake of niche technologies, and to facilitate deep structural changes in incumbent regimes to create room for their penetration.

Incumbent electricity regimes have deep-rooted economic and socio-political influence (Yang and Sharma 2020). Major changes to the regime will therefore have widespread ramifications extend into these realms, affecting a diverse range of policy issues, such as energy security and affordability, industrialization, and social welfare. For example, while Vietnam needs grid expansion to further advance its solar uptake, project developers need to overcome issues that may conflict with extant rules governing public projects, such as land acquisition from local communities, deforestation caused by land clearing, and raising purchase prices.

Such major changes to incumbent regimes thus require careful planning and consideration of these cross-cutting issues. They cannot happen simply because of strong political will overcoming incumbent resistance. Given the substantial complexity involved in regime change, the transition process is often considered as "messy, conflictual, and highly disjointed" (Meadowcroft 2009), and developing coherent and effective policy frameworks to drive its progress, informed by an appreciation of the aforementioned underlying complexity, will take a long time. Addressing climate change, however, requires rapid transitions toward a clean electricity future in the next two or three decades, including considerable utility-scale solar expansion.

Transition policy framework thus should recognize the need to reconcile the dichotomy of a usually prolonged electricity transition and the present need to achieve rapid transitions to help save the world from the climate crisis. One way to achieve such a reconciliation is to focus initial efforts on promoting renewable energy technologies that already play an important role in the energy mix, which could reduce immediate demand for major changes to electricity regimes and hence ensure a quick start to transitions. It could also buy time for policymakers and energy planners to work out how to reconfigure incumbent regimes.

References

ACE (2022) The 7th ASEAN energy outlook: 2020–2040. Jakarta

ADB (2022) Asian development outlook 2022. Manila

Asia News Network (2022) Vietnam won't add wind, solar power in 2022. The Phnom Penh Post

Barroco J, Herrera M (2019) Clearing barriers to project finance for renewable energy in developing countries: a Philippines case study. Energy Policy 135:111008. https://doi.org/10.1016/j.enpol.2019.111008

Bergek A, Jacobsson S, Carlsson B, Lindmark S, Rickne A (2008) Analyzing the functional dynamics of technological innovation systems: a scheme of analysis. Res Policy 37:407–429. https://doi.org/10.1016/j.respol.2007.12.003

Bridle R, Suharsono A, Mostafa M (2019) Indonesia's coal price cap: a barrier to renewable energy deployment. Canada

Do TN, Burke PJ, Baldwin KGH, Nguyen CT (2020a) Underlying drivers and barriers for solar photovoltaics diffusion: the case of Vietnam. Energy Policy 144:111561. https://doi.org/10.1016/j.enpol.2020.111561

Do TN, Burke PJ, Baldwin KGH, Nguyen CT (2020b) Underlying drivers and barriers for solar photovoltaics diffusion: the case of Vietnam. Energy Policy2 144:111561

Do TN, Burke PJ, Nguyen HN, Overland I, Suryadi B, Swandaru A, Yurnaidi Z (2021) Vietnam's solar and wind power success: policy implications for the other ASEAN countries. Energy Sustain Dev 65:1–11. https://doi.org/10.1016/j.esd.2021.09.002

Dorband II, Jakob M, Steckel JC (2020) Unraveling the political economy of coal: insights from Vietnam. Energy Policy 147:111860. https://doi.org/10.1016/j.enpol.2020.111860

Fathoni HS, Boer R, Sulistiyanti (2021) Battle over the sun: resistance, tension, and divergence in enabling rooftop solar adoption in Indonesia. Global Environmental Change 71:102371. https://doi.org/10.1016/j.gloenvcha.2021.102371

Foong HM (2022) Supporting the development of ASEAN economies: understanding the strong power demand growth in the region. Singapore

Gallagher KS, Bhandary R, Narassimhan E, Nguyen QT (2021) Banking on coal? Drivers of demand for Chinese overseas investments in coal in Bangladesh, India, Indonesia and Vietnam. Energy Res Soc Sci 71:101827. https://doi.org/10.1016/j.erss.2020.101827

Geels FW (2002) Technological transitions as evolutionary reconfiguration processes: a multi-level perspective and a case-study. Res Policy 31:1257–1274. https://doi.org/10.1016/S0048-7333(02)00062-8

Geels FW (2005) Technological transitions and system innovations: a co-evolutionary and socio-technical analysis. Edward Elgar Publishing, United Kingdom

Geels FW (2014) Regime resistance against low-carbon transitions: introducing politics and power into the multi-level perspective. Theory Cult Soc 31:21–40. https://doi.org/10.1177/0263276414531627

Geels FW (2018) Low-carbon transition via system reconfiguration? A socio-technical whole system analysis of passenger mobility in Great Britain (1990–2016). Energy Res Soc Sci 46:86–102. https://doi.org/10.1016/j.erss.2018.07.008

Hamdi E (2021) Indonesia wants to go greener, but PLN is stuck with excess capacity from coal-fired power plants. United States

Handayani K, Anugrah P, Goembira F, Overland I, Suryadi B, Swandaru A (2022) Moving beyond the NDCs: ASEAN pathways to a net-zero emissions power sector in 2050. Appl Energy 311:118580. https://doi.org/10.1016/j.apenergy.2022.118580

Huang P (2021) When government-led experimentation meets social resistance? A case study of solar policy retreat in Shenzhen, China. Energy Res Soc Sci 75:102031. https://doi.org/10.1016/j.erss.2021.102031

IEA (2015) Energy policies beyond IEA countries: Indonesia. Paris

IEA (2022) Vietnam—country profile. Paris

IESR (2023) Indonesia energy transition outlook. Jakarata

International Trade Administration (2022) Vietnam—country commercial guide. United States

IRENA (2017) Renewable energy prospects: Indonesia. Abu Dhabi

IRENA (2021) Renewable energy statistics 2021. Abu Dhabi

IRENA (2022a) Renewable energy outlook for ASEAN: towards a regional energy transition. Abu Dhabi

IRENA (2022b) Renewable energy statistics 2022. Abu Dhabi

Jayaraman K, Paramasivan L, Kiumarsi S (2017) Reasons for low penetration on the purchase of photovoltaic (PV) panel system among Malaysian landed property owners. Renew Sustain Energy Rev 80:562–571. https://doi.org/10.1016/j.rser.2017.05.213

Kennedy SF (2018) Indonesia's energy transition and its contradictions: emerging geographies of energy and finance. Energy Res Soc Sci 41:230–237. https://doi.org/10.1016/j.erss.2018.04.023

Koerner SA, Siew WS, Salema AA, Balan P, Mekhilef S, Thavamoney N (2022) Energy policies shaping the solar photovoltaics business models in Malaysia with some insights on Covid-19 pandemic effect. Energy Policy 164:112918. https://doi.org/10.1016/j.enpol.2022.112918

Köhler J, Geels FW, Kern F, Markard J, Onsongo E, Wieczorek A, Alkemade F, Avelino F, Bergek A, Boons F, Fünfschilling L, Hess D, Holtz G, Hyysalo S, Jenkins K, Kivimaa P, Martiskainen M, McMeekin A, Mühlemeier MS, Nykvist B, Pel B, Raven R, Rohracher H, Sandén B, Schot J, Sovacool B, Turnheim B, Welch D, Wells P (2019) An agenda for sustainability transitions research: state of the art and future directions. Environ Innov Soc Transit 31:1–32. https://doi.org/10.1016/j.eist.2019.01.004

Lee N, Flores-Espino F, Oliveira R, Roberts B, Bowen T, Katz J (2020) Exploring renewable energy opportunities in select Southeast Asian countries. Denver

Maulidia M, Dargusch P, Ashworth P, Ardiansyah F (2019) Rethinking renewable energy targets and electricity sector reform in Indonesia: a private sector perspective. Renew Sustain Energy Rev 101:231–247. https://doi.org/10.1016/j.rser.2018.11.005

McCawley P (1971) The Indonesian electric supply industry. Canberra

Meadowcroft J (2009) What about the politics? Sustainable development, transition management, and long term energy transitions. Policy Sci 42:323–340. https://doi.org/10.1007/s11077-009-9097-z
MEMR (2022) Handbook of energy & economic statistics of Indonesia. Jakarta
Minh Do T, Sharma D (2011) Vietnam's energy sector: a review of current energy policies and strategies. Energy Policy 39:5770–5777. https://doi.org/10.1016/j.enpol.2011.08.010
PLN (2022) Company profile. Jakarta
Potisat T, Tongsopit S, Aksornkij A, Moungchareon S (2017) To buy the system or to buy the service: the emergence of a solar service model in Thailand. Renewable Energy Focus 21:1–10. https://doi.org/10.1016/j.ref.2017.06.002
Rababah HE, Ghazali A, Mohd Isa MH (2021) Building integrated photovoltaic (BIPV) in Southeast Asian countries: review of effects and challenges. Sustainability 13:12952. https://doi.org/10.3390/su132312952
Rip A, Kemp R (1998) Technological change. In: Rayner S, Malone EL (eds) Human choice and climate change. Battelle Press, Columbus, pp 327–399
Roberts C, Geels FW, Lockwood M, Newell P, Schmitz H, Turnheim B, Jordan A (2018) The politics of accelerating low-carbon transitions: towards a new research agenda. Energy Res Soc Sci 44:304–311. https://doi.org/10.1016/j.erss.2018.06.001
Sang X (2021) Vietnam plans to cut 500 million kWh of photovoltaic power generation in 2021. Seetao
Setyawati D (2020) Analysis of perceptions towards the rooftop photovoltaic solar system policy in Indonesia. Energy Policy 144:111569. https://doi.org/10.1016/j.enpol.2020.111569
Smith A, Raven R (2012) What is protective space? Reconsidering niches in transitions to sustainability. Res Policy 41:1025–1036. https://doi.org/10.1016/j.respol.2011.12.012
Smith A, Voß J-P, Grin J (2010) Innovation studies and sustainability transitions: the allure of the multi-level perspective and its challenges. Res Policy 39:435–448. https://doi.org/10.1016/j.respol.2010.01.023
Steen M, Weaver T (2017) Incumbents' diversification and cross-sectorial energy industry dynamics. Res Policy 46:1071–1086. https://doi.org/10.1016/j.respol.2017.04.001
The White House (2022) Indonesia and international partners secure groundbreaking climate targets and associated financing. Washington, DC
Ting MB, Byrne R (2020) Eskom and the rise of renewables: regime-resistance, crisis and the strategy of incumbency in South Africa's electricity system. Energy Res Soc Sci 60:101333. https://doi.org/10.1016/j.erss.2019.101333
Tongsopit S, Moungchareon S, Aksornkij A, Potisat T (2016) Business models and financing options for a rapid scale-up of rooftop solar power systems in Thailand. Energy Policy 95:447–457. https://doi.org/10.1016/j.enpol.2016.01.023
Turnheim B, Geels FW (2012) Regime destabilisation as the flipside of energy transitions: lessons from the history of the British coal industry (1913–1997). Energy Policy 50:35–49. https://doi.org/10.1016/j.enpol.2012.04.060
Turnheim B, Geels FW (2013) The destabilisation of existing regimes: confronting a multi-dimensional framework with a case study of the British coal industry (1913–1967). Res Policy 42:1749–1767. https://doi.org/10.1016/j.respol.2013.04.009
Turnheim B, Sovacool BK (2020) Forever stuck in old ways? Pluralising incumbencies in sustainability transitions. Environ Innov Soc Transit 35:180–184. https://doi.org/10.1016/j.eist.2019.10.012
Vu K (2022) Vietnam slashes power usage at its largest solar farm. Reuters
Wang P, Yang M, Mamaril K, Shi X, Cheng B, Zhao D (2021) Explaining the slow progress of coal phase-out: the case of Guangdong-Hong Kong-Macao Greater Bay Region. Energy Policy 155:112331. https://doi.org/10.1016/j.enpol.2021.112331
World Bank (2022a) World development indicators. Wasington, DC
World Bank (2022b) Vietnam—overview. Washington, DC

Yang M, Sharma D (2020) The spatiality and temporality of electricity reform: a comparative and critical institutional perspective. Energy Res Soc Sci 60:101327. https://doi.org/10.1016/j.erss.2019.101327

Yang M, Sharma D, Shi X (2022) Policy entry points for facilitating a transition towards a low-carbon electricity future. Frontiers of Engineering Management. https://doi.org/10.1007/s42524-022-0214-4

Chapter 12
Impact of Policy on Solar PV Supply for ASEAN and Beyond

Rohan Best, Rabindra Nepal, and Han Phoumin

Abstract This paper assesses the role of renewable energy policy in solar photovoltaic energy supply. Cross-country findings are based on cross-sectional regressions and panel analysis including fixed effects and multiple approaches to give robust standard errors for within-group and cross-sectional dependence, showing that a composite renewable energy policy index has a significant influence at lags of up to six years on changes in solar energy supply per capita. There are also key results for more specific renewable energy policy types, with carbon pricing and such incentives as feed-in tariffs having the most robust impact on solar use. Association of Southeast Asian Nations (ASEAN) member states could benefit from further focus on renewable energy policies, given their relatively low solar photovoltaic supply per capita levels and renewable policy scores alike. The analysis suggests that expanded implementation of carbon pricing in ASEAN member states is an opportunity not to be missed.

1 Introduction

Large-scale deployment of solar photovoltaic (PV) energy has great potential to aid countries around the world (International Energy Agency 2022) in achieving climate goals including net-zero emissions (United Nations 2022), as a significant part of the necessary transition from fossil fuels and toward renewables. This transition may also achieve net savings based on renewables becoming increasingly inexpensive

R. Best (✉)
Macquarie University, North Ryde, Balaclava Road, Sydney, Australia
e-mail: rohan.best@mq.edu.au

R. Nepal
University of Wollongong, Wollongong, Australia
e-mail: rnepal@uow.edu.au

H. Phoumin
Economic Research Institute for ASEAN and East Asia, Jakarta, Indonesia
e-mail: han.phoumin@eria.org

H. Phoumin et al. (eds.), *Large-Scale Development of Renewables in the ASEAN*, Economics, Law, and Institutions in Asia Pacific,
https://doi.org/10.1007/978-981-99-8239-4_12

over time (Way et al. 2022). The speed of this transition will determine whether modern civilization succeeds in limiting global warming to two degrees Celsius above pre-industrial levels.

Association of Southeast Asian Nations (ASEAN) member states and other Asian countries have a crucial part to play in meeting these global climate and energy goals, as they are likely to experience increasing energy demand to drive their growing economies. Fossil fuels also have a significant share of their current energy mixes, as with most other countries (IEA 2022). This creates a degree of lock-in, where past fossil fuel use can be reflected in continued use thereof (Best and Burke 2020).

At least some ASEAN member states have potentially abundant renewable resources, especially solar energy, owing to these countries having abundant land and sun exposure (Wang et al. 2021). ASEAN member states in general have significant untapped potential where renewable alternatives to fossil fuels are concerned (IRENA and ACE 2022).

Policy support is likely to be crucial for large-scale deployment of solar energy in ASEAN, both to overcome the incumbency of legacy energy sources and to address instances of higher costs for installing solar energy infrastructure. A range of studies at the global level have considered various policies that can influence solar deployment and renewables uptake more broadly. Best and Burke (2018) found a significant impact for carbon pricing on solar deployment. Polzin et al. (2015) found that while feed-in tariffs (FiTs) are effective for less mature technologies, renewable portfolio standards seem to be more effective for mature technologies. Baldwin et al. (2017) found that policy success for renewable energy development varies among income brackets.

This study focuses on numerous research questions which are relevant for continued expansion of solar PV energy in ASEAN. First, we consider ASEAN's average progress in solar PV deployment and how it compares to global averages, after which we consider what explains the differences in solar PV supply across countries. A key factor is policy impacts on solar PV per capita, and in this context, we consider which policies are most influential. This analysis allows us to assess ASEAN member states' solar energy uptake compared to other countries. A forward-looking perspective can then be taken to consider which policies are most important for promoting deployment of solar PV in ASEAN.

This paper starts with a more comprehensive discussion of prior literature and the policy context in Sect. 2, covering a range of drivers of solar and other renewables. This includes economic and financial aspects, i.e., income or capital. Physical endowments have also been assessed, for both fossil fuels and renewables. A broad range of potential policies are considered extending across planning, technology-specific incentives, systemic characteristics, and broad-based climate policies. Section 3 describes and justifies methodologies with respect to the many important variables that frequently emerge in limited historical time series. Section 4 provides results on drivers of solar PV supply per capita in a cross-country sample, emphasizing the importance of institutional contexts and policy levers. Section 5 concludes by summarizing the main findings and providing suggestions for policy directions in ASEAN.

2 Literature Review and Policy Context

A growing number of studies have investigated drivers of energy mix differences among countries. Burke (2010) considered the impact of income, Best (2017), Brunnschweiler (2010), and Pfeiffer and Mulder (2013) investigated financial sector development influences, Best and Burke (2018) considered select policies and perceptions, and Escoffier et al. (2021) studied oil price influences. Aguirre and Ibikunle (2014) found that policies with voluntary participation might have a negative relationship with renewables uptake.

More recently, there has been increased focus on explanatory variables pertaining to institutions. Chen et al. (2021) used a panel threshold model to investigate institutional impact. Abban and Hasan (2021) assessed the impact of political ideology. Sweidan (2021) found that openness fosters energy transition. Best and Burke (2017) found that government effectiveness is an important indicator of electrification outcomes in developing countries.

A systematic literature review of determinants of renewable energy deployment by Bourcet (2020) includes recommendations for future research. One is for consideration of the share of renewable energy as a dependent variable. Dogan et al. (2021) also suggest that the choice between share and levels is important. Broader sets of explanatory variables also have potential for inclusion in research. Natural endowments are one type of explanatory variable that is not used by many studies. Socioeconomic explanatory variables might also be considered further, as well as methods which account for path dependency (Bourcet 2020).

This study includes more explanatory variables than prior studies, including different policy variables and an ASEAN indicator variable. Policy variables cover energy and climate aspects with greater breadth than prior research. Rarely used natural endowments serve as controls, helping to avoid endogeneity concerns from other energy-related explanatory variables. A further value-added aspect is more recent data than that used in prior studies.

A range of renewable energy policies might be important for solar energy uptake. The Regulatory Indicators for Sustainable Energy (RISE) include seven main categories of renewable energy policies (ESMAP 2020), as follows:

- A legal framework for renewable energy (REP1);
- Planning for renewable energy expansion (REP2);
- Incentives and regulatory support for renewable energy (REP3);
- Attributes of financial and regulatory incentives (REP4);
- Network connection and use (REP5);
- Counterparty risk (REP6); and
- Carbon pricing and monitoring (REP7).

Legal framework for renewable energy (REP1) refers to such issues as allowance for private-sector ownership of renewable energy generation and official renewable energy targets. The latter may be legally binding, as well as linked to international commitments and explicit strategies for fulfillment.

Planning for renewable energy expansion (REP2) may involve separate plans for electricity, heating and cooling, and transport. An important aspect is whether institutions in each country are tasked with monitoring and reporting upon such plans. This work can be a precursor to the institutions adjusting said plans or targets.

Incentives and regulatory support for renewable energy (REP3) include financial and regulatory support, which may comprise support through feed-in tariffs (FiTs) for electricity, grid access, heating and cooling, and transport. Many studies have considered the impact of FiTs on renewable energy uptake, with varying findings. A global study on large-scale renewables use by Baldwin et al. (2017) found that FiTs had a substantial impact. Smith and Urpelainen (2014) also found that FiTs can effectively support renewable electricity generation. Best and Burke (2018), however, did not find that FiTs had a substantial impact on large-scale solar PV supply when controlling for other key variables. FiTs are also available for small-scale household solar energy production, albeit tending to benefit wealthy households most (Best et al. 2021). Other policy support includes upfront support, such as capital subsidies and grants, which have been shown to be effective for both households and small businesses in numerous countries (Best et al. 2019; Best and Trück 2020; de Groote and Verboven 2019), as well as rebates and tax credits or reductions.

Attributes of financial and regulatory incentives (REP4) refer to auctions or fixed tariffs. Auctions have tended to be used for large-scale renewable energy generation, where the chief motivation is the pursuit of cost-effectiveness through competition, in that generation contracts are awarded to lower bids. Other provisions must be taken into account to ensure that bidders can complete their promised renewable energy installations, including pre-qualification of bidders and tariff indexing. Fixed production tariffs, by contrast, may apply to payments for small-scale producers of renewable energy, at least for given periods of time. Considerations here relate to such aspects as contract lengths, quantity restrictions, and again, potential indexing of tariffs. Reverse auctions might also be cost-effective for rooftop solar installations (Mayr et al. 2014).

Network connection (REP5) refers to procedures including connecting renewable electricity to the grid. Grid use is a further crucial aspect that policy frameworks must address, possibly by considering rules for buying and selling electricity through the grid. Allocation of costs for connections is another important policy aspect. Ancillary services are also important for ensuring a reliable supply of electricity over time when renewable electricity features in the electricity mix.

Counterparty risk (REP6) is an important part of policy frameworks in multiple dimensions. Government guarantees may provide underwriting relating to payments by special purpose entities for renewable energy power purchase agreements (PPAs). The financial positions of large utilities can also be audited and made available to give greater confidence in their ability to pay electricity generators and operate crucial energy services for customers.

Carbon pricing (REP7) is often discussed as a cost-effective policy approach to reducing emissions (Aldy and Stavins 2012), and an increasing amount of empirical evidence suggests that it is indeed effective (Andersson 2019; Bayer and Aklin 2020; Best et al. 2020). A central component is the transition from fossil fuels to renewable energy, which carbon pricing promotes despite the challenge of lock-in of fossil fuel influence (Best and Burke 2020). An important component of broader attempts to price emissions is to have monitoring and verification systems in place.

3 Methodology and Data

Equation (1) shows the structure of the cross-sectional regressions for this research:

$$S_c = \alpha + \beta P_c + \delta R_c + \gamma A_c + \psi E_c + \lambda N_c + \varepsilon_c \tag{1}$$

The solar dependent variable (S) relates to PV energy supply per capita for each country (c) for which data are available. More precisely, it is the per capita supply of solar PV energy in 2020 minus the equivalent per capita supply in a prior year, over a range of prior years with lags of one to eight years.

The range of explanatory variables are shown on the right of Eq. (1). P is the key policy variable. R is the regulatory quality index. A is a binary variable for ASEAN member states. Economic variables (E) include the private credit variable and log GDP per capita. The natural endowments vector (N) includes variables for solar and wind endowments, and coal, oil, and natural gas reserves. The constant (α) and the error term (ε) are also shown.

Equation (2) gives the framework for the fixed-effects panel investigations:

$$S_{c,t} = \alpha + \beta P_{c,t} + \delta R_{c,t} + \psi E_{c,t} + I_c + I_t + \varepsilon_{c,t} \tag{2}$$

Equation (2) has some similarities to Eq. (1). The dependent variable is again the change in solar PV supply per capita. In the panel context, one-year changes are used to allow for a larger sample size. Key explanatory variables are also the same, including the renewable energy policy score (P), regulatory quality (R), and economic (E) variables of private credit and log GDP per capita. ASEAN and endowment variables are not included, given that there is little or no variation over time. Fixed effects are also included in Eq. (2) for countries (c) and the time dimension (t). The time fixed effects are for years. As the policy data start in 2010, and the dependent variable has a one-year change, there are binary variables for eight fixed effects from 2012 to 2019, relative to a base year 2011. The constant and error are again shown, with the error term having country and time subscripts.

Cross-country data are from such sources as the following. Solar PV energy supply for each country is from the World Energy Statistics and Balances of the International Energy Agency (IEA 2022). Population for per capita measure is taken from the World Development Indicators (WDI) of the World Bank (2021), as is gross domestic product (GDP) per capita in 2017 international dollars with purchasing power parity. Another key economic variable of the ratio of private credit to GDP is derived from the Global Financial Development Database (World Bank 2022). The policy variables are extracted from the World Bank Regulatory Indicators for

Sustainable Energy (ESMAP 2020), including the summary variable for renewable energy policies and seven components thereof. The general institutional context within which policies are made is also part of the analysis, with a regulatory quality index being extracted from the Worldwide Governance Indicators (Kaufmann et al. 2010).

Variables used in our research include per capita measures and indexes or scores which are not directly influenced by the size of a country's economy or population. As described above, the dependent variable is based on solar PV energy supply per capita. Fossil fuel reserves are also converted into per capita terms and then transformed using the inverse hyperbolic sine transformation so that zero values are not dropped when taking the log of zero. The GDP variable is also in per capita terms. The capital stock variable of private credit is expressed as a ratio to GDP to give an indication of the size of financial stocks relative to ongoing economic production. The policy score variables are an index with values on a scale from 1 to 100. The regulatory quality variable is a normalized index over a standard distribution. Natural wind and solar endowments are not per capita measures, given that these are abundant and renewable resources. The solar measure is the log of the global horizontal irradiance, while the wind measure is the log of the total wind resource, including both offshore and onshore measures (Breyer and Gerlach 2010; Lu et al. 2009). A binary variable is also used to identify ASEAN member states as part of investigating whether there are further explained influences which other explanatory variables do not capture.

The prior literature indicates no single methodology or model type (Bourcet 2020). Methods used include cross-sectional, fixed-effects panel regressions, Bayesian Model Averaging, Fully Modified Ordinary Least Squares, quantile regression, autoregressive distributed lag, and methods relating to Granger Causality (Bourcet 2020; Li et al. 2022; Saadaoui and Chtourou 2022). Appropriate models vary depending on specific research questions and available data. This paper takes a number of approaches, including cross-sectional and panel regressions. There are multiple aspects to the rationale for using cross-sectional regressions. One is that solar PV growth is quite recent, and thus, the time series of past values is limited. Such key explanatory variables are similarly limited, in part because energy policies have only recently been introduced, with data in some instances dating back only as far as 2010 (ESMAP 2020), partially signifying the difficulty in measuring and constructing comparable variables across countries when different countries take different approaches.

Another factor is that some explanatory variables, such as global horizontal irradiance and wind resources, generally do not vary over time, as is also true of other variables, such as regulatory quality as one measure of broader institutional frameworks. This suggests that changes in important variables might not show a strong relationship with changes in solar energy supply per capita.

A further issue for panel estimations is that cross-sectional dependence tests require a sufficient number of intervals, which this paper has, albeit barely. We therefore do both cross-sectional and panel regressions. In this context, we employ fixed-effects panels to account for unobserved heterogeneity through country and year fixed effects.

Another attribute is that the dependent variable is the change in solar PV energy supply per capita, incorporating some temporal dimensions. This change is the difference between solar PV energy supply in the most recent data for 2020, and the lagged value from a prior year. A range of prior years are used in Sect. 4 to give an indication of the length of lags, which are likely to result from policy development to subsequent growth in solar PV energy supply.

Table 1 has descriptive statistics for seven ASEAN member states and 86 other countries in 2019. Average solar PV energy supply per capita is lower for ASEAN member states than other countries. Renewable energy policy scores are lower for ASEAN member states on average, for these countries for which data from 2019 is available for every variable. ASEAN member states have relatively larger stocks of private credit relative to GDP than other countries, which may help drive capital intensive transitions to renewables (Best 2017). Average regulatory quality is lower in the ASEAN member states in Table 1 than other countries. ASEAN member states also have lower endowments on average in four of five natural resources, the sole exception being global horizontal irradiance (GHI). This may suggest that solar energy is an appropriate option for ASEAN member states to promote in their energy mixes.

Figure 1 displays growth of solar PV share from 2011 to 2019 with averages for ASEAN and other countries. The line for the world excluding ASEAN shows strong growth, which appears to accelerate over time. By contrast, the ASEAN growth rate is lower and less stable. The initial gap between ASEAN and other countries in 2011 has widened by 2019, such that the solar PV share for ASEAN member states is substantially lower than that for other countries.

Figure 2 shows each country for which data is available for renewable energy policy scores and solar PV share of total energy supply. The nine ASEAN member states for which data is available tended to have relatively low solar shares in 2020, with all but Vietnam and Cambodia below the best fit line.

Figure 3 shows average renewable energy policy scores for each of the seven policy components in ASEAN member states and other countries as of 2019. It is evident that ASEAN member states have lower scores in most cases, with the carbon pricing variable (REP7) being most pronounced. Two exceptions are attributes of incentives (REP4), which is close to the average, and counterparty risk (REP6), where the ASEAN average is higher.

4 Results

Table 2 shows the impact of the composite renewable energy policy score on the change in solar PV supply per capita. The explanatory variables have lag lengths which are specified in the column headings, and the change in solar PV supply is for the interval starting with the lagged year and ending in 2020. For example, a lag of four means that the explanatory variables are from 2016 and the change in solar supply per capita is for 2016 to 2020.

Table 1 ASEAN versus non-ASEAN descriptive statistics

	ASEAN			Non-ASEAN		
Variable	Minimum	Mean	Maximum	Minimum	Mean	Maximum
Solar p.c	0.00	0.00	0.01	0	0.01	0.05
REP1	0.20	0.66	0.80	0.20	0.86	1
REP2	0.18	0.59	0.97	0.04	0.68	1
REP3	0.24	0.44	0.60	0.06	0.57	1
REP4	0.08	0.47	0.83	0	0.51	1
REP5	0.06	0.36	0.70	0	0.57	1
REP6	0.17	0.62	0.83	0	0.63	1
REP7	0	0.14	0.50	0	0.53	1
Pr. Cred	0.29	0.85	1.38	0.06	0.55	1.64
GDP p.c	4388.80	12,110.19	28,421.46	1097.95	23,539.47	89,966.45
ASEAN	1	1	1	0	0	0
REP sum	0.24	0.53	0.69	0.20	0.64	0.97
Reg. qual	− 0.75	− 0.14	0.56	− 1.50	0.19	1.88
GHI	1665	1837.29	1939	1055	1784.67	2261
Wind	265	590	1375	250	6403.43	143,000
Gas	0	0.01	0.03	0	0.06	2.72
Oil	0	0.00	0.02	0	0.04	1.12
Coal	0	0.02	0.10	0	0.10	2.12

These statistics are based on a sample of 93 countries for which data for each variable is available, including 7 ASEAN member states and 86 other countries
Key:
Solar p.c.: solar photovoltaic energy supply per capita
REP: renewable energy policy score, as described in Sect. 2
Private credit: ratio of private credit to GDP
GDP per capita: gross domestic product per capita in 2017 constant international dollars in purchasing power parity terms
ASEAN is a binary variable for member states of the Association of Southeast Asian Nations
REP sum.: summary variable for renewable energy policy scores
Reg. qual: regulatory quality
GHI: global horizonal irradiance
Wind: total wind resource
Gas, oil, and coal: inverse hyperbolic sine transformations of fossil fuel reserves per capita

The renewable energy policy score has positive and significant coefficients for lags of two to four years, with the coefficient magnitudes tending to increase from lag 1 to lag 4, and decreasing thereafter. These results suggest that a four-year lag may represent the interval necessary for renewable energy policy scores to have their maximum impact on average. It would be reasonable that such impact grows over time, as there are likely to be lags between policy implementation and impact, but there would be greater scope for other changes that might overshadow historical

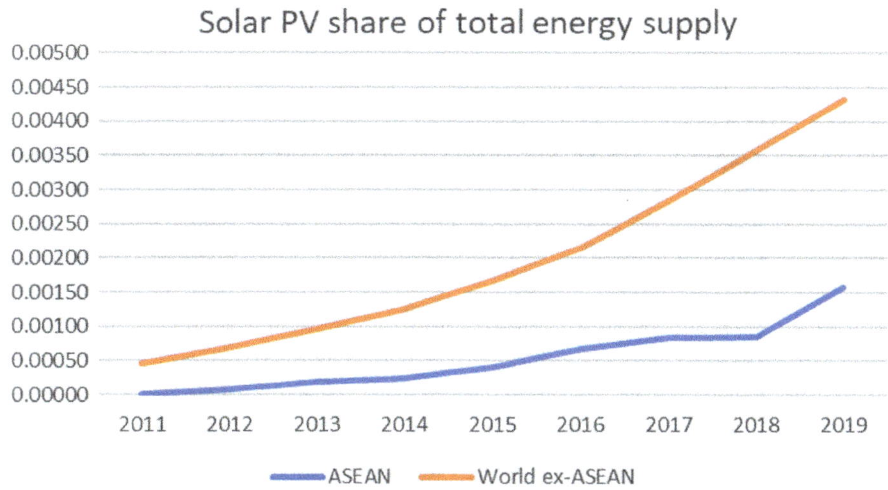

Fig. 1 Solar PV share of total energy supply, 2011–2019. *Source* IEA (2022)

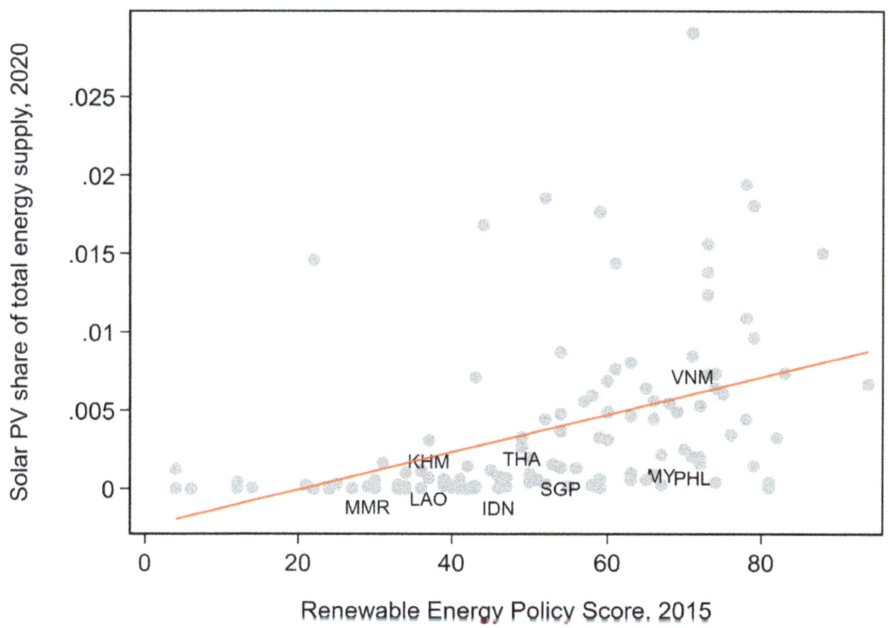

Fig. 2 Scatter plot showing the relation between renewable energy policy score in 2015 and solar PV share of total energy supply in 2020, with one dot representing each of 116 countries represented for which data for both variables is available, including nine ASEAN member states. Malaysia is shown as "MY" immediately to the left of the Philippines. *Source* IEA (2022), ESMAP (2020)

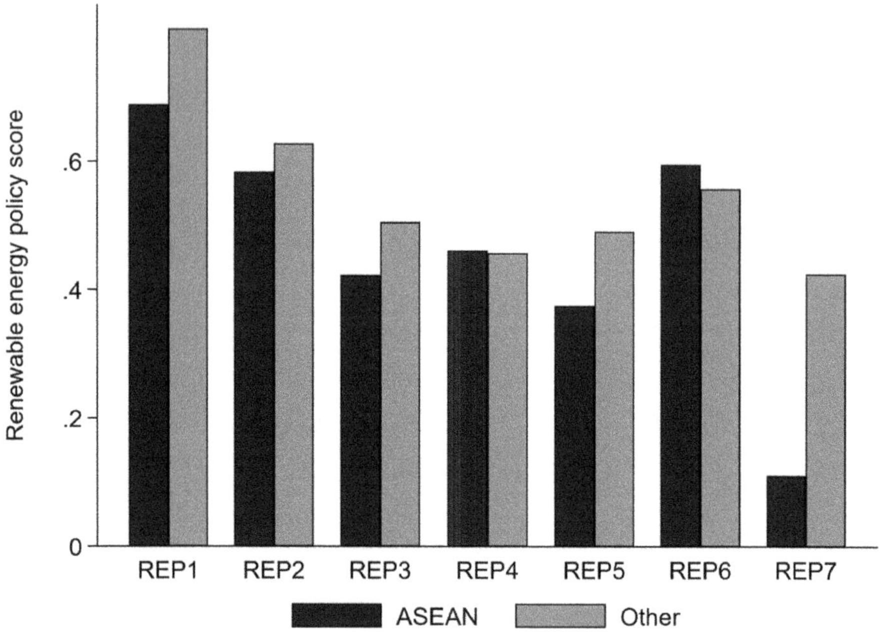

Fig. 3 Policy breakdown showing average scores for seven renewable energy policy (REP) components for nine ASEAN member states and 128 other countries for which data is available for 2019. *Source* ESMAP (2020). Key: *REP1* Legal framework; *REP2* Planning; *REP3* Incentives; *REP4* Attributes of incentives; *REP5* Network connection and use; *REP6* Counterparty risk; *REP7* Carbon pricing and emissions monitoring

policy changes in some instances, over intervals of five years or longer. These might include broader institutional changes beyond renewable energy policies, as well as economic, socio-political, or technological changes.

The regulatory quality indicator also has a significant positive influence on solar PV supply per capita, across all lags and apparently growing generally in magnitude as the lag length increases. A positive impact of regulatory quality on solar energy use is intuitive, given that renewables integration is complicated and there are many related rules which can have an impact. This variable accounts for broader institutional influences beyond specific renewable energy policies. The importance of broader governance characteristics in driving energy outcomes was also found by Best and Burke (2017), in that government effectiveness was important for electrification outcomes in developing countries, such as grid access.

There are also significant positive coefficients for the coal reserves variable in explaining the change in subsequent solar energy supply per capita. It is possible that having larger coal reserves, which is the fossil fuel with the highest carbon intensity, might be related to attempts to promote renewable energy such as solar PV to compensate for high local emissions where coal is used. Most of the coefficients for oil reserves are insignificant, while gas reserves show some significant negative

Table 2 Results for change in solar PV supply per capita up to 2020

	Lag = 1	Lag = 2	Lag = 3	Lag = 4	Lag = 5	Lag = 6	Lag = 7	Lag = 8
Renew. policy	0.003	0.010**	0.013**	0.014**	0.012	0.008	0.001	– 0.009
	(0.002)	(0.005)	(0.006)	(0.007)	(0.007)	(0.006)	(0.006)	(0.007)
Regulat. qual	0.001*	0.002*	0.003**	0.003**	0.004**	0.004***	0.005***	0.006***
	(0.001)	(0.001)	(0.001)	(0.001)	(0.002)	(0.002)	(0.002)	(0.002)
Private credit	0.001	0.001	0.001	0.002	0.002	0.003	0.004	0.008***
	(0.001)	(0.002)	(0.002)	(0.003)	(0.003)	(0.003)	(0.003)	(0.003)
Log GDP p.c	0.000	0.000	– 0.000	0.000	0.001	0.001	0.002	0.002
	(0.000)	(0.001)	(0.001)	(0.001)	(0.001)	(0.001)	(0.001)	(0.001)
Gas reserves	– 0.005***	– 0.002	– 0.002	– 0.003	– 0.003	– 0.003	– 0.004	– 0.005*
	(0.001)	(0.003)	(0.002)	(0.002)	(0.002)	(0.002)	(0.003)	(0.002)
Oil reserves	0.010***	0.005	0.006	0.006	0.004	0.003	0.001	– 0.002
	(0.002)	(0.008)	(0.007)	(0.008)	(0.007)	(0.007)	(0.007)	(0.006)
Coal reserves	0.005***	0.009***	0.010**	0.011**	0.010*	0.009	0.009	0.017***
	(0.002)	(0.003)	(0.004)	(0.005)	(0.005)	(0.008)	(0.008)	(0.005)
Observations	93	92	93	94	94	98	98	95
R-squared	0.500	0.464	0.474	0.481	0.459	0.442	0.431	0.539

The column names show lags for the explanatory variables and the length of the growth interval for the dependent variable. For example, lag = 3 means that explanatory variables from the three years prior to 2020 are used to explain the change in solar PV supply per capita between 2017 and 2020. Each column is a separate regression. For statistical significance, *** = 1%, ** = 5%, * = 10%. Coefficients are not shown for the constant, the ASEAN binary variable, log of global horizontal irradiance for solar exposure, or the log of the wind resource. Nor are these variables lagged. The inverse hyperbolic sine (IHS) transformation is used to give a log transformation for fossil fuel reserves without omitting countries with no such reserves

coefficients. This last is consistent with the idea that gas can substitute for renewable energy in some cases as countries seek to lower emissions. While natural gas has been suggested as a stopgap fuel when transitioning from extensive coal use to renewables, this bridge narrative has also been described as hindering renewable energy transitions (Kemfert et al. 2022). Countries can leapfrog straight to renewables (van Benthem 2015), or gas and renewables can also theoretically complement one another, as fast-start gas generators can fill in when intermittent renewables are offline.

The economic variables tend to be insignificant in most cases in Table 2. One exception is that private credit has a significant positive coefficient in the final column for eight-year lags. The intuition that the link between private credit and renewable energy would be subject to such a substantial lag is based on the long intervals involved between financial system changes, subsequent financing of energy projects, construction and installation of renewable power generation, and grid integration. The importance of private credit or other variables for financial capital stocks was also found in prior analyses of wind and other renewables (Brunnschweiler 2010; Best 2017), which tend to show stronger links between financial capital and renewable energy use. One reason for the insignificance in most columns of Table 2 might be a correlation between the regulatory quality variable and the economic variables, i.e., countries with better regulatory quality tend to have better economic outcomes. This is reflected in a relatively high variance inflation factor for a lag of 1, for example, for the regulatory quality variable of 5.5, even though the average variance inflation factor of 2.8 is below common thresholds of 5 or 10.

Table 3 removes the regulatory quality variable, due to its relatively high variance inflation factor in Table 2. The renewable energy policy score variable again has significant positive coefficients for lags 2–4, and each lag from 1 to 6 is now significantly positive at the 5% level. The renewable policy variable may therefore be incorporating the influence of regulatory quality in Table 3, which may suggest that regulatory quality helps promote favorable renewable policies, which in turn drive solar energy adoption.

The economic variables are significant more often in Table 3 than in Table 2. For example, the private credit variable is significantly positive in four columns in Table 2, with magnitudes again showing a rising trend. The corresponding magnitudes in Table 3 also have larger point estimates than in Table 2, as is also true of the log of gross domestic product per capita variable. These significant positive coefficients for the economic variables, when the regulatory quality variable is omitted, signify the strong positive correlation between economic and institutional variables, with the influence of strong institutions probably visible to some extent through the economic variables in Table 3.

Fossil fuel reserve variables are similar in Table 3 relative to Table 2. For example, coal reserves again show significant positive coefficients at most lags. Fossil fuels not being affected substantially by omission of the regulatory quality variable is consistent with a weak relationship between fossil fuel reserves and institutional variables. Prior research has noted that fossil fuel reserves are an important control variable when understanding related policies, and fossil fuel reserves have the added advantage that they can be largely considered as exogenous (Best and Zhang 2020).

Table 4 splits the renewable energy policy score into separate variables for its seven components. The control variables are not shown to save space as they match Table 3 and produce similar outcomes.

Four of the seven policy components have a statistically significant impact on solar PV supply per capita in Table 4. These tend to be from policy types which may have a more immediate impact than others. For instance, incentives and carbon pricing may affect solar PV supply over a shorter term than legal frameworks and planning, which might take longer than the one-year lag to have an impact. The strongest links between renewable energy polices and solar energy use appear to be from attributes of these aforementioned incentives and carbon pricing, with significant coefficients at the 5% level. The carbon pricing coefficient is also significant at the 5% level if the

Table 3 Results for change in solar PV supply p.c. up to 2020, omitting regulatory quality

	Lag = 1	Lag = 2	Lag = 3	Lag = 4	Lag = 5	Lag = 6	Lag = 7	Lag = 8
Renew. policy	0.004**	0.011**	0.016**	0.017**	0.016**	0.011**	0.004	− 0.004
	(0.002)	(0.005)	(0.006)	(0.007)	(0.007)	(0.005)	(0.005)	(0.006)
Private credit	0.002*	0.002	0.003	0.003	0.004	0.006**	0.007**	0.011***
	(0.001)	(0.002)	(0.002)	(0.003)	(0.003)	(0.003)	(0.003)	(0.003)
Log GDP p.c	0.001*	0.002**	0.002**	0.002**	0.003***	0.003***	0.004***	0.005***
	(0.000)	(0.001)	(0.001)	(0.001)	(0.001)	(0.001)	(0.001)	(0.001)
Gas reserves	− 0.005***	− 0.003	− 0.002	− 0.003	− 0.003	− 0.003	− 0.004	− 0.004*
	(0.001)	(0.003)	(0.002)	(0.002)	(0.002)	(0.003)	(0.002)	(0.002)
Oil reserves	0.010***	0.004	0.005	0.004	0.002	0.001	− 0.001	− 0.004
	(0.002)	(0.008)	(0.007)	(0.008)	(0.008)	(0.007)	(0.007)	(0.007)
Coal reserves	0.006***	0.010***	0.012***	0.013***	0.012**	0.011	0.011	0.018***
	(0.002)	(0.003)	(0.004)	(0.005)	(0.005)	(0.008)	(0.008)	(0.005)
Observations	93	92	93	94	94	98	98	95
R-squared	0.482	0.444	0.446	0.450	0.430	0.402	0.384	0.485

The column names show the lags for the explanatory variables and the length of the growth period for the dependent variable. For example, lag = 3 means that the explanatory variables from three years prior to 2020 are used to explain the change in solar PV supply per capita from 2017 to 2020. Each column is a separate regression. For statistical significance, *** = 1%, ** = 5%, * = 10%. Coefficients are not shown for the constant ASEAN binary variable, the log of global horizontal irradiance for solar exposure, or the log of the wind resource. Nor are these variables lagged. The inverse hyperbolic sine (IHS) transformation is used to give a log transformation for fossil fuel reserves per capita without omitting countries with no such reserves

Table 4 Results for the change in PV supply p.c. up to 2020 for 7 policy components

Renewable policy component	Coefficient	Standard error
Legal framework	0.001	0.002
Planning	0.002	0.001
Incentives and regulatory support	0.003*	0.001
Attributes of incentives	0.003**	0.001
Network connection and use	0.002*	0.001
Counterparty risk	0.000	0.001
Carbon pricing and monitoring	0.001**	0.001

Each row is a separate regression. Control variables are not shown but match Table 3. The lag is one year. There are 93 observations in each case and the R-squared values range from 0.46 to 0.49. For statistical significance, ** = 5%, * = 10%

Table 5 Results for change in PV supply p.c. up to 2020 with lag1 and larger sample size

Renewable policy component	Coefficient	Standard error
Legal framework	0.001	0.001
Planning	0.002	0.001
Incentives and regulatory support	0.003***	0.001
Attributes of incentives	0.002***	0.001
Network connection and use	0.002**	0.001
Counterparty risk	0.002**	0.001
Carbon pricing and monitoring	0.002***	0.001

Each row is a separate regression. Control variables of log GDP per capita and an ASEAN binary variable are not shown. This concise control set allows for a larger sample size than that in Table 4. There are 117 observations in each case and the R-squared values range from 0.18 to 0.24. For statistical significance, *** = 1%, ** = 5%

dependent variable is changed to be the share of solar PV energy rather than change in per-capita level.

Table 5 repeats the analysis with only two control variables, log GDP per capita and the binary ASEAN variable, to increase the sample size, in contrast with the larger control set in Table 4, as is evident in Table 3. Table 5 is similar to Table 4 in some respects. The legal framework and planning coefficients, while positive, are not statistically significant. One difference is that the counterparty risk variable is statistically significant in Table 5, as are the other four coefficients, while some coefficient magnitudes tend to be slightly larger than in Table 4. There is also stronger statistical significance, including at the 1% level for incentives and carbon pricing in Table 5. The similarity of coefficient magnitudes when control variables are omitted in Table 5, along with the greater statistical significance, may motivate consideration of a panel structure with a larger sample and fewer control variables.

Table 6 Panel results for one-year change in PV supply p.c. 2010–2020

	Coefficient	Standard error
Renewable policy score	0.004***	0.001
Regulatory quality	0.001	0.001
Private credit	0.001	0.001
Log GDP p.c	0.001	0.001

Year and country fixed effects are included. There are 1048 observations. The R-squared is 0.099. The dependent variable is the one-year change in solar PV supply per capita. The explanatory variables are lagged by one year. Coefficients for the constant and year fixed effects are not shown. For statistical significance, *** = 1%

Table 6 shows fixed effects panel results, with the dependent variable again being the one-year change in solar PV supply per capita. The control set includes country and year fixed effects to account for time-invariant and commonly varying factors respectively. Other assessed variables include regulatory quality and economic variables of private credit and log GDP per capita. Controls are no longer included for the ASEAN binary variable, fossil fuel reserves, or renewable endowments of solar exposure or wind resources, as these variables are mostly fixed over time.

The panel results in Table 6 show the renewable energy policy score having a significant positive impact on the one-year change in solar PV supply per capita, with statistical significance at the 1% level. These panel results appear to show a stronger link between renewable energy policy scores and the solar PV dependent variable than the prior tables. The control variables in Table 6 have positive coefficients, consistent with prior tables, although each of the controls is statistically insignificant.

Cross-sectional dependence could theoretically be an issue for panel-data analysis where common shocks and unobserved components become part of the error term (de Hoyos and Sarafidis 2006). The risk of cross-sectional dependence may have risen in recent years with increasing regional and global integration. This paper addresses the risk of cross-sectional dependence in the following ways. First, we have the dependent variable being a change in solar energy supply rather than a level. Given that the impact of such cross-sectional dependence can be more concerning in dynamic settings, this reduces the inclination to have a dynamic panel structure involving a lagged dependent variable. A second approach is use of Driscoll and Kraay (1998) standard errors, which can be useful in cases where the unobserved common factors are uncorrelated with the individual regressors. With this alternative approach to calculating standard errors, the renewable energy policy variable is still statistically significant at the 1% level. Finally, we use multiple tests. Post-estimation tests including the Pesaran and Friedman test (Friedman 1937; Pesaran 2004), can be conducted to ensure that cross-sectional dependence is not a major issue. For numerous concise models with enough common observations to perform these tests, the null hypothesis of no cross-sectional dependence is not rejected.

5 Conclusion and Policy Recommendations

The study uses a cross-country comparison of different energy and climate policy types to show the effectiveness of renewable energy policies in driving solar energy uptake, which may help focus policymaker attention on the most effective policies. While ASEAN member states tend to have lower renewable energy policy scores and lower uptake of solar PV per capita, the results suggest successful solar uptake may follow from progress in policy. There are no significant coefficients for the ASEAN variable in any of the tables, suggesting that there might not be any additional and systematic barriers to uptake for ASEAN member states other than the control variables shown.

The results help reveal how long the lags are when assessing the impact of renewable energy policies on solar PV supply. The strongest links tend to be evident for lags of 2–4 years. There are also significant relationships in the range of 1–6 years when using a more concise control set, which probably means that general institutional quality is being partly reflected in more specific renewable energy policy scores. The largest coefficient for renewable energy policy score is for a four-year lag. It is reasonable that this composite variable works at a substantial lag, as it includes some variables which would not have immediate impact, including the legal framework and planning components.

The results also revealed which types of policies can be crucial for social PV supply. While the strongest relationships were found with incentives and carbon pricing, it is likely that a general institutional foundation is necessary to support more targeted policies. This is evident in the renewable energy policy score variable becoming larger and more significant when excluding the regulatory quality variable, implying that the impact of renewable energy policy scores are partially correlated with the general institutional environment. The legal framework and planning components may also similarly correlate with the general institutional environment, although no statistical significance of this was evident in the results.

The results may suggest a roadmap for ASEAN member states. Carbon pricing being currently low in ASEAN, and linked to greater solar PV supply per capita, may motivate promoting this policy, which may in turn have short- and medium-term impact on solar energy adoption. There is also mounting evidence that carbon pricing has an intended impact in promoting transitions from fossil fuels to renewables (Best and Burke 2018, 2020), which may be reassuring to ASEAN member states. These countries might also scale up financial and regulatory incentives for solar PV supply per capita, as the results show a robust influence of this policy component on solar PV supply per capita as well. Historically, ASEAN member states have on average had lower incentive variable scores than other countries.

References

Abban AR, Hasan MZ (2021) Revisiting the determinants of renewable energy investment—new evidence from political and government ideology. Energy Policy151. https://doi.org/10.1016/j.enpol.2021.112184

Aguirre M, Ibikunle G (2014) Determinants of renewable energy growth: a global sample analysis. Energy Policy 69:374–384. https://doi.org/10.1016/j.enpol.2014.02.036

Aldy JE, Stavins RN (2012) The promise and problems of pricing carbon: theory and experience. J Environ Dev 21(2):152–180. https://doi.org/10.1177/1070496512442508

Andersson JJ (2019) Carbon taxes and CO_2 emissions: Sweden as a case study. Am Econ J Econ Pol 11(4):1–30

Baldwin E, Carley S, Brass JN, MacLean LM (2017) Global renewable electricity policy: a comparative policy analysis of countries by income status. Journal of Comparative Policy Analysis: Research and Practice 19(3):277–298. https://doi.org/10.1080/13876988.2016.1166866

Bayer P, Aklin M (2020) The European Union emissions trading system reduced CO_2 emissions despite low prices. Proc Natl Acad Sci USA 117(16):8804–8812. https://doi.org/10.1073/pnas.1918128117

Best R (2017) Switching towards coal or renewable energy? The effects of financial capital on energy transitions. Energy Economics 63:75–83. https://doi.org/10.1016/j.eneco.2017.01.019

Best R, Burke PJ (2017) The importance of government effectiveness for transitions toward greater electrification in developing countries. Energies 10(9). https://doi.org/10.3390/en10091247

Best R, Burke PJ (2018) Adoption of solar and wind energy: the roles of carbon pricing and aggregate policy support. Energy Policy 118:404–417. https://doi.org/10.1016/j.enpol.2018.03.050

Best R, Burke PJ (2020) Energy mix persistence and the effect of carbon pricing. Aus J Agri & Res Econ 64(3):555–574. https://doi.org/10.1111/1467-8489.12370

Best R, Burke PJ, Nishitateno S (2019) Evaluating the effectiveness of Australia's small-scale renewable energy scheme for rooftop solar. Energy Economics84. https://doi.org/10.1016/j.eneco.2019.104475

Best R, Burke PJ, Jotzo F (2020) Carbon pricing efficacy: cross-country evidence. Environ Resource Econ 77(1):69–94. https://doi.org/10.1007/s10640-020-00436-x

Best R, Chareunsy A, Li H (2021) Equity and effectiveness of Australian small-scale solar schemes. Ecological Economics 180(March 2020):106890. https://doi.org/10.1016/j.ecolecon.2020.106890

Best R, Trück S (2020) Capital and policy impacts on Australian small-scale solar installations. Energy Policy 136(November 2019):111082. https://doi.org/10.1016/j.enpol.2019.111082

Best R, Zhang QY (2020) What explains carbon-pricing variation between countries? Energy Policy 143(January):111541. https://doi.org/10.1016/j.enpol.2020.111541

Bourcet C (2020) Empirical determinants of renewable energy deployment: a systematic literature review. Energy Economics 85. https://doi.org/10.1016/j.eneco.2019.104563

Breyer C, Gerlach A (2010) Population density and area weighted solar Irradiation: global overview on solar resource conditions for fixed tilted, 1-axis and 2-axis PV systems

Brunnschweiler CN (2010) Finance for renewable energy: an empirical analysis of developing and transition economies. Environ Dev Econ 15(3):241–274. https://doi.org/10.1017/S1355770X1000001X

Burke PJ (2010) Income, resources, and electricity mix. Energy Economics 32(3):616–626. https://doi.org/10.1016/j.eneco.2010.01.012

Chen C, Pinar M, Stengos T (2021) Determinants of renewable energy consumption: importance of democratic institutions. Renewable Energy 179:75–83. https://doi.org/10.1016/j.renene.2021.07.030

de Groote O, Verboven F (2019) Subsidies and time discounting in new technology adoption: evidence from solar photovoltaic systems. American Economic Review 109(6):2137–2172. https://doi.org/10.1257/aer.20161343

de Hoyos RE, Sarafidis V (2006) Testing for cross-sectional dependence in panel-data models. The Stata Journal 6(4)
Dogan E, Inglesi-Lotz R, Altinoz B (2021) Examining the determinants of renewable energy deployment: does the choice of indicator matter? Int J Energy Res 45(6):8780–8793. https://doi.org/10.1002/er.6413
Driscoll JC, Kraay AC (1998) Consistent covariance matrix estimation with spatially dependent panel data, vol 80, issue 4. https://about.jstor.org/terms
Escoffier M, Hache E, Mignon V, Paris A (2021) Determinants of solar photovoltaic deployment in the electricity mix: do oil prices really matter? Energy Economics97. https://doi.org/10.1016/j.eneco.2020.105024
ESMAP (2020) Regulatory indicators for sustainable energy. https://rise.esmap.org/
Friedman M (1937) The use of ranks to avoid the assumption of normality implicit in the analysis of variance. J Am Stat Assoc 32(200):675–701. https://doi.org/10.1080/01621459.1937.10503522
IEA (2022) IEA world energy statistics and balances; extended world energy balances. International Energy Agency. https://www.oecd-ilibrary.org/
International Energy Agency (2022) Renewable energy market update 2022. www.iea.org/t&c/
IRENA and ACE (2022) Renewable energy outlook for ASEAN towards a regional energy transition. International Renewable, Energy Agency, Abu Dhabi; and ASEAN Centre for Energy, Jakarta. www.irena.org
Kaufmann D, Kraay A, Mastruzzi M (2010) The worldwide governance indicators. Policy research working paper 5430
Kemfert C, Präger F, Braunger I, Hoffart FM, Brauers H (2022) The expansion of natural gas infrastructure puts energy transitions at risk. Nat Energy 7(7):582–587. https://doi.org/10.1038/s41560-022-01060-3
Li X, Ozturk I, Syed QR, Hafeez M, Sohail S (2022) Does green environmental policy promote renewable energy consumption in BRICST? Fresh insights from panel quantile regression. Economic Research-Ekonomska Istrazivanja. https://doi.org/10.1080/1331677X.2022.2038228
Lu X, McElroy MB, Kiviluoma J (2009) Global potential for wind-generated electricity. Proc Natl Acad Sci 106(27):10933–10938. https://doi.org/10.1073/pnas.0904101106
Mayr D, Schmidt J, Schmid E (2014) The potentials of a reverse auction in allocating subsidies for cost-effective roof-top photovoltaic system deployment. Energy Policy 69:555–565. https://doi.org/10.1016/j.enpol.2014.01.029
Pesaran MH (2004) General diagnostic tests for cross section dependence in panels
Pfeiffer B, Mulder P (2013) Explaining the diffusion of renewable energy technology in developing countries. Energy Economics 40:285–296. https://doi.org/10.1016/j.eneco.2013.07.005
Polzin F, Migendt M, Täube FA, von Flotow P (2015) Public policy influence on renewable energy investments-A panel data study across OECD countries. Energy Policy 80:98–111. https://doi.org/10.1016/j.enpol.2015.01.026
Saadaoui H, Chtourou N (2022) Do institutional quality, financial development, and economic growth improve renewable energy transition? Some evidence from Tunisia. Journal of the Knowledge Economy. https://doi.org/10.1007/s13132-022-00999-8
Smith MG, Urpelainen J (2014) The effect of feed-in tariffs on renewable electricity generation: an instrumental variables approach. Environ Resource Econ 57(3):367–392. https://doi.org/10.1007/s10640-013-9684-5
Sweidan OD (2021) Is the geopolitical risk an incentive or obstacle to renewable energy deployment? Evidence from a panel analysis. Renewable Energy 178:377–384. https://doi.org/10.1016/j.renene.2021.06.089
United Nations (2022) The closing window climate crisis calls for rapid transformation of societies. https://www.unep.org/emissions-gap-report-2022
van Benthem AA (2015) Energy leapfrogging. J Assoc Environ Resour Econ 2(1):93–132. https://doi.org/10.1086/680317

Wang H, Yang X, Lou Q, Xu X (2021) Achieving a sustainable development process by deployment of solar PV power in ASEAN: A SWOT analysis. Processes 9(4). https://doi.org/10.3390/pr9040630

Way R, Ives MC, Mealy P, Farmer JD (2022) Empirically grounded technology forecasts and the energy transition. Joule 6(9):2057–2082. https://doi.org/10.1016/j.joule.2022.08.009

World Bank (2021) World development indicators. https://databank.worldbank.org/source/world-development-indicators

World Bank (2022) Global financial development database. https://www.worldbank.org/en/publication/gfdr/data/global-financial-development-database

www.ingramcontent.com/pod-product-compliance
Ingram Content Group UK Ltd.
Pitfield, Milton Keynes, MK11 3LW, UK
UKHW022212051026
15624UKWH00001B/134

9789819982387